STATISTICAL PHYSICS AND CHAOS IN FUSION PLASMAS

NONEQUILIBRIUM PROBLEMS IN THE PHYSICAL SCIENCES AND BIOLOGY

Editors: **I. Prigogine and G. Nicolis**
Université Libre de Bruxelles
Brussels, Belgium

Volume I: **G. NICOLIS, G. DEWEL, AND J. W. TURNER,**
Order and Fluctuations in Equilibrium and Nonequilibrium Statistical Mechanics: XVIIth International Solvay Conference on Physics

Volume II: **C. W. HORTON, JR., L. E. REICHL, AND V. G. SZEBEHELY,**
Long-Time Prediction in Dynamics

Volume III: **C. W. HORTON, JR. and L. E. REICHL,**
Statistical Physics and Chaos in Fusion Plasmas

STATISTICAL PHYSICS AND CHAOS IN FUSION PLASMAS

Edited by

C. W. HORTON, JR. and L. E. REICHL

A WILEY-INTERSCIENCE PUBLICATION
JOHN WILEY & SONS
New York • Chichester • Brisbane • Toronto • Singapore

Copyright © 1984 by John Wiley & Sons, Inc.

All rights reserved. Published simultaneously in Canada.

Reproduction or translation of any part of this work beyond that permitted by Section 107 or 108 of the 1976 United States Copyright Act without the permission of the copyright owner is unlawful. Requests for permission or further information should be addressed to the Permissions Department, John Wiley & Sons, Inc.

Library of Congress Cataloging in Publication Data:

Main entry under title:
  Statistical physics and chaos in fusion plasmas.

   (Nonequilibrium problems in the physical sciences and biology, ISSN 0275-9292; v. 3)

     "A Wiley-Interscience publication."
     Based on papers presented at a workshop held at the University of Texas at Austin, Dec. 1982.
       Includes index.
       1. Plasma confinement—Congresses. 2. Statistical physics—Congresses. 3. Chaotic behavior in systems—Congresses. 4. Nonlinear theories—Congresses.
     I. Horton, C. W. (Claude Wendell), 1942.
     II. Reichl, L. E. III. Series. IV. Title: Fusion plasmas.
QC718.5.C65S73   1984   621.48'4   83-19649
ISBN 0-471-88310-7

Printed in the United States of America

10 9 8 7 6 5 4 3 2 1

CONTRIBUTORS

R. BALESCU — Campus Plaine U.L.B., Boulevard du Triomphe, 1050 Brussels, Belgium

J. R. CARY — Institute for Fusion Studies, University of Texas, Austin, Texas 78712

P. Y. CHEUNG — Department of Physics, University of California, Los Angeles, California 90024

E. G. D. COHEN — Rockefeller University, 1230 York Avenue, New York, N.Y. 10021

R. DE FAINCHTEIN — Center for Studies in Statistical Mechanics, University of Texas, Austin, Texas 78712

P. H. DIAMOND — Institute for Fusion Studies, University of Texas, Austin, Texas 78712

G. D. DOOLEN — Los Alamos National Laboratory, Los Alamos, New Mexico 87545

D. F. DU BOIS — Los Alamos National Laboratory, Los Alamos, New Mexico 87545

C. GREBOGI — Plasma and Fusion Energy Studies, University of Maryland, College Park, Maryland 20742

J. M. GREENE — General Atomic, P.O. Box 81608, San Diego, California 92138

B. HAFIZI — Science Applications Inc., Plasma Research Institute, Boulder, Colorado 80302

T. HATORI — Institute of Plasma Physics, Nagoya University, Nagoya, Japan

C. W. HORTON, JR. — Institute for Fusion Studies, University of Texas, Austin, Texas 78712

Y. H. ICHIKAWA — Institute of Plasma Physics, Nagoya University, Nagoya, Japan

T. KAMIMURA — Institute of Plasma Physics, Nagoya University, Nagoya, Japan

C. F. F. KARNEY — Princeton Plasma Physics Laboratory, P.O. Box 451, Princeton, New Jersey 08544

T. R. KIRKPATRICK — Rockefeller University, 1230 York Avenue, New York, N.Y. 10021

M. KONO	Research Institute for Applied Mechanics, Kyushu University 87, Hakozaki Higashi-ku, Fukuoka 812, Japan
J. A. KROMMES	Princeton Plasma Physics Laboratory, P.O. Box 451, Princeton, New Jersey 08544
Y. KURAMOTO	Research Institute for Fundamental Physics, Kyoto University, Kyoto, Japan
R. G. LITTLEJOHN	Department of Physics, University of California, Los Angeles, California 90024
W. H. MATTHAEUS	Department of Physics, College of William & Mary, Williamsburg, Virginia 23185
S. W. MCDONALD	University of Maryland, College Park, Maryland 20742
J. MEISS	Institute for Fusion Studies, University of Texas, Austin, Texas 78712
J. H. MISGUICH	Association Euratom-CEA sur la Fusion, BP6 92260 Fontenay-aux-Roses, France
D. MONTGOMERY	Department of Physics, College of William & Mary, Williamsburg, Virginia 23185
H. MORI	Department of Physics, Kyushu University, Fukuoka, Japan
E. OTT	Department of Physics & Astronomy, University of Maryland, College Park, Maryland 20742
T. PETROSKY	Center for Studies in Statistical Mechanics, University of Texas, Austin, Texas 78712
L. E. REICHL	Center for Studies in Statistical Mechanics, University of Texas, Austin, Texas 78712
H. A. ROSE	Los Alamos National Laboratory, Los Alamos, New Mexico 87545
G. SCHMIDT	Stevens Institute of Technology, Castle Point Station, Hoboken, New Jersey 07030
M. SUZUKI	Department of Physics, University of Tokyo, Kunkyo-ku, Tokyo 113, Japan
J. SWIFT	Department of Physics, University of Texas, Austin, Texas 78712
T. TANIKAWA	Department of Physics, University of California, Los Angeles, California 90024
P. W. TERRY	Institute for Fusion Studies, University of Texas, Austin, Texas 78712
R. B. WHITE	Princeton Plasma Physics Laboratory, P.O. Box 451, Princeton, New Jersey 08544

A. WOLF	Department of Physics, University of Texas, Austin, Texas 78712
A. Y. WONG	Department of Physics, University of California, Los Angeles, California 90024
N. YAJIMA	Research Institute Applied Mechanics, Kyushu University, Fukuoka 812, Japan
J. A. YORKE	University of Maryland, College Park, Maryland 20747

PREFACE

The subject of this book is the statistical properties of nonlinear dynamical systems with emphasis on their relation to the problem of plasma confinement. That aspect of plasma confinement, which is related to properties of nonlinear Hamiltonian systems, has been studied for many years. Such systems have two facets. On the one hand, they behave like integrable systems and can sustain the propagation of coherent waves (solitons) and, on the other hand, they can decay into chaos. It is not surprising that presently a question of active investigation from several different perspectives is the search for theoretical descriptions containing both the integrable aspects of the problem and the chaos or stochasticity. It is a remarkable fact, as the reader will find in the study of this volume, that the new mathematical methods grounded in physics, such as scaling, renormalization, and fractal structure, allow quantitative theories to be constructed for a variety of important conservative systems. Perhaps the unique contribution of this volume is to show how with the aid of these new mathematical methods theories for the structure of magnetic fields, the confinement of charged particles, the description of correlations in the Vlasov–Poisson system, and the derivation of a (Markovian) kinetic equation are currently being developed.

The physical problems mentioned in the preceding paragraph involve conservative systems with volume-preserving flows and maps. A second realm of dynamics occurs in dissipative systems where flows and maps are volume contracting. Although the distinction in some plasma systems can be blurred by the choice of the fluid versus the kinetic description, the mathematical consequences from the choice tend to be sharp. Dissipative systems, which share many common characteristics in their behavior, include neutral fluid turbulence, reaction diffusion systems of chemistry, Langmuir-ion acoustic turbulence described by the Zakharov model, and the problem of drift wave turbulence in an inhomogeneous plasma. In these systems problems of great current interest are the chaotic attractor and its basin of attraction, the period doubling and other sequences of bifurcations, and the role of coherent dissipative structures. The importance of the chaotic attractor for general systems and its rich variety of properties in only now being realized although its existence for simple maps and flows has been known for 20 years or so. Chapters in this volume explore the fractional dimensionality of the attractor subspace in the system's phase space as well as the fractal boundary of the boundary of the attractor. These studies show that such previously mysterious phenomena as intermittonery can be studied within the context of the fractal boundary of the attractor.

A number of traditional topics in the study of nonlinear dynamics have been treated in this volume. Recent developments in the theory of solitons are reviewed with emphasis on the so-called exotic solitons. An application of fractal analysis to the study of phase transitions is presented. A new method of computing Lyapunov exponents from experimental data is developed, and a method for calculating the splitting of periodic orbits in conservative systems using Hamilton's principle is developed.

Finally, one unique aspect of this book is the collection of works on the problem of long-lived correlations in phase space of the Vlasov–Poisson system. In plasma physics these long-lived correlations are called clumps and for some time they have remained a rather mysterious phenomenon. In this volume, a powerful, new diagrammatic perturbation theory is presented for a first-principles calculation of the clump kinetic equation. In addition, in this volume, for the first time there are some detailed calculations and experimental considerations that purport to describe the practical consequences of these correlations.

In conclusion the editors believe that the reader of this volume will obtain a new and deeper understanding of the rich interplay between the statistical and near-integrable behavior that occurs in nonlinear physics. The contributions included in this volume grew out of a workshop on Statistical Physics and Chaos in Fusion Plasmas held at the University of Texas at Austin in December 1982. The editors are grateful for the support given that workshop by the Institute for Fusion Studies and the Center for Studies in Statistical Mechanics at the University of Texas.

C. W. HORTON, JR.
L. E. REICHL

Austin, Texas
January 1984

CONTENTS

1. CONSERVATIVE DYNAMICS

Renormalization and the Breakup of Magnetic Surfaces   3
   by J. M. Greene

Statistical Properties of Nonlinear Area-Preserving Maps   21
   by Y. H. Ichikawa, T. Kamimura, and T. Hatori

Long-Time Correlations in Stochastic Systems   33
   by C. F. F. Karney

Magnetic Field Lines, Hamiltonian Mechanics, and Symmetries and Invariants   43
   by R. G. Littlejohn

Searching for Integrable Systems   51
   by J. R. Cary

Hamilton's Principle and the Splitting of Periodic Orbits   57
   by G. Schmidt

Resonance Behavior of the Perturbed Toda Lattice   65
   by R. de Fainchtein and L. E. Reichl

Recent Developments of Soliton Research in Plasma Physics   79
   by Y. H. Ichikawa and N. Yajima

2. DISSIPATIVE DYNAMICS

Onset of Chaos in Continuous Media: Case of Reaction–Diffusion Systems   93
   by Y. Kuramoto

Progress in Computing Lyapunov Exponents from Experimental Data 111
    by A. Wolf and J. Swift

Fractal Basin Boundaries in Nonlinear Dynamical Systems 127
    by C. Grebogi, S. W. McDonald, E. Ott, and J. A. Yorke

Evolution from Coherence to Turbulence in Plasmas 131
    by A. Y. Wong, P. Y. Cheung, and T. Tanikawa

Coherence in Chaos and the Zakharov Model 155
    by G. D. Doolen, D. F. DuBois, H. A. Rose, and B. Hafizi

3. KINETIC THEORY

Kinetic Theory of Fluctuations in Fluids Far from Equilibrium 171
    by E. G. D. Cohen and T. R. Kirkpatrick

Kinetic Description of a Chaotic Motion in a Classical Conservative System with Two Degrees of Freedom 189
    by T. Y. Petrosky

Collisional Diffusion in a Torus with Imperfect Magnetic Surfaces 209
    by R. B. White

Applications of Fractal Analysis to Phase Transitions and Other Phenomena 225
    by M. Suzuki

4. TURBULENCE THEORY

Time Evolution of Vorticity Field and Turbulent Diffusion 235
    by H. Mori

Topics in the Theory of Statistical Closure Approximations for Plasma Physics 241
    by J. A. Krommes

Statistical Description of Drift Wave Turbulence 253
    by C. W. Horton, Jr.

Solitons in Turbulent Flow   273
    by J. D. Meiss

Dynamic Alignment and Selective Decay in MHD   285
    by W. H. Matthaeus and D. Montgomery

5. CLUMPS IN PLASMA PHYSICS

A Diagrammatic Approach to the Theory of Clumps in Turbulent Plasmas   295
    by R. Balescu and J. H. Misguich

Global Analysis of Nonlinear Transient Phenomena Near the Instability Point   311
    by M. Suzuki

Clump Kinetics in Turbulent Plasmas   327
    by M. Kono

Impact of Clumps on Plasma Stability and the Nature of Turbulence in a Saturated State   335
    by P. W. Terry and P. H. Diamond

Author Index   353

Subject Index   359

STATISTICAL PHYSICS AND CHAOS IN FUSION PLASMAS

Part 1

CONSERVATIVE DYNAMICS

RENORMALIZATION AND THE BREAKUP OF MAGNETIC SURFACES

J. M. Greene
GA Technologies Inc., San Diego, California

1. INTRODUCTION

Containment of field lines is a fundamental problem for magnetically controlled fusion. Transport along field lines is much faster than transport across field lines, so that containment of field lines is the first step toward containment of plasma in toroidal devices. Thus this problem has been studied for many years.

A critical point in its solution was the recognition by Feigenbaum[2] that renormalization provided a fruitful way of studying period doubling in one-dimensional maps. This was very far from the problem of interest to fusion, but some years later Kadanoff[5] and co-workers showed that these techniques could be applied to problems of more-direct relevance, namely, the existence of containing magnetic surfaces. MacKay[7] then carried Kadanoff's program to conclusion.

The purpose of this chapter is to gather these results together and to set them into a fusion context. Sections 2 and 3 discuss the magnetic-containment problem. First, magnetic field lines in axisymmetric systems are treated, and then the effect of nonaxisymmetric perturbations is introduced. Period doubling in a magnetic-field-line context is discussed in Section 4, primarily as an introduction to renormalization in such systems. The breakup of magnetic surfaces is treated in Section 5 and Section 6 contains conclusions.

2. MAGNETIC GEOMETRY

In this section, the general nature of magnetic geometry will be outlined briefly.

First, consider axisymmetric toroidal systems. This case can be treated precisely numerically, so most calculations of field lines have been in such systems. In general, for the solutions of interest, magnetic field lines wind around the tori, also called magnetic surfaces, to which they are confined. These tori form a nested set, with a closed line, the magnetic axis, at the center. An important geometric quantity is the winding number on each torus. This will be called the safety factor for historical reasons. It will be denoted by q and will be defined by

$$q = \frac{\text{Revolutions around the symmetry axis}}{\text{Revolutions around the magnetic axis}} \quad (1)$$

A convenient way to picture these tori is to consider a cross section through the system. Then the tori are represented by closed curves. Successive traverses of a field line through this surface will be represented by a series of dots. The orbits of individual field lines can be reduced to a scheme for predicting successive intersections. Thus, the problem of following field lines can be reduced to a map of the cross section onto itself.

We can make the identification of field lines with orbits of a dynamical system in a three-dimensional phase space. Thus, the problem treated in this chapter is of more general interest. Choosing a relevant example from a multitude of possibilities, particle-guiding center motion in a tokamak can be treated by the methods described here. However, here the physical intuition will be restricted to the very pictorial field-line problem.

The map of the cross section generated by the magnetic field lines is flux preserving; that is, a bundle of field lines is mapped into another bundle with the same flux. This is equivalent to the area conservation of Poincaré's invariant for Hamiltonian systems.

The structure of this map is straightforward for axisymmetric systems. The safety factor varies continuously from some limiting value at the magnetic axis. On the tori for which it is rational, field lines close on themselves after a finite number of iterations of the map. The numerator of the fractional representation of the rational will be called the period of the closed line. When the safety factor is irrational, the field line covers the entire torus densely. This distinction between rational and irrational numbers is nonphysical, but it is clear how it comes about. Without setting a limit to the length of closed field lines, we are considering a closed line to be closed no matter how long it is. If the period is very long, the line covers a surface quite well, and tends asymptotically to dense covering in the limit. Thus, a given finite length of magnetic line can be treated as either part of a closed line or part of a magnetic surface with irrational safety factor, and the choice depends on the longest line length that can be usefully treated as closed. Thus both halves of the asymptotic long-length-line concept of rational and irrational safety factors can play a role in treating finite line segments. In this respect it is a typical result of the point of view taken in this chapter.

The field-line orbits of perturbed, nonaxisymmetric systems can also be reduced to a map. As long as the system is generally toroidal, a suitable cross section can be found. If the perturbation is not too large, the map of this cross section onto itself will be recognizably similar to the unperturbed map.

Having reduced the problem of magnetic-field-line orbits to a map, it makes sense to consider the general subject of maps. Renormalization techniques have made it posible to make statements about all maps, including those generated by field lines. Thus, the field-line problem will not be lost in the generality of the map problem. The maps that will be treated are area preserving. However, no more than a change of metric is required to make them flux preserving.

3. TWO-DIMENSIONAL MAPS

This section will discuss the qualitative behavior induced by nonaxisymmetric perturbations of toroidal systems, in terms of specific maps. First consider the map

$$x' = -y + px - (1-p)x^2$$
$$y' = x - px' + (1-p)x'^2 \qquad (2)$$

This transforms a point (x, y) to a point (x', y'), and is area preserving. That is, the Jacobian of the transformation from (x, y) to (x', y') is unity. This map can be denoted by T, $(x', y') = T(x, y)$. Thus T stands for a pair of functions. The parameter p is a control parameter that can be thought of as related to the amount of nonaxisymmetric perturbation. While this figure will be discussed as if it were generated by field-line orbits, it was in fact introduced by Hénon as the nonlinear map that was simplest from the standpoint of computing orbits. Since it is not derived from a magnetic system, there is a limit to which its analogy to such systems can be carried. However, the map it produces does have common features with the magnetic field plots of White.[8]

Figure 1 shows several orbits for this map, for $p = 0.24$. Each orbit is a series of points generated by Eq. (2) from an initial point. The corresponding magnetic line can be imagined as connecting successive points by looping around in and out of the paper. First consider the orbit at the center of the figure. Equation (2) transforms the point $(0, 0)$ into itself, and the corresponding orbit is represented by a small circle. This corresponds to the magnetic axis. Such simple closed lines also exist in nonaxisymmetric systems, as can be proven from various fixed-point theorems.

Magnetic surfaces with sufficiently irrational safety factor survive under small finite nonaxisymmmetric perturbation; that is, there are some closed toroidal surfaces to which a field line is restricted, and which it winds

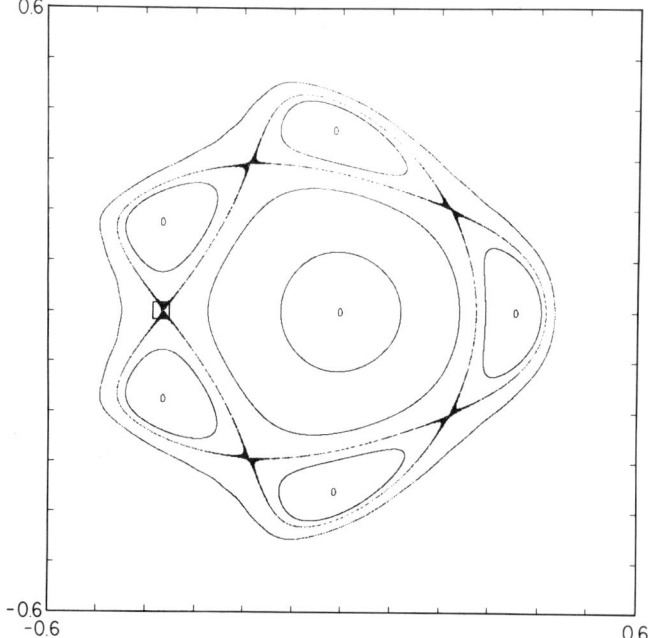

Figure 1. Some illuminating orbits for the map of Eq. (2).

around and covers densely. This was proven about 20 years ago by Arnol'd and Moser, following a suggestion of Kolmogorov. Hence, they are also called KAM orbits, or KAM curves. Two such curves are shown in Fig. 1 near the magnetic axis.

Nonaxisymmetry has a more-profound effect on surfaces with a rational safety factor. The magnetic lines that exist in the vicinity of one such surface are shown in Fig. 1, forming a fivefold island structure near a surface where the safety factor is $q = 5$. The perturbation is resonant here, so that in some regions lines are coherently pushed out and in other regions they are pulled in across the unperturbed magnetic surface. On the borders of these regions a few closed lines survive. One such lies at the center of the island structure illustrated in Fig. 1. It visits the center of each island successively before closing on itself. Lines of force that are swept off a resonant surface find themselves caught in a flow with a different safety factor, and thus fall out of resonance. As a result, these lines form the island KAM structures that surround the island's central closed line.

In principle, there is an island structure corresponding to every rational safety factor, but only one has the appropriate scale to be visible in this figure. Thus, surfaces with rational and irrational safety factors respond quite differently to perturbation—the former transforming into islands, while the latter retain their character. The distinction between rational and irrational numbers is more extreme than that considered in the previous section, but the same principles apply.

In addition to the closed lines and magnetic surfaces, an additional type of magnetic line completes the inventory of possible behaviors. This is represented by the fuzzy or stochastic orbit that outlines the island chain of Fig. 1. It is shown in more detail in Fig. 2, which is an enlargement of the region in the box on the left of Fig. 1. The dark center of this picture is a single orbit, and the points that compose it occur virtually randomly within the band. Corresponding magnetic lines wander through and fill a volume. An analytic function that could reproduce this in detail is beyond comprehension, so it is reasonably clear that the problem of representing field lines in closed form over indefinitely long lengths will never be solved.

On the other hand, there are possibilities for making reasonable statements about asymptotic behavior in the limit of infinitely long length lines. For example, it may be adequate for many purposes to know that a given orbit is asymptotically a perfect mess. More importantly, the existence of magnetic surfaces is an asymptotic problem, since only in the limit are surfaces densely covered and defined by a field line. Existence is important because, by uniqueness and continuity, lines in a region bounded by magnetic surfaces must always remain there. To the extent that magnetic surfaces can be proven to exist, stability or containment properties of nearby lines are known. Knowledge of bounding magnetic surfaces is useful because diffusion of a line of force is generally fairly rapid through most of its accessible region. Thus, even moderately long lines can display

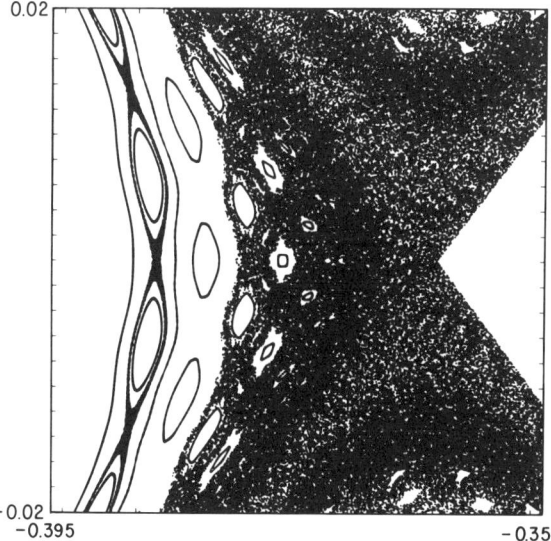

Figure 2. More orbits of the map of Fig. 1. The frame of this figure is the small box on the left of Fig. 1.

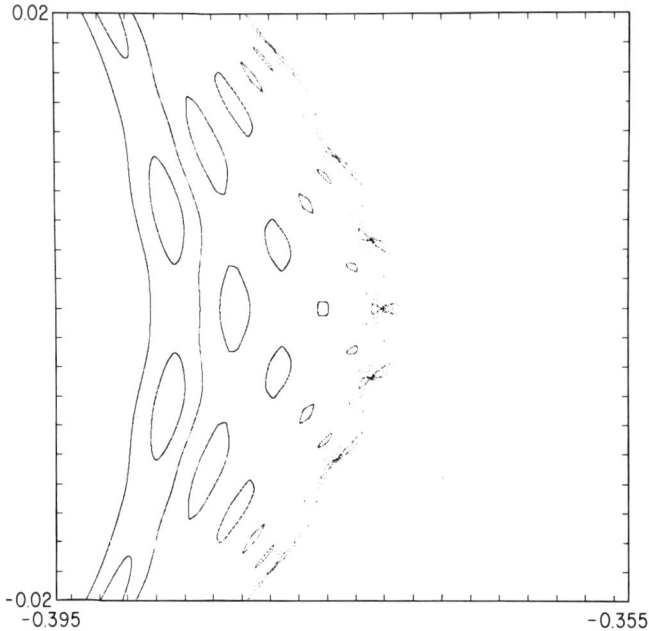

Figure 3. This is the same as Fig. 2, but with the stochastic orbits omitted for clarity.

behaviors that are best represented by concepts, such as "stochastic" or "KAM," that are only defined precisely in an asymptotic limit.

The nature of maps can be explored by a further examination of the orbits of the map of Eq. (1). Figure 3 shows some orbits exactly corresponding to those of Fig. 2, but with the stochastic orbits omitted. The islands shown, from left to right, have safety factors 61/12, 66/13, 71/14, 76/15, 81/16, and 86/17. The different chains are best distinguished along the diagonals on which they are aligned. These island chains encircle the magnetic axis conforming with the chain illustrated in Fig. 1, along with the separating KAM surfaces. The existence of the fivefold island chain distorts and perturbs nearby surfaces, with the level of nonaxisymmetric perturbation steadily increasing toward the right as the island is approached. The effect of this will be described in the next few paragraphs.

Returning to Fig. 2, we see that the less-perturbed island chain on the left is isolated from its neighbors by intervening KAM surfaces, while the more-perturbed islands are engulfed by the main stochastic orbit. This shows that KAM surfaces do not survive under large perturbation, since they would dam the stochastic sea if they did exist. Thus, the problem of proving the existence of a given surface in a given geometry is nontrivial. Toward the right, the stochastic orbit appears dense. Hence, the most-perturbed islands disappear entirely. Thus, under increasing perturbation there is a progression from interleaved island and KAM orbits, to isolated islands in a stochastic sea, and on to all-encompassing stochasticity. Quantifying this sequence will be the task of the remainder of this chapter. The second of these transitions—the vanishing of islands—is discussed in Section 4, while the first is treated in Section 5. It is particularly important to understand the level of perturbation that leads to the destruction of the KAM surfaces that otherwise limit the extent of the stochastic orbit shown in Fig. 2. KAM surfaces are the dikes that contain plasma in toroidal fusion devices.

To better understand this, note that the island structure reproduces the tokamak structure on a smaller scale. In three dimensions, the island is a tube winding helically around the main magnetic axis. The periodic orbit at the center of an island is its magnetic axis. In the dominant island chain of Fig. 1, this axis is approximately five times longer than that of the main tokamak, but it is otherwise similar. In particular, it is surrounded by KAM curves, and these have an associated safety factor, relative to their magnetic axis. Note that there are different levels of safety factor. The island chain of Fig. 1 lies in the region of $q = 5$ around the main magnetic axis, but has a safety factor $q_i = 12.67$ at its center. The safety factor at the center of an island is distinguished by the subscript i. In its definition, the numerator of Eq. (1) is replaced by "Revolutions in multiples of the period of the island's magnetic axis."

The magnitude of the perturbation resonant at the unperturbed surface affects the thickness of the island and also the safety factor at its magnetic axis. In fact, these quantities can be used as a local measure of the magnitude of the perturbation. They are directly related in perturbation theory.[3,6] The quantity q_i varies inversely with the size of the perturbation, and is infinite in the axisymmetric limit. The major result of the work described in this chapter is that the perturbation required to reduce the safety factor at the center of an island to the value 6 is very nearly the same as the perturbation required to destroy all the KAM surfaces that separate the island from its neighbors. That is, the transition from interleaved-islands-and-KAM-surfaces to islands-isolated-in-a-wide-ranging-stochastic-sea occurs when the island's central safety factor passes through 6. For example, in Fig. 3, the central safety factor of the island chain at the left is $q_i = 6.77$, for the next to the right it is $q_i = 5.68$, and the next, $q_i = 4.65$. The significance of the value 6 is described in Section 5.

In Section 4 the magnitude of the perturbation required to destroy the island structure, as seen in the right part of Fig. 2, is discussed.

4. PERIOD DOUBLING

This section will focus on one particular island structure, say the fivefold island chain at the $q = 5$ surface in Fig. 1. The question that will be

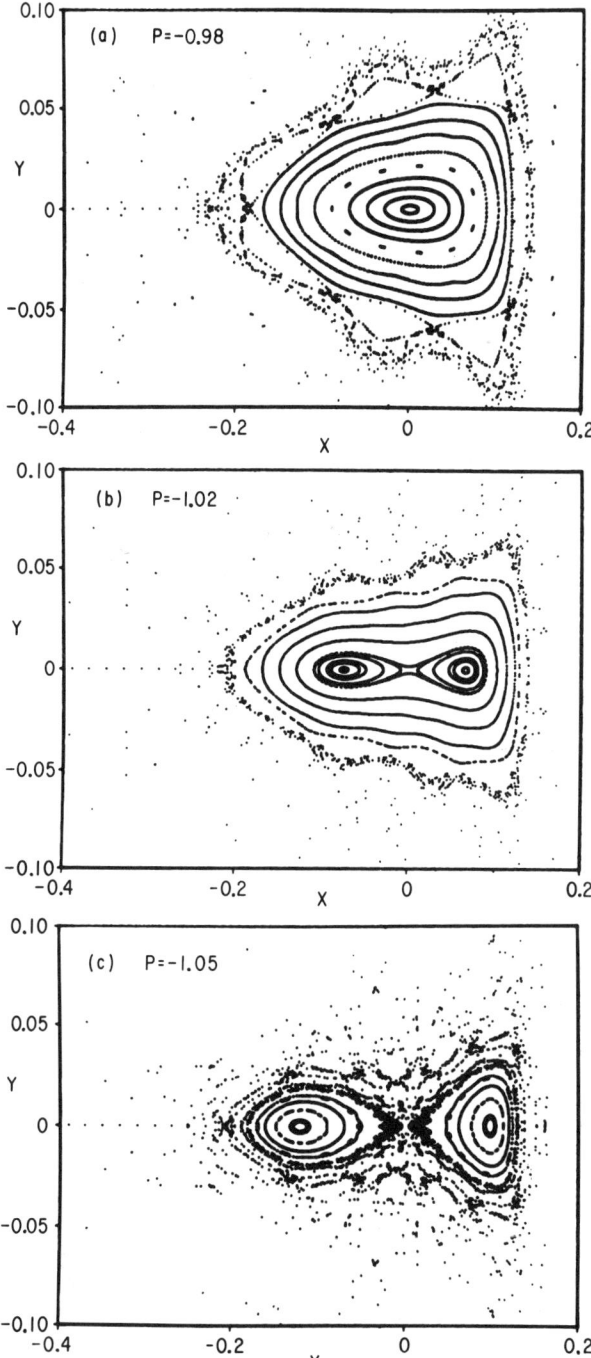

Figure 4. A typical period-doubling sequence, showing behavior near a magnetic axis under increasing perturbation.

addressed is that of how, with increasing perturbation, this island completely disintegrates and merges with the adjoining stochastic sea. This gives some insight into the origin of the sea, and perhaps also into its properties, but the primary reason for this section is to provide an introduction to renormalization. The original papers for this section are Greene et al.[4] and Collet et al.[1]

With increasing perturbation the central q value of this island steadily drops. When it reaches a value $q_i = 2$, a 10-fold island structure arises, winding around the magnetic axis at the center of the fivefold island. The usual situation is that the magnetic axis becomes unstable, and the stability is passed on to a magnetic axis of twice the length, as shown in Fig. 4. This is period doubling. It has also occurred in the 86/17 island chain toward the right of Fig. 3. This kind of instability is suppressed by symmetry in most plots of magnetic field lines.

The 10-fold island is born with an infinite central q, but with increasing perturbation this rapidly drops. It too undergoes period doubling when the central q hits 2. Successive period doublings accumulate geometrically at a finite perturbation strength, with intervals between doublings decreasing geometrically.

The next step is to fix the perturbation at the critical value at which the period doubling in a given island accumulates, and examine field-line plots in the immediate vicinity of some of the longer lines. In appropriate

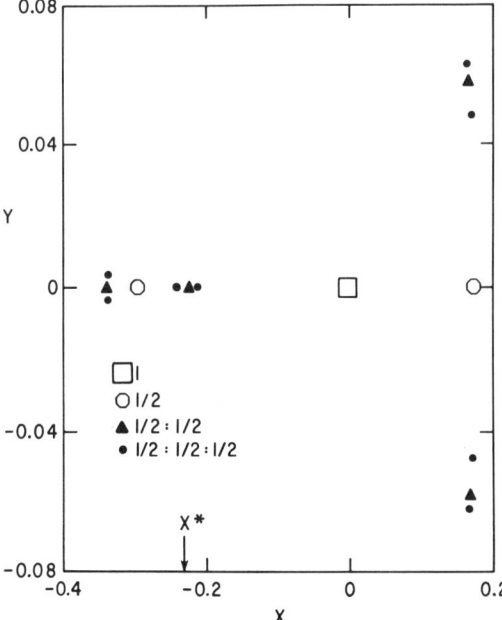

Figure 5. Some orbits of the universal period-doubling map, T^*.

coordinates, one obtains a picture like Fig. 5. This does not look like usual field-line plots since there are no stable magnetic lines, but positions of unstable closed lines have been plotted. The orbit denoted by the large square has bifurcated to yield the orbit denoted by the octagon, that in turn has produced the orbits given by the triangles and then the one given by the small circles.

The remarkable thing about this picture is that it is self-similar, around the point $(x^*, 0)$; that is, magnification by appropriate factors yields the same figure again. The large anchor shape with the hooks vertically on the right is duplicated by a smaller anchor with the hooks given by the dots on the left. When more orbits are plotted, this is repeated ad infinitum.

Kenneth Wilson, together with Kadanoff and others, showed us in the work leading to his recent Nobel Prize, that such self-similar pictures are specified by a small amount of information. A few bits of information, plus self-similarity, can yield masses of data. Thus, Fig. 5 leads to a powerful method for studying the nature of period doubling.

We will now proceed to make algebra out of this picture, and the name of the technique for doing this is renormalization. The picture says that two maps are equivalent, the given map and one obtained by scaling this map. We break this into two steps, first producing a functional operator that generates a new map by scaling any given map, and then looking for equivalence between these two maps.

In Fig. 5, we expect to show equivalence between the orbit denoted by the square and the orbit denoted by the left octagon. However, the latter orbit has a period twice that of the former. Thus, we wish to renormalize length along the field line by a factor of 2; that is, the newly generated map, which we will denote by T', should be related to the result of two successive iterations of the given map, T, which we denote as T^2. Furthermore, the newly generated map should be renormalized across the field lines, so that it matches the given map. From numerical computations, the appropriate factors are $\alpha = -4.0180\ldots$ for the scaling in x, and $\beta = 16.3638\ldots$ for the scaling in y. Putting this together, we define an operator N that generates a pair of functions, T', from a pair of functions by T by

$$T' = N(T) \tag{3}$$

where

$$N(T) \equiv BT^2 B^{-1}, \tag{4}$$

and the perpendicular scaling is linear, so that it can be represented by the matrix

$$B \equiv \begin{pmatrix} \alpha & 0 \\ 0 & \beta \end{pmatrix}, \tag{5}$$

with α and β as previously given. This can also be given in a completely explicit form. If the map T from (x, y) to (x', y') is represented by

$$\begin{aligned} x' &= f(x, y) \\ y' &= g(x, y) \end{aligned} \tag{6}$$

then the map T' is represented by

$$x' = \alpha f[f(x/\alpha, y/\beta), g(x/\alpha, y/\beta)]$$
$$y' = \beta g[f(x/\alpha, y/\beta), g(x/\alpha, y/\beta)]. \quad (7)$$

Examining successively longer orbits in the immediate neighborhood of a chosen point of a given map is achieved by repeated applications of the operator N. Clearly, maps that are invariant under N have a special place in the theory, and one of these is the map illustrated in Fig. 5. We denote this map by T^*,

$$T^* = N(T^*). \quad (8)$$

To better understand this equation, consider the linearization of the operator N, dN, in the neighborhood of T^*; that is, consider the effect of N on maps that are very close to T^*. This is an eigenvalue problem, yielding those deviations from T^* that are form invariant under N, but that change in magnitude. Since the space of maps on which N operates is infinite dimensional, there are an infinity of eigenvalues. The fundamental fact of renormalization theory is that all but a small number of these are less than unity, so that T^* is an attractor in almost all directions.

With that background, return to Eq. (8). Given one solution to this equation, there is a two-parameter family of solutions with differing scales in the x and y directions. From another point of view, x and y can be measured in any units in Fig. 5. As with the scale invariance of linear eigenvalue problems, this leads to Eq. (8) being formally over determined. It can only be solved for particular values of the scaling factors α and β. Thus Eq. (8) determines α and β as well as T^*.

Actually, there are five unstable eigenvalues of the linearization dN. For example, if the scaling is not centered on $(x^*, 0)$ in Fig. 5, this point exponentially falls off the maps resulting from successive applications of N (with growth rates α and β). In all, four of the unstable eigenvectors can be projected out by appropriate choice of the alignment and origin of the coordinates. Thus there is essentially only one unstable eigenvalue, which is called the relevant eigenvalue. To project this out, a control parameter, such as p of Eq. (2), is adjusted to a value appropriate for the particular orbits of the map under study.

These considerations lead to the picture given in Fig. 6. This sketch represents the infinite-dimensional space of all two-dimensional area-preserving maps; that is, each point in the space represents a map, including a choice of coordinates and especially the origin of the scaling of the operator N. We assume that this is picked to be the point of accumulation of the period-doubled orbits that branch from a chosen orbit.

Applying N successively to a given initial map yields a succession of maps that is an orbit in this function space. The map T^* is illustrated in the center of the picture. The horizontal curve passing through it represents a surface with one fewer dimension than the full function space. That is, it represents all area-preserving maps that satisfy one additional condition. The condition is that the control parameter has the value such that these maps do not excite the relevant eigenvector. Thus every map in this multidimensional surface is attracted into T^* under repeated applications of N. This will be called the surface of critical maps, since the magnetic line near the origin for each of these maps is just at its accumulation of period doublings.

The vertical line in Fig. 6 represents a one-parameter family of maps that are precisely repelled by N. Near T^* it is tangent to the relevant, unstable eigenvector.

Variation of a parameter, such as p of Eq. (2), produces a continuous sequence of maps such as is illustrated in Fig. 6 by the dashed line. Consider the map in this sequence that lies on the intersection of the dashed line and the surface of critical maps passing through T^*. Applying N to this map many times, that is, examining it on successively smaller scales, leads inevitably to T^*; that is, given proper choice of coordinates and parameter, locally all area-preserving maps look the same. This is called universality.

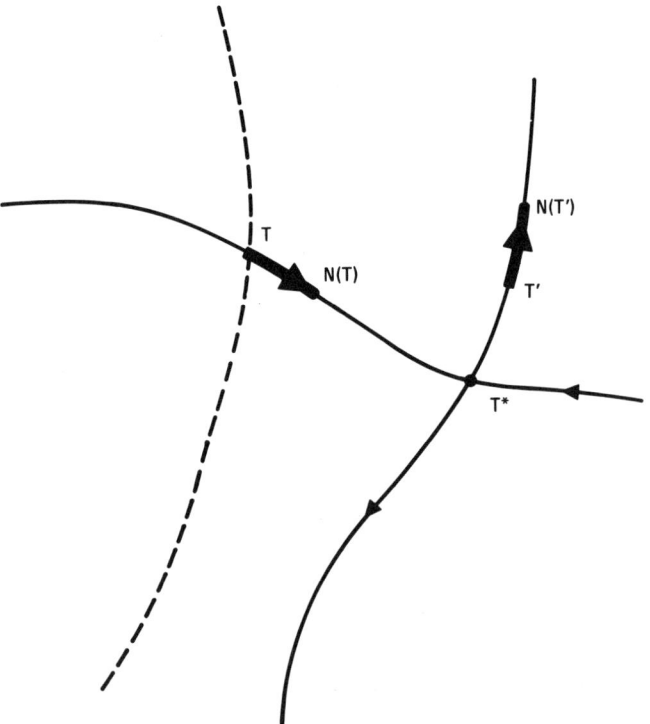

Figure 6. A sketch representing the infinite-dimensional function space of all area-preserving maps, and the flow in this space produced by the operator N.

Note that it is the property of the limit of many applications of N, and, therefore, a property of asymptotically long field lines.

The significance of T^* for Fig. 2 is that all orbits that have bifurcated out of the central orbit, for example, the orbit represented by the square in Fig. 5, are unstable. With each successive period doubling, stability is passed to a longer orbit. Thus, at the accumulation value, all finite-period orbits in the bifurcation tree are unstable. It is very difficult to prove absolutely that every single periodic orbit of a map is unstable, but numerical computations show that such stable orbits are at most negligibly important in T^*. Thus, the unstable orbits of T^* are like pins in a pinball game, and a typical line bounces from one to another randomly. This is the way the stochastic orbit of Fig. 2 approximately fills in the most highly perturbed region of the map.

It follows that the surface of critical maps of Fig. 6 represents the second transition of the previous section, from isolated islands to effectively uniform stochastic motion. In the final section of this chapter another characterization of this surface will be given, which can be implemented on a computer.

5. DESTRUCTION OF MAGNETIC SURFACES

The problem of the breakup of a magnetic surface has many features in common with the problem of period doubling. Now one chooses a magnetic surface. The most-interesting choice is the most-rugged one that separates two stochastic orbits, such as the pair illustrated on the right and left of Fig. 2. Then the destruction of this surface heralds the merging of the two orbits on either side, and, hence, a major change in the containment properties of the system.

The most-rugged surface is one that is in some sense the farthest from nearby islands, so that it is least distorted in conforming to their shapes. This is a problem in number theory. The answer is given in terms of the continued-fraction representation of the safety factor. Thus, we express

$$q = a_0 + \cfrac{1}{a_1 + \cfrac{1}{a_2 + \cdots}} \qquad (9)$$

or, in shorthand,

$$q = [a_0, a_1, a_2, \ldots] \qquad (10)$$

The coefficients a_n are called partial quotients. They are positive integers. An irrational number has an infinite sequence of partial quotients, but a rational number has a finite sequence. Some picture of the relations associated with continued fractions can be deduced from Fig. 3. The safety factors of the islands shown have the continued-fraction representations [5, 12], [5, 13], ..., [5, 17]. The larger the partial quotient, the closer to the nearby big island, and the bigger the perturbation.

The result of these considerations is that the most-interesting surfaces have irrational safety factors whose continued-fraction expansion has only ones beyond some level. These will be called noble numbers. For example, a good candidate for the most-rugged surface in the vicinity of rightmost surface shown in Fig. 3 is one with the safety factor [5, 12, 1, 1, ...].

We need one other result from the theory of continued fractions. The most important islands close to the chosen surface have safety factors given by truncating the continued-fraction representation of its safety factor. If part of the sequence of successively closer islands is given by

$$q_{n-1} = [a_0, \ldots, a_{n-1}]$$
$$q_n = [a_0, \ldots, a_{n-1}, a_n] \quad (11)$$
$$q_{n+1} = [a_0, \ldots, a_{n-1}, a_n, a_{n+1}]$$

then the periods of these islands, which are the numerators of the fractional representation of the safety factor, are related by

$$P_{n+1} = a_{n+1} P_n + P_{n-1} \quad (12)$$

For noble numbers, with $a_n = 1$, this yields a Fibonacci sequence.

So, after choosing a magnetic surface, one adjusts a perturbation parameter until the surface has structure on all space scales. This is the critical value. Examining relatively long orbits in a small neighborhood in well-chosen coordinates, one obtains Fig. 7. The critical KAM surface is

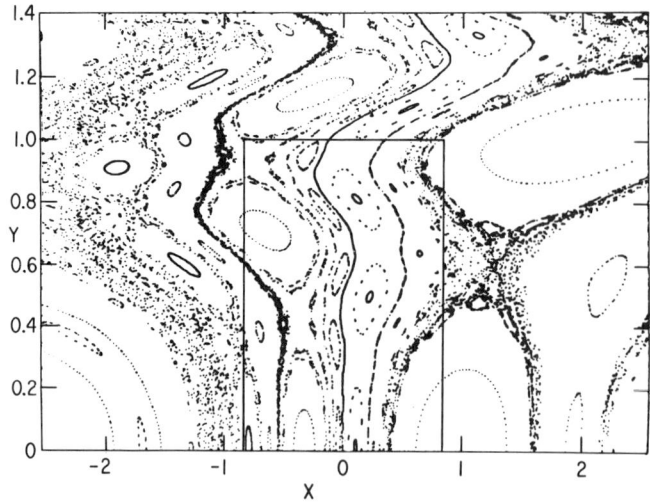

Figure 7. Orbits of the universal map, T^*, of magnetic surface destruction.

the one passing through the point $(0, 0)$. Again, this picture is self-similar. Every feature within the outer frame is repeated within the inner frame.

Reducing this picture to renormalization equations proceeds much as before. The scale factors of the renormalization perpendicular to the magnetic field lines are $\alpha = -1.4148 \cdots$ in y and $\beta = -3.0668 \cdots$ in x.

The scaling in the length along the field line is more complicated. Since the chosen safety factor was a noble number, the periods of successive orbits that are equivalent under the spatial scaling have a Fibonacci relation. Specifically, the period of the orbit at the center of the island at $(-0.3, 0)$ is the sum of the periods of the islands centered at $(1., 0)$ and $(-3., 0)$. This three term recursion relation for the periods [Eq. (12)] can be reduced to a two term relation in two variables,

$$\bar{P}_{n+1} = P_n$$
$$P_{n+1} = P_n + \bar{P}_n \tag{13}$$

Note that the recursion relation for periods in the previous section was also of the two-term form,

$$P_{n+1} = 2P_n \tag{14}$$

Hence, the equivalent renormalization relation is most elementarily expressed in terms of two maps, T and U, as

$$\begin{pmatrix} U' \\ T' \end{pmatrix} = K \begin{pmatrix} U \\ T \end{pmatrix} \tag{15}$$

where

$$K \begin{pmatrix} U \\ T \end{pmatrix} = \begin{pmatrix} BTB^{-1} \\ BTUB^{-1} \end{pmatrix} \tag{16}$$

Equations (13) and (16) correspond if P is the period of a given orbit in T and $\bar{P}$ is the period of the equivalent orbit in U.

Again, the critical map is the equilibrium solution of this equation,

$$\begin{pmatrix} U^* \\ T^* \end{pmatrix} = K \begin{pmatrix} U^* \\ T^* \end{pmatrix} \tag{17}$$

We must make one additional restriction. For the interesting solution, the maps T^* and U^* must commute. This is a physically reasonable restriction since, by the construction of Fig. 7, they are both powers of the same map. It is a necessary restriction because the linearization of K, dK, has unstable eigenvalues whose eigenmaps are noncommuting.

The calculation now proceeds much as in the case of period doubling. There were many interesting nitty gritty problems that had to be overcome, as described in MacKay.[7] Figure 6 can be used as a sketch of the relevant function space. Again, there is only one relevant unstable eigenvalue. The surface of critical maps separates the maps in which the chosen KAM surface exists from those in which it does not. Hence, it represents the transition from a banded island and KAM structure to an isolated island in a stochastic sea, which was discussed in Section 3.

In Section 6 a further characterization of this surface will be given.

6. CONCLUSIONS

In this chapter we have shown how renormalization yields a solution of the problem of finding the transitions, under increasing perturbation, from ordered behavior to chaotic behavior. These transitions appear in Fig. 2 as different orbits in different regions and in Fig. 6 as surfaces in an abstract function space. This section will endeavor to relate these two pictures.

The problem is to decide whether a given map in a given region is above or below the critical surface in Fig. 6. Using a computer to study a sequence of longer lines on a smaller scale, according to the rules of either the operator N or K, yields maps that are successively closer to the vertical line of maps that are repelled from T^* by N, or K. Thus, roughly, locally, maps can be given a one-dimensional order along this line. We can call the ordering parameter the local perturbation strength.

We want a good way of estimating this parameter. Two possibilities strike the eye in Fig. 7. One measure of size of the perturbation is the width of the islands. However, the islands are fuzzy at the edge, and have a rather indefinite width. Furthermore, this width is not completely coordinate invariant.

A better number to use is the safety factor at the center of the island. This is completely independent of coordinates. Furthermore, the principal islands of (U^*, T^*) are transformed into each other by K, so they all have the same central safety factor, which happens to be $q_i = 5.9988 \cdots$.

This is a reasonable description of the surface of critical maps of Fig. 6. If islands have a central q greater than 6, there are nearby KAM surfaces, and containment. If the central safety factor is smaller, there is an enclosing stochastic sea. This central-safety-factor description of the surface of critical maps is only precise close to (U^*, T^*), but it is asymptotically accurate in the local limit.

A similar criterion can be found for the critical surface of period doubling. Here there are no islands, and, hence, no central safety factor. However, there are still closed lines where the islands used to be. One can calculate the instability of these lines, that is, the rate at which nearby lines diverge. if the distance increases at a rate less, in absolute value, then $\lambda_c = -2.057\ldots$ per period, the map is below the surface of critical maps, and there are probably some small islands nearby. Otherwise, there probably are none.

Thus, renormalization provides a computationally efficient way of determining the containment of magnetic field lines. The finite computations can be tied to a mathematical theory of asymptotic behavior, so that they can give completely credible results.

ACKNOWLEDGMENTS

It was a great pleasure to be associated with Dr. MacKay during the period these ideas were becoming clear. The contributions of all the Princeton Plasma Physics community during this time are gratefully acknowledged. The assistance of Dr. Karney with all aspects of computing was especially appreciated.

This work was supported by the Department of Energy, Contract DE-AT03-76ET51011.

REFERENCES

1. P. Collet, J.-P. Eckmann, and H. Koch. "On Universality for Area Preserving Maps of the Plane," *Physica* **3D**, 457-467 (1981).
2. M. J. Feigenbaum, "Quantitative Universality for a Class of Nonlinear Transformations," *J. Stat. Phys.* **19**, 25-52 (1978).
3. J. M. Greene, "The Calculation of KAM Surfaces," in *Nonlinear Dynamics*, R. H. G. Helleman, ed. (New York Academy of Sciences, New York, 1980), pp. 80-89.
4. J. M. Greene, R. S. Mackay, F. Vivaldi, and M. J. Feigenbaum. "Universal Behaviour in Families of Area-Preserving Maps," *Physica* **3D**, 468-486 (1981).
5. L. P. Kadanoff, "Scaling for a Critical Kolmogorov-Arnol'd-Moser Trajectory." *Phys. Rev. Lett.* **47**, 1641-1643 (1981).
6. A. J. Lichtenberg, "Determination of the Transition between Adiabatic and Stochastic Motion," in *Intrinsic Stochasticity in Plasmas*, D. Gresillon, ed. (Editions de Physique, Orsay, France, 1980), pp. 13-40.
7. R. S. MacKay, "Renormalization in Area Preserving Maps," Princeton University Thesis, 1982 [University Microfilms International, Ann Arbor, Michigan (order no. 830 1411)].
8. R. B. White, "Collisional Diffusion in a Torus with Imperfect Magnetic Surfaces." (In Part 3 of this volume) (Wiley, New York, 1984).

STATISTICAL PROPERTIES OF NONLINEAR AREA-PRESERVING MAPS

Yoshi H. Ichikawa, T. Kamimura, and T. Hatori
Institute of Plasma Physics, Nagoya University, Nagoya, Japan

Abstract. *Numerical observations have been carried out on diffusion processes associated with a general class of the radial twist maps. Detailed observations on orbits of the standard map reveals that at the integer values of the stochastic parameter particles are trapped into the accelerating modes, and thus they contribute to enhancement of the stochastic diffusion. Intermittent transitions between chaotic and periodic orbits are shown to take place with large probability at the half-integer values of the stochastic parameter. Although the present analysis remains qualitative, it will be sufficient to indicate that the standard map possesses a peculiar property owing to its periodic structure. Therefore, one needs to pay special attention in reducing general nonlinear area-preserving maps to the standard map by a local linearization approximation.*

1. INTRODUCTION

Recent advances in fusion plasma confinement experiments have attracted intense theoretical investigations on the long-time behavior of plasma particles in the systems. As one of the powerful approaches available, analysis of nonlinear dynamical maps has been applied to the confinement analysis of energetic α particles in tokamak plasmas,[3] the radial transport of ions in tandem-mirror devices,[2] and the stochastic heating of the plasma by external radio-frequency waves.[5] Basing our investigations on a nonlinear tandem-mirror map, we have carried out direct numerical observation of the radial diffusion of collisionless guiding centers in the tandem-mirror devices.[4]

In order to explore the statistical properties of nonlinear area-preserving maps, we present results of numerical observation of the diffusion process for a general class of the radial twist maps in Section 2. A peculiar feature of the diffusion process of the standard map is emphasized. Thus, we are led to study the details of the orbital behavior of particles in the standard map, in Section 3. We show explicitly that the enhancement of the diffusion process in the specific range of the stochastic parameter is associated with accelerator modes. In Section 4, we illustrate explicitly the intermittent transition between chaotic orbits and periodic orbits, which appears to be responsible for suppressing the diffusion process at the half-integer of the stochastic parameter. Section 5 presents concluding remarks.

2. STOCHASTIC DIFFUSION IN NONLINEAR MAPS

We have expressed the complicated tandem-mirror map as a squared map of the element map T defined by

$$T: \quad Y_{n+1} = Y_n + A \sin(2\pi X_n)$$
$$X_{n+1} = X_n + F(Y_{n+1}) \quad (1a)$$

with

$$F(Y) = \tfrac{1}{2} Y^2 - \tfrac{1}{2} \quad (1b)$$

The constant term $-\tfrac{1}{2}$ of Eq. (1b) is included because the end plugs are twisted by $\pi/2$ with respect to each other. If the end plugs were not twisted, the radial transport could be described by the squared map of Eq. (1a) with

$$F(Y) = \tfrac{1}{2} Y^2 \quad (2)$$

Very frequently, one reduces a general radial twist map [Eq. (1a)] into the standard map by a linearization approximation, for which $F(Y)$ is given by

$$F(Y) = Y \quad (3)$$

In order to investigate the statistical properties of the area-preserving nonlinear maps, we have carried out numerical observations of evolution of the mean square displacement of particles after N time steps:

$$\langle \Delta Y_N^2 \rangle = \langle [Y_N(i) - Y_0(i)]^2 \rangle \quad (4)$$

where the average $\langle \ \rangle$ is taken over the initial ensemble of particles. As an initial ensemble, we took 4000 particles equally spaced in the interval $0 < X < 1$ along the line of $Y = 0$. Figure 1 illustrates the results of such observations for the standard map [Eq. (1a) with Eq. (3)] for values of the stochastic parameter $A = 0.7$ and $A = 1.1$, respectively. Since $\langle \Delta Y_N^2 \rangle$ increases monotonically in proportion to the time step N in Fig. 1a, we can

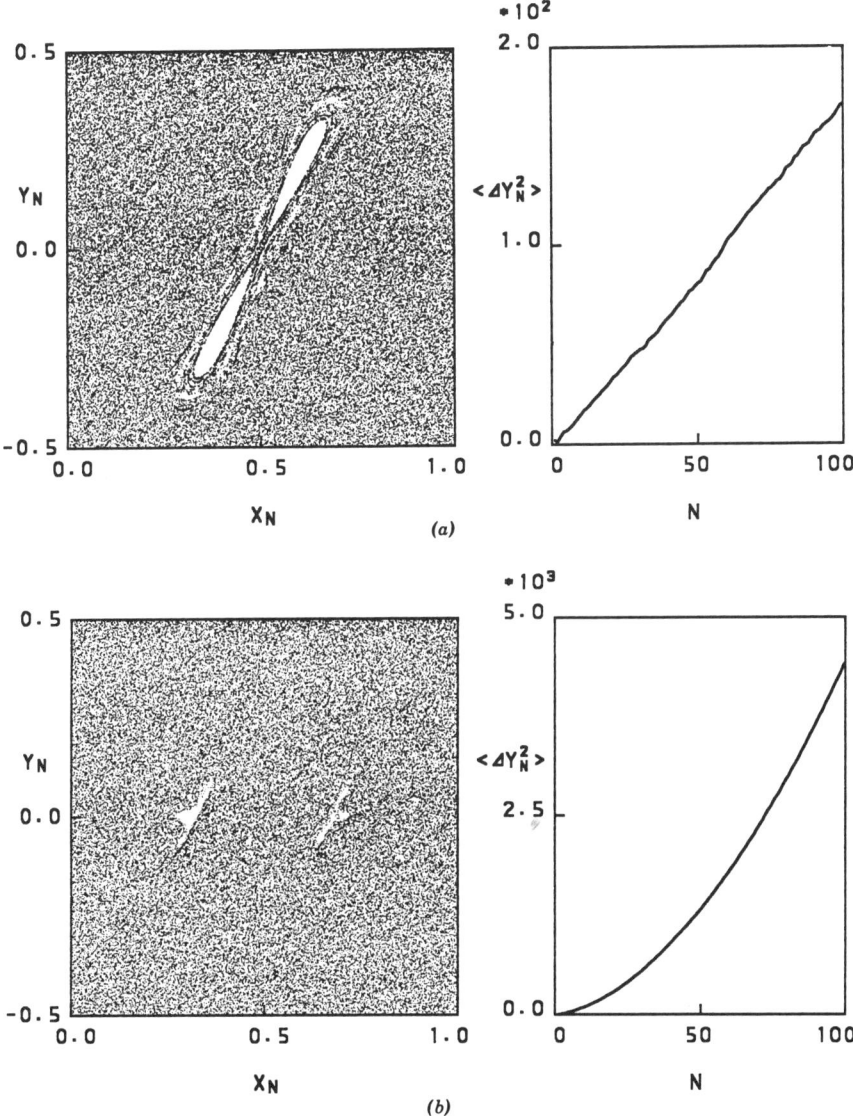

Figure 1. (a) Spreading of 4000 particles in the X–Y space and $\langle \Delta Y_N^2 \rangle$ after $N = 100$ steps for the values of $A = 0.7$. (b) Spreading of 4000 particles in the X–Y space and $\langle \Delta Y_N^2 \rangle$ after $N = 100$ steps for the value of $A = 1.1$. A dot-printer failed to reproduce highly populated regions, which appear as white islands.

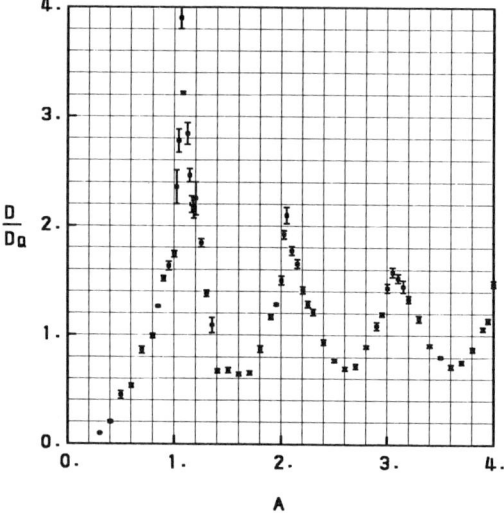

Figure 2. The normalized diffusion rate for various values of A in the standard map.

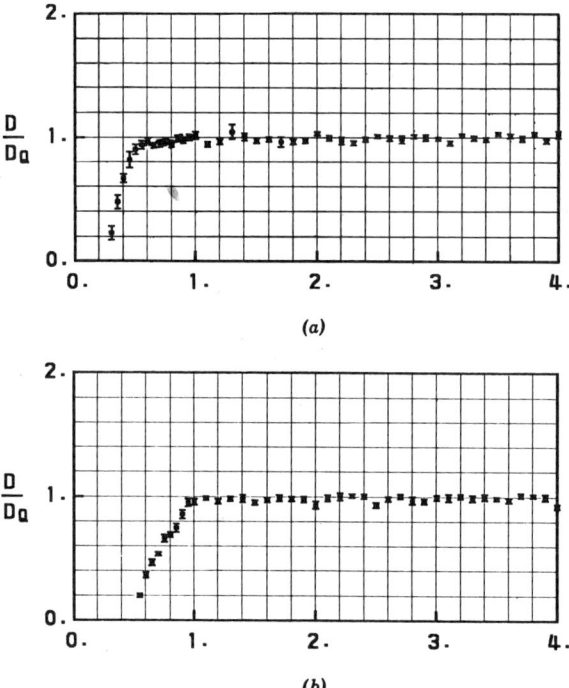

Figure 3. (a) Normalized diffusion rate for the radial twist map with $F(Y) = \frac{1}{2} Y^2$. (b) Normalized diffusion rate for the radial twist map with $F(Y) = \frac{1}{2} Y^2 - \frac{1}{2}$.

determine the average diffusion rate for the stochastic motion of particles, defined as

$$D = \lim_{N \to \infty} \frac{1}{2N} \langle \Delta Y_N^2 \rangle \qquad (5)$$

However, rapid increase of the mean square displacement for the value of $A = 1.1$ is due to the accelerator modes, the detailed structure of which will be discussed in Section 3. Eliminating the particles situated at the accelerating fixed points, we could obtain a diffusionlike spread of the particles for the value of $A = 1.1$. A similar treatment has been applied in the range of $2 < A < 2.2$.

We summarize in Fig. 2 the diffusion rate normalized relative to the quasilinear diffusion rate $D_Q = A^2/4$, for the standard map. Similarly, Fig. 3 illustrates the observed diffusion rate of the radial twist maps with Eqs. (1b) and (2). Comparing Figs. 3a and 3b, we confirm that the $\pi/2$-twist of the end plugs suppresses the radial transport of particles for the stochastic parameter A less than the unity.

The standard map's relevant statistical properties have been examined extensively by Chirikov,[1] Rechester et al.,[8] and many others. In particular, the oscillatory deviation from the quasilinear diffsion rate is accounted for by the sum of the Bessel function, which is legitimate asymptotically in the range of large stochastic parameter.

Applying the Fourier-space-path method to the general radial twist maps, we find that the diffusion rate defined in the limit of $N \to \infty$ is reduced to the quasilinear diffusion rate. Therefore, we note that our approximation [Eq. (24) in Ref. 4] is indeed appropriate for the observed diffusion rate in Fig. 3a, but it is too crude to express the diffusion rate for the standard map. This also explains the remarkable similarity between Figs. 6 and 7 of our previous work.[4]

3. ACCELERATOR MODES OF THE STANDARD MAP

As observed in Section 2, at a certain value of the stochastic parameter A in the standard map, the mean square displacement increases quadratically as the time steps N. This suggests that the accelerator modes are predominant.

Since Fig. 1b indicates that a large number of particles are accelerating in the Y direction, we now examine the detailed properties of accelerating particles. For this, we followed tracks of $10^3 \times 10^3$ particles uniformly distributed over the domain of $0 < X < 1$ and $0 < Y < 1$. We registered the initial positions of particles when their displacement in the Y direction continues to increase in the successive time steps. For $A = 1.00$, after 80 time steps, there are only two particles for which ΔY is increasing. These

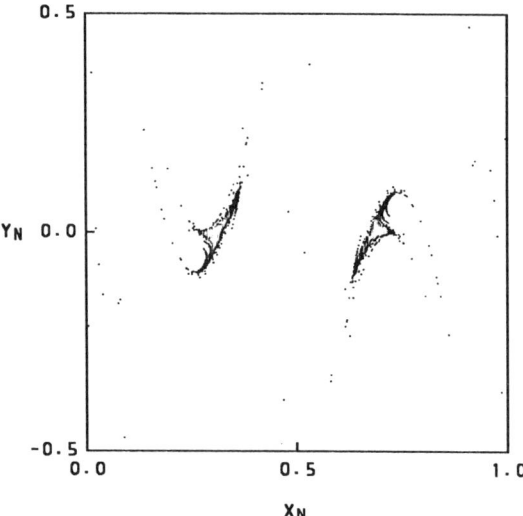

Figure 4. The initial positions of particles whose displacement in the Y direction continues to increase during the $N = 200$ time steps; $A = 1.10$. As with Fig. 1b, a dot-printer failed to reproduce the highly populated central region properly.

are just the two stable fundamental accelerating points determined by the conditions

$$A \sin(2\pi X_a) = 1 \qquad (6a)$$

$$-4 < 2\pi A \cos(2\pi X_a) < 0 \qquad (6b)$$

On the other hand, for $A = 1.10$, we find 3926 particles are accelerating continuously for 200 time steps. Their $X - Y$ space distribution is shown in Fig. 4. Similar patterns are obtained for other values of A. In order to identify the properties of these accelerating islands, we followed orbits of the particles whose initial position was in the interval of $0.25 < X < 0.375$ at $Y = 0$ for $A = 1.100$. Figures 5a and 5b illustrate the evolution of Y displacement of particles that started at $(X_0 = 0.2875, Y_0 = 0)$ and $(X_0 = 0.3000, Y_0 = 0)$. Following their paths in $X - Y$ space, we confirmed that they are the period-3 accelerating points. Figure 5a suggests that $(X_0 = 0.2875, Y_0 = 0)$ is a stable period-3 accelerating point, while $(X_0 = 0.3000, Y_0 = 0)$ is an unstable one. Figure 5b indicates that the orbit is trapped back again into the accelerating island at the time steps of $\sim 0.30 \times 10^4$. Figure 5c shows that the chaotic orbit started at $(X_0 = 0.3625, Y_0 = 0)$ is trapped into the accelerating mode at the time steps of $\sim 0.75 \times 10^4$. Thus, we expect that this occasional trapping of particles into the unstable accelerating orbit will enhance the stochastic diffusion processes associated with the standard map.

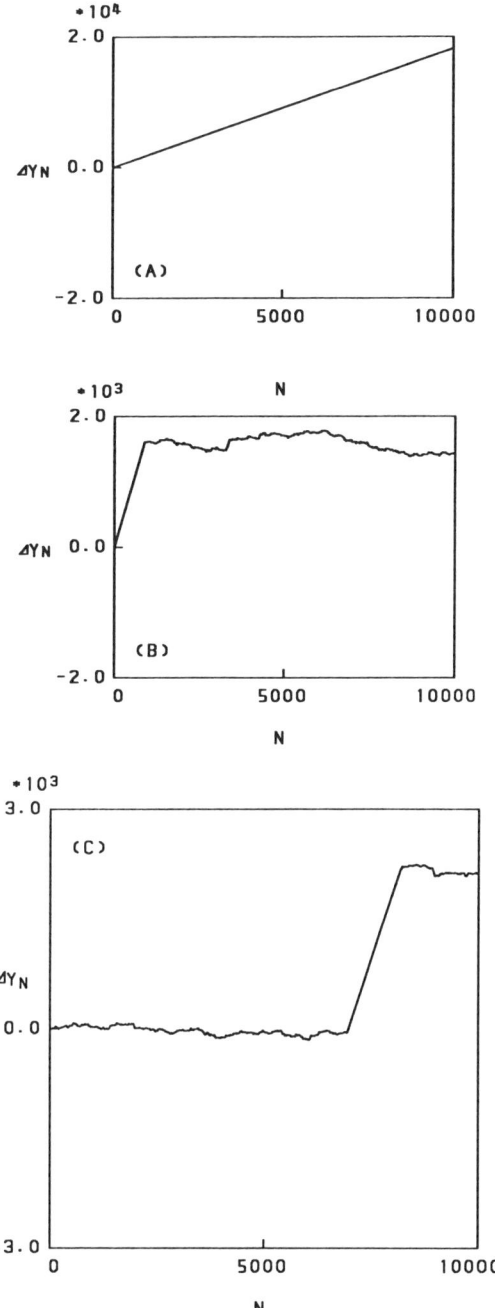

Figure 5. (a) A stable period-3 accelerating orbit. (b) An unstable period-3 accelerating orbit. (c) Trapping of a chaotic orbit, whose initial position was ($X_0 = 0.3625$, $Y_0 = 0$), into the accelerating orbit; $A = 1.100$.

The structure of the period-3 accelerating islands is determined by the constant-phase condition of the 3-step map,

$$A[\sin(2\pi X_1) + \sin(2\pi X_2) + \sin(2\pi X_3)] = l \quad (7a)$$

$$3Y_1 + 3A\sin(2\pi X_1) + 2A\sin(2\pi X_2) + A\sin(2\pi X_3) = m \quad (7b)$$

where l and m are integers, and

$$X_2 = X_1 + Y_1 + A\sin(2\pi X_1) \quad (8a)$$

$$X_3 = X_2 + Y_2 + A\sin(2\pi X_1) + A\sin(2\pi X_2) \quad (8b)$$

the stability condition is given as

$$-4 < 3(\alpha_1 + \alpha_2 + \alpha_3) + 2(\alpha_1\alpha_2 + \alpha_2\alpha_3 + \alpha_3\alpha_1) + \alpha_1\alpha_2\alpha_3 < 0 \quad (9)$$

where

$$\alpha_i = 2\pi A \cos(2\pi X_i) \quad (i = 1, 2, 3) \quad (10)$$

Thus, we could construct the period-3 accelerator island as shown in Fig. 6. The shape of accelerating island observed in Fig. 4 agrees with that of Fig. 6.

By increasing the stochastic parameter A slightly to $A = 1.110$, the stable period-3 accelerator points disappear, and the stable fundamental accelerator island grows in size, which becomes unstable at $A = 1.185 \cdots$. Then, as a result of bifurcation of the fundamental accelerating point, there appears the period-2 accelerator modes at $A \simeq 1.2$.

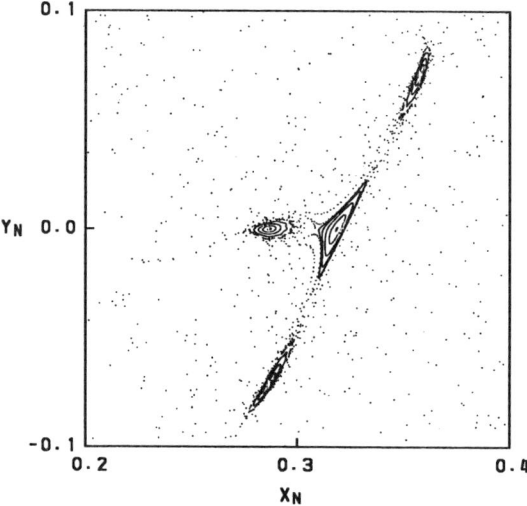

Figure 6. Period-3 accelerator islands for $A = 1.100$. Stable period-3 accelerator points at $(X_1 = 0.2865, Y_1 = 0)$, $(X_2 = 1.3577, Y_2 = 1.0712)$, and $(X_3 = 3.2865, Y_3 = 1.9288)$ and unstable period-3 accelerator points at $(X_1 = 0.3094, Y_1 = 0.0)$, $(X_2 = 1.3337, Y_2 = 1.0242)$, and $(X_3 = 3.3094, Y_3 = 1.9758)$ are determined from Eqs. (7)–(9).

4. INTERMITTENT TRANSITION BETWEEN CHAOTIC AND PERIODIC ORBITS

When the stochastic parameter A is increased to 1.5, we found two stable islands in the regions of $(0.24 < X < 0.25,\ Y \simeq 0)$ and $(0.31 < X < 0.32,\ -0.02 < Y < +0.02)$. Traces of orbits that make up these islands confirm that they are not the accelerating island, but are the period-4 fixed points. Figure 7 illustrates one of these orbits, starting at $(X_0 = 0.2440,\ Y_0 = 0)$. Now, the most interesting observation of the manifestation of these periodic fixed points is shown in Figs. 8a and 8b, which shows the temporal evolution of the Y displacement of particles starting at $(X_0 = 0.12,\ Y_0 = 0.15)$ and $(X_0 = 0.13,\ Y_0 = 0.15)$, respectively. All of these figures indicate that intermittent transition between chaotic orbits and periodic orbits occurs very frequently. Trapping into the periodic orbits prevents the stochastic diffusion of particles.

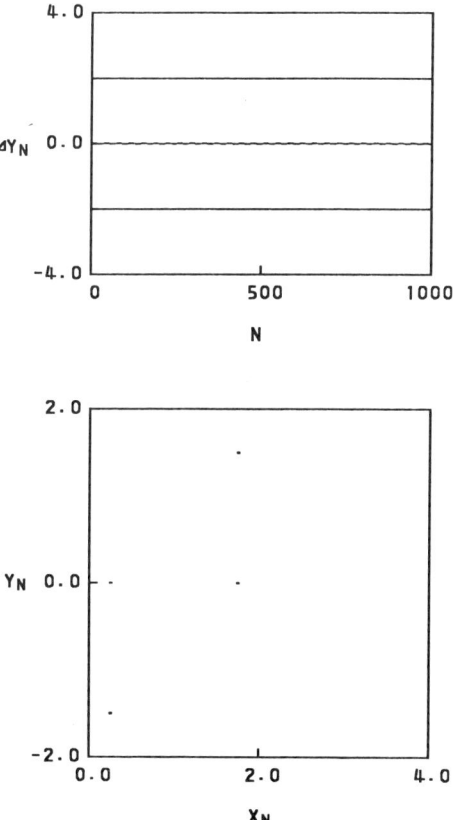

Figure 7. Period-4 fixed points for the value of $A = 1.5$.

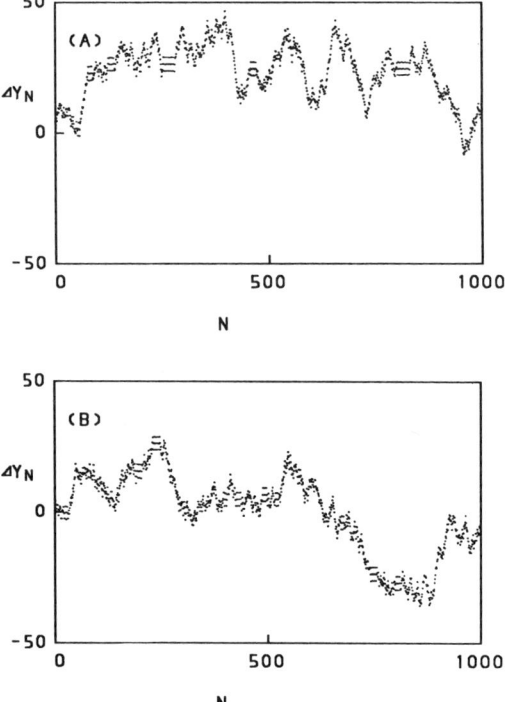

Figure 8. Intermittent transition between chaotic orbits and the period-4 fixed point orbit.

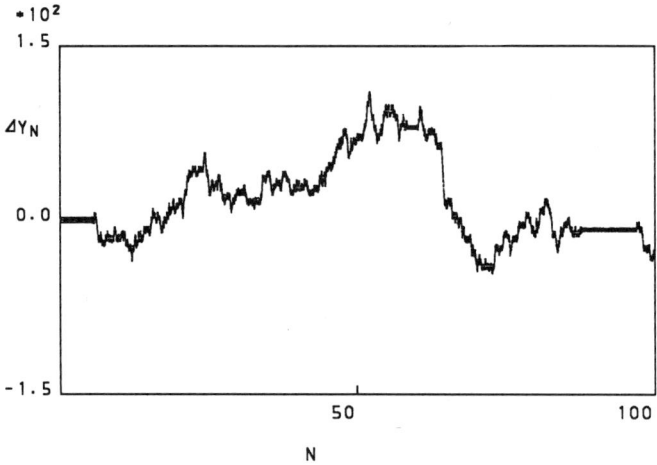

Figure 9. Intermittent transition between chaotic orbits and the periodic fixed point orbits for the value of $A = 0.5$.

Figure 9 is another illustration of the intermittent transition between the chaotic orbits and the periodic orbit for $A = 0.5$. Although we cannot predict quantitative effects at the present stage of investigation, we expect that these intermittent transitions between the chaotic orbits and the periodic orbits suppress the stochastic diffusion of particles in the standard map.

5. CONCLUDING REMARKS

In the present analysis we have shown that the period-3 accelerator mode effects are predominant on the enhanced diffusion in the standard map. We have observed that particles have a finite probability to get trapped into these period-3 accelerator islands, and are accelerated for a certain period of the time. Hence, the resulting diffusion rate is much enhanced compared to the quasilinear diffusion rate.

Karney et al.[6] have examined the effects of the accelerator modes on the diffusion process in the standard map. Recently, Karney[7] analyzed the long-time correlation in the stochastic regime of the standard map, and discussed the trapping probability of particles into the accelerator islands. Although the relationship of the present analysis with his investigation is not clear at the present stage, we believe that our findings of the detailed structure of the accelerator modes will provide new insights.

REFERENCES

1. B. V. Chirikov, *Phys. Rep.* **52**, 263 (1979).
2. R. H. Cohen, *Intrinsic Stochasticity in Plasmas*, G. Laval and D. Gresillion, Eds. (Editions de Physique, Orsay, 1979), p. 181.
3. R. J. Goldston, R. B. White, and A. H. Boozer, *Phys. Rev. Lett.* **47**, 647 (1981).
4. Y. H. Ichikawa, T. Kamimura, and C. F. F. Karney, *Physica* **6D**, 233 (1983).
5. C. F. F. Karney, *Phys. Fluids* **22**, 2188 (1979).
6. C. F. F. Karney, A. B. Rechester, and R. B. White, *Physica* **4D**, 425 (1982).
7. C. F. F. Karney, *Physica* **8D**, 360 (1983).
8. A. B. Rechester, M. N. Rosenbluth, and R. B. White, *Phys. Rev. A* **23**, 2664 (1981).

LONG-TIME CORRELATIONS IN STOCHASTIC SYSTEMS

Charles F. F. Karney

Plasma Physics Laboratory, Princeton University, Princeton, New Jersey

Abstract. In recent years there has been considerable interest in understanding the motion in Hamiltonian systems when phase space is divided into stochastic and integrable regions. This paper studies one aspect of this problem, namely, the motion of trajectories in the stochastic sea when there is a small island present. The results show that the particle can be stuck close to the island for very long times. For the standard mapping, where accelerator modes are possible, it appears that the mean squared displacement of particles in the stochastic sea may increase faster than linearly with time indicating nondiffusive behavior.

INTRODUCTION

Many important problems in physics are described by Hamiltonians of two degrees of freedom. Examples are the motion of a charged particle in electrostatic waves, the motion of a charged particle in various magnetic-confinement devices, the acceleration of a particle bouncing between a fixed and an oscillating wall, the wandering of magnetic field lines, etc. In such systems, there is usually a range of parameters (normally when the coupling between the two degrees of freedom is large), where the motion in nearly the whole of phase space is stochastic. Such behavior is seen for instance in the standard mapping[2]

$$r_t - r_{t-1} = -k \sin \theta_{t-1} \qquad \theta_t - \theta_{t-1} = r_t$$

When k is large, most of phase space is stochastic. However, there may still be small islands present; stochastic trajectories can wander close to these islands and remain there for a long time leading to unexpectedly long correlations. The effect of these correlations can be dramatic. The simplest approximation for the diffusion coefficient,

$$\mathcal{D} = \lim_{t \to \infty} \frac{\langle (r_t - r_0)^2 \rangle}{2t}$$

(where the average is over some appropriate ensemble), is given by assuming that the phase θ is a random variable in the equation for r. This gives the "quasilinear" result $\mathcal{D} = \mathcal{D}_{ql} = \frac{1}{4}k^2$. However, a numerical determination[6] of the diffusion at $k = 6.6$, where the random-phase approximation might be expected to be accurate, gave $\mathcal{D}/\mathcal{D}_{ql} \sim 80$. At this value of k there is an island ("accelerator mode") present in the stochastic sea. This leads to long-time correlations in the acceleration of the particle and an enhanced diffusion coefficient.

In this paper, we examine more closely the effect these islands have on a stochastic trajectory. As far as determining the effect on the correlation function, this involves determining how "sticky" the island is. Given that the stochastic trajectory comes within a certain distance of the boundary of the island, how long do we expect it to stay close to the island? This approach is inspired by work of Channon and Lebowitz[1] on the correlations of a trajectory in the stochastic band trapped between two KAM surfaces in the Hénon map. Similar work has been carried out on the whisker map by Chirikov and Shepelyansky[4]; this work is being extended by B. V. Chirikov and F. Vivaldi. The work reported herein is described in more detail in Ref. 5.

DERIVATION OF MAPPING

Far into the stochastic regime for a general mapping, the islands which appear via tangent bifurcations are very small and exist only for a small interval in parameter space. This allows us to approximate them by a Taylor expansion in both phase and parameter space about the tangent bifurcation point retaining only the leading terms. This was carried out by Karney et al.,[6] where the resulting mapping was reduced to a canonical form

$$Q: \quad y_t - y_{t-1} = 2(x_{t-1}^2 - K) \qquad x_t - x_{t-1} = y_t$$

Here K is proportional to $k - k_{tang}$ (k_{tang} is the parameter value where the tangent bifurcation takes place) and x and y are related to the original phase-space coordinates by a smooth transformation. The quadratic mapping Q represents an approximation of the general mapping close to the point of tangent bifurcation. For $0 < K < 1$, this mapping has stable (elliptic) and unstable (hyperbolic) fixed points at $(x, y) = (\mp\sqrt{K}, 0)$, respectively. The elliptic fixed point is usually surrounded by integrable trajectories (KAM curves) that define a stable region (the island) in which the motion is bounded. An example of island structure is shown in Fig. 1 for $K = 0.1$ (the value of K at which extensive numerical calculations have been carried out).

Referring to the islands shown in Fig. 1, consider a particle which at $t = 0$ is close to, but outside, the islands. Initially, the particle will stay close to the islands; however, as we let $t \to \pm\infty$, we find $(x, y) \to (\infty, \pm\infty)$. It is just such trajectories that we are interested in, because they correspond to particles in the stochastic region of the general mapping approaching the islands, staying there for some time (and contributing to long-time correlations), and then escaping back to the main part of the stochastic region.

What we need is some way of bringing these particles back to the vicinity of the island. We do this by defining an $L \times L$ square around the island. This square spans the region $x_{\min} \leq x < x_{\max}$ and $-\frac{1}{2}L \leq y < \frac{1}{2}L$, where $L = x_{\max} - x_{\min}$. Whenever an orbit leaves this square at (x_t, y_t), we pick a new initial condition

$$(x_0, y_0) = (x_t - mL, y_t - nL)$$

with m and n being integers chosen so that (x_0, y_0) lies inside the square. We also record the length t of the previous orbit segment. This procedure defines the periodic quadratic map Q^*, which can be shown to sample the orbits close to the island in the same way that the general mapping does. Examples of the orbits of Q^* are shown in Fig. 2 for the same parameters as for Fig. 1.

One useful way of looking at Q^* is as a magnification of a small region near a tangent bifurcation in the general mapping. The difference is that

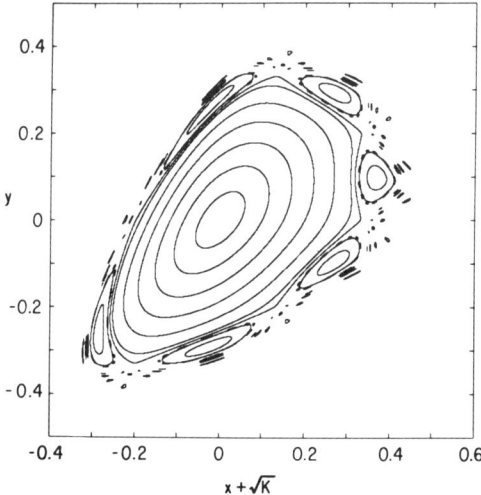

Figure 1. Some islands of the quadratic map Q for $K = 0.1$.

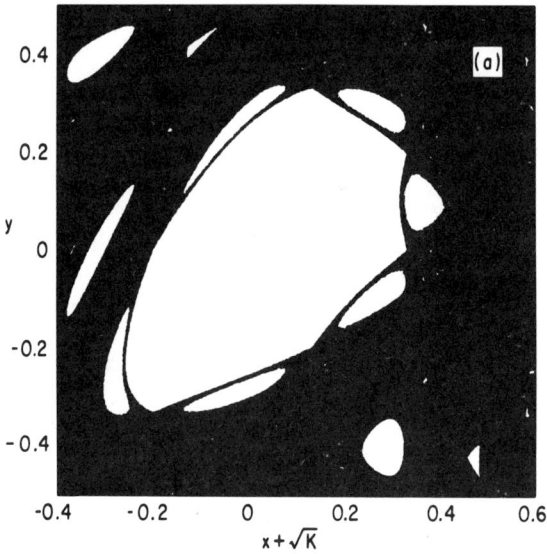

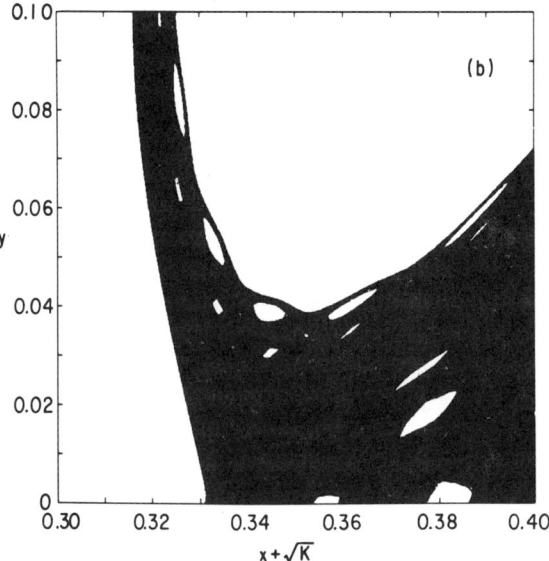

Figure 2. (*a*) Stochastic trajectories for periodic quadratic map Q^* for $K = 0.1$. (*b*) An enlargement of a portion of (*a*). Here $x_{min} + \sqrt{K} = -0.4$, $x_{max} + \sqrt{K} = 0.6$.

once the trajectory leaves the vicinity of the islands, it is immediately reinjected on the other side of the islands. In the general map, the trajectory will spend some long time, which depends on the ratio of the size of the islands to the total accessible portion of phase space, in the stochastic sea before coming back to the vicinity of the islands.

Assuming that the long-time behavior of stochastic orbits is dominated by the region close to the islands, there are two advantages to reducing the problem to a study of Q^*. First, since Q^* describes the behavior of most islands far into the stochastic regime, the properties of many mappings may be treated by looking at a special mapping Q^* that depends only on a single parameter K. The second advantage is that the properties of orbits close to the islands may be studied much more efficiently because there is no need to follow orbits while they spend a long and uninteresting time far from the islands.

TRAPPING STATISTICS

The prescription for numerically determining the stickiness of the island system in Q is to compute a long orbit in the stochastic region of Q^*. The orbit is divided into segments at those points where it leaves the $L \times L$ square. The main results of the calculation are then the *trapping statistics* f_t, which are proportional to the number of orbit segments that have a length of t. Suppose that the total length of the orbit is T and N_t is the number of segments of length t. If T is so large that we can ignore partial segments at the ends of the orbit, then we have $\sum tN_t = T$; the total number of segments is $N = \sum N_t$. The trapping statistics are defined by $f_t = N_t/T$ and are therefore normalized so that $\sum tf_t = 1$. The mean length of the orbits is given by $\alpha = 1/\sum f_t \; (= T/N)$. The probability that a particular segment has length t is $p_t = \alpha f_t \; (= N_t/N)$. If an arbitrary point is chosen in the orbit, then tf_t is the probability that this point belongs to a segment of length t and f_t is the probability that it belongs to the beginning, say, of a segment of length t.

The survival probability

$$P_t = \sum_{\tau=t+1}^{\infty} p_\tau$$

is the probability that an orbit beginning in a segment at $t = 0$ is still trapped in the same segment at time t. Note that $P_0 = 1$ as required. This is the quantity studied by Channon and Lebowitz[1] and Chirikov and Shepelyansky.[4] The correlation function

$$C_\tau = \sum_{t=\tau}^{\infty} (t-\tau) f_t = \sum_{t=\tau}^{\infty} P_t/\alpha$$

is the probability that a particle is trapped in the same segment at two times τ apart. Again, we have $C_0 = 1$.

There is another way of interpreting C_τ: Consider a drunkard who executes a one-dimensional random walk with velocity $v = dr/dt = \pm 1$. The direction of each step is chosen randomly, while the durations of the steps are chosen to be the lengths of consecutive trapped segments of Q^*. Then for integer τ, C_τ is just the usual correlation function for such a process, that is, $\langle v_t v_{t+\tau} \rangle_t$. The behavior of this random-walk process is similar to the behavior of an orbit in the general mapping when two accelerator modes with opposite values of the acceleration are present. (This is the case with the first-order accelerator modes for the standard mapping.)

A diffusion coefficient may be defined by

$$D = \tfrac{1}{2} C_0 + \sum_{\tau=1}^{\infty} C_\tau = \sum \tfrac{1}{2} t^2 f_t.$$

This gives the diffusion rate for the drunkard in the random-walk problem above. It is also related to the diffusion coefficient for the general mapping.

RESULTS FOR $K = 0.1$

We have measured f_t for K between 0 and 1.3 at intervals of 0.05, and at most of the values of K a slow algebraic decay of f_t is seen. A representative case is $K = 0.1$, whose trapping statistics are given in Fig. 3a, which illustrates the slow decay for very long times $t \sim 10^7$. Also given in Fig. 3 are P_t, C_τ, and $\alpha \equiv -d \log C_\tau / d \log \tau$ (thus locally $C_\tau \sim \tau^{-\alpha}$). This last plot shows the power at which C_τ decays, varying between about $\tfrac{1}{4}$ and $\tfrac{3}{2}$.

A glance at Fig. 2 shows the origin of this behavior. The central island is surrounded by a chain of sixth-order islands. Around each of these islands are several other sets of islands. This picture repeats itself at deeper and deeper levels. A particle which manages to penetrate into this maze can get stuck in it for a long time.

For $\tau \lesssim 10^4$, Fig. 3d gives $\alpha \approx \tfrac{1}{4}$. Correspondingly we have $P_t \sim t^{-p}$, where $p = 1 + \alpha \approx \tfrac{5}{4}$. This is close to the asymptotic ($t \to \infty$) result found by Chirikov and Shepelyansky[4] for the whisker map, in which $\langle p \rangle \approx 1.45$. However, in our case, α shows some strong variations beyond $\tau \approx 10^4$ where C_τ "steps down" (e.g., between 10^4 and 3×10^5). This means that the asymptotic form of C_τ is very difficult to determine numerically.

The diffusion coefficient D is given by the summation of C and is approximately 6400. The error in this estimate of D depends on the asymptotic form for C_τ. On the basis of the numerical results, we cannot rule out the possibility that as $\tau \to \infty$, C_τ decays with $\alpha \le 1$. In that case, D would be infinite!

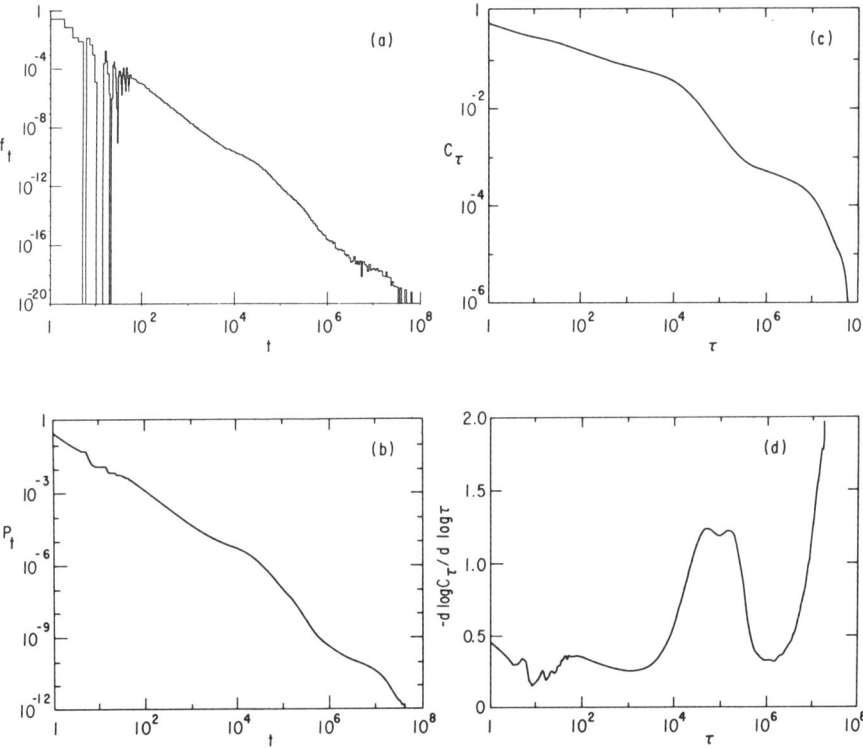

Figure 3. (a) The trapping statistics f_t for $K = 0.1$. (b), (c), and (d) show P_t, C_τ, and $d \log C_\tau / d \log \tau$.

If D is indeed infinite, we would wish to know how a group of particles spreads with time. We again consider the drunkard's walk based on Q^*, which was introduced earlier. The second moment of r is related to the correlation function by

$$S_t \equiv \langle (r_t - r_0)^2 \rangle = tC_0 + 2 \sum_{\tau=1}^{t} (t - \tau) C_\tau$$

This is plotted in Fig. 4a, using the data of Fig. 3. For $t \lesssim 10^4$, S_t grows somewhat faster than $t^{3/2}$ (see Fig. 4b) and even until $t \approx 10^7$, S_t is growing significantly faster than linearly. Beyond 10^7, the numerical data show a convergence to a linear rate; but this is merely because no segments longer than about 6×10^7 were observed. For $t \to \infty$, S_t grows as $t^{2-\alpha}$, assuming that the exponent α at which C_τ decays asymptotically is less than 1. If the diffusion coefficient is estimated from $D_t = \frac{1}{2} S_t / t$, then D_t grows with t as shown in Fig. 4c.

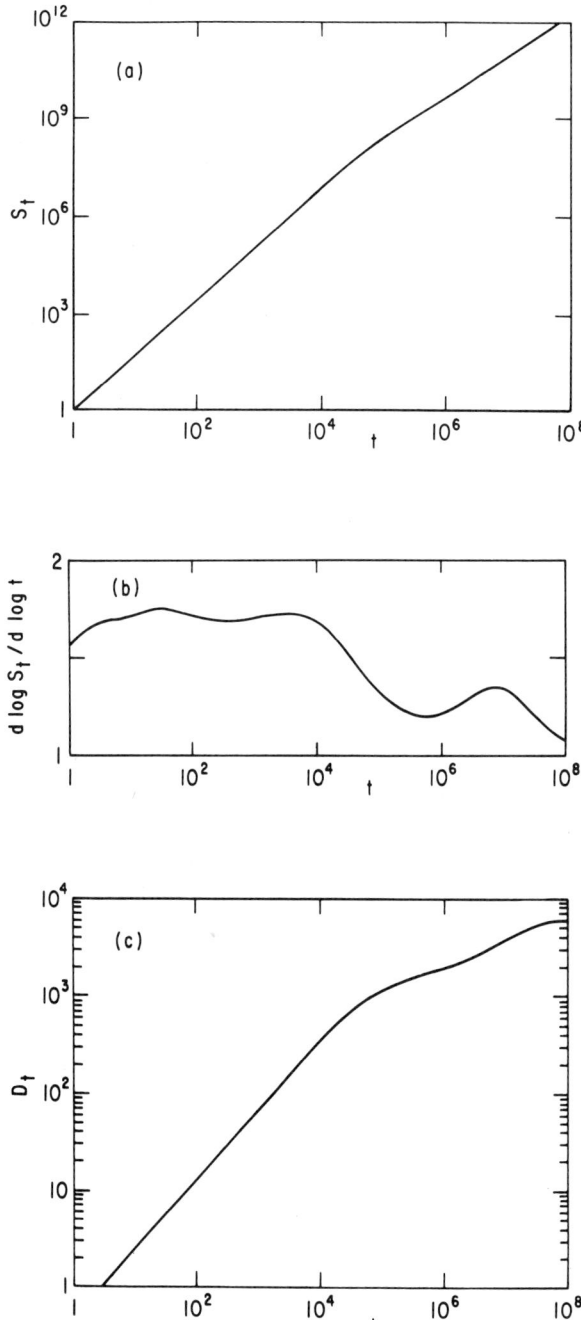

Figure 4. (a) The variance S_t for the case given in Fig. 3. (b) and (c) show $d \log S_t / d \log t$ and $D_t = \frac{1}{2} S_t / t$.

DISCUSSION

We can apply these results to the determination of the correlation function of a general mapping. Suppose the correlation function is defined by

$$\mathscr{C}(\tau) = \langle h(\mathbf{x}(t))h(\mathbf{x}(t+\tau))\rangle_t$$

where h is some smooth function of the position in phase space $\mathbf{x}$. Then the contribution of an island located at $\mathbf{x}_0$ to $\mathscr{C}(\tau)$ is[5]

$$\mathscr{C}_{is}(\tau) = h^2(\mathbf{x}_0)(B/A)C_\tau$$

where A is the total area of the stochastic component of the general mapping and B is the portion of that area which is near the island. (More precisely, when we regard the $L \times L$ square of Q^* as being a magnification of a small area of the general mapping, then B measures the area of the stochastic component within this small area.) Similar relations connect D and S_t and the corresponding quantities for the general mapping.

In the case of an accelerator mode in the standard mapping, K is related to the parameter k by $k^2 = (2\pi n)^2 + 16K$, where n is an integer. For $K = 0.1$, we take 6400 as a lower bound for D. We find that the contribution to the diffusion coefficient is increased over its quasilinear value by a factor of at least $360/n^2$. Thus for $n = 1$ or $k \approx 6.41$, the islands completely dominate the diffusion. The first-order accelerator modes continue to have such a large effect at least until $k \approx 100$. If D is in fact infinite, even arbitrarily small accelerator modes will eventually dominate the motion, and Fig. 4 can be used to estimate the time at which the accelerator modes become important.

In summary, small islands within the stochastic sea lead to correlations in the stochastic orbits for extremely long times. When the islands are accelerator modes, this may cause the particles to behave nondiffusively, that is, the mean squared displacement of the particles may increase faster than linearly with time.

In order to provide a definitive answer to this question, the asymptotic behavior of C_τ must be determined. Because the asymptotic regime starts at such a large τ (greater than 10^7), it appears its properties cannot be studied by the numerical method used in this paper. What is needed is a better analytical understanding of the behavior of trajectories close to the border between integrability and stochasticity. Chirikov[3] has made some useful steps in this direction.

ACKNOWLEDGMENTS

This work was supported by the U.S. Department of Energy under Contract DE-AC02-76-CHO3076. I first became interested in this problem at the previous workshop on this subject held in Kyoto in November 1981. I began working on it while at the Institute of Plasma Physics, Nagoya University, participating in the U.S.–Japan Fusion Cooperation Program. Some of the work was also carried out while at the Aspen Center for Physics. I would like to thank J. M. Greene, R. S. MacKay, and F. Vivaldi for stimulating discussions. B. V. Chirikov also provided some enlightening comments.

REFERENCES

1. S. R. Channon and J. L. Lebowitz, "Numerical Experiments in Stochasticity and Homoclinic Oscillation," in *Nonlinear Dynamics, Annals of the New York Academy of Sciences* **357**, 108–118 (1980).
2. B. V. Chirikov, "A Universal Instability of Many-Dimensional Oscillator Systems," *Phys. Rept.* **52**, 263–379 (1979).
3. B. V. Chirikov, "Chaotic Dynamics in Hamiltonian Systems with Divided Phase Space," in *7th Sitges Conference on Dynamical Systems and Chaos*, Barcelona (Institute of Nuclear Physics Preprint INP 82-132, Novosibirsk, 1982).
4. B. V. Chirikov and D. L. Shepelyansky, "Statistics of Poincaré Recurrences and the Structure of the Stochastic Layer of a Nonlinear Resonance" (in Russian), in *9th International Conference on Nonlinear Oscillations*, Kiev (Institute of Nuclear Physics Preprint INP 81-69, Novosibirsk; English translation: Princeton Plasma Physics Laboratory Report PPPL-TRANS-133, 1983).
5. C. F. F. Karney, "Long-Time Correlations in the Stochastic Regime," *Physica* **8D**, 360 (1983).
6. C. F. F. Karney, A. B. Rechester, and R. B. White, "Effect of Noise on the Standard Mapping," *Physica* **4D**, 425–438 (1982).

MAGNETIC FIELD LINES, HAMILTONIAN MECHANICS, AND SYMMETRIES AND INVARIANTS

Robert G. Littlejohn

Department of Physics, University of California, Los Angeles, California

It is clear from the other chapters in this volume that the theory of maps is still generating a great deal of interest in the plasma-physics community. It seems that many people got interested in map theory in the first place by tracing magnetic field lines around tokamaks and stellarators, so perhaps it is appropriate here to note more about the connection between magnetic field lines and Hamiltonian mechanics.

I will not attempt to give an adequate historical accounting of all the threads of ideas that have arisen over the years concerning magnetic-field-line flow, but I will try to give a kind of unified and general treatment of field-line flow as a Hamiltonian system, which should pull all these threads together.

It turns out that the way to do this is to modify somewhat both the usual description of magnetic-field-line flow and that of Hamiltonian mechanics. In the case of Hamiltonian mechanics, the modification consists of using noncanonical coordinates in phase space.

From a mathematical standpoint, this is a rather simple thing to do. Mathematicians tend to eschew coordinates anyway. But it is a step with a number of useful consequences for plasma physics. In addition to magnetic-field-line flow, the subject of this chapter, there are also guiding-center theory[6–8] and the subject of gyrokinetic equations,[4] which are clarified by the use of noncanonical coordinates.

Hamiltonian mechanics, whether canonical or noncanonical, provides a number of useful features that are not present in non-Hamiltonian systems. For example, there are variational principles, which can be used for a variety of purposes (see Ref. 5 and the chapter by Schmidt in this volume); there are covariance properties or transformation laws that can be used to great advantage; there is Liouville's theorem, which imposes powerful constraints on the asymptotic behavior of orbits; there is Noether's theorem, which relates symmetries to invariants; there is KAM theory; and there are systematic perturbation theories.[1] Many of these features have never adequately been explored for magnetic-field-line flow.

We must also modify the usual description of magnetic-field-line flow, in this case by employing a variational principle. Let $\mathbf{B} = \nabla \times \mathbf{A}$ be a magnetic field, and let $\mathbf{x} = \mathbf{x}(\lambda)$ be a field line, where λ is an arbitrary parameter describing position along the field line. Then it turns out that the field-line equations are contained implicitly in the variational principle,

$$\delta \int \mathbf{A}(\mathbf{x}) \cdot \frac{d\mathbf{x}}{d\lambda} d\lambda = 0 \tag{1}$$

in which it is understood that the variations satisfy $\mathbf{A} \cdot \delta \mathbf{x} = 0$ at the endpoints. The Euler–Lagrange equations stemming from Eq. (1) are simply

$$(\nabla \times \mathbf{A}) \times \frac{d\mathbf{x}}{d\lambda} = 0 \tag{2}$$

which shows that $\mathbf{x} = \mathbf{x}(\lambda)$ is indeed a field line.

It is interesting to note that the integral of Eq. (1) would be the magnetic flux if the path $\mathbf{x} = \mathbf{x}(\lambda)$ were closed. We are not restricted to closed paths, but for the sake of giving the integral a name, we may call it the flux functional $\Phi = \Phi[\mathbf{x}(\lambda)]$, for any kind of path. Then the field line is characterized by $\delta \Phi = 0$. To put this into words, we may say that the physical field line is characterized by the fact that first-order variations in the path cause only second-order variations in the flux.

One of the main practical advantages of Lagrangian or variational principles is the fact that they allow arbitrary changes of coordinates (i.e., they are generally covariant). This comes about because the Lagrangian is a scalar. In the case of Lagrangians in mechanics, we have $L = L(\mathbf{x}, d\mathbf{x}/dt)$; here we have $L = L(\mathbf{x}, d\mathbf{x}/d\lambda) = \mathbf{A}(\mathbf{x}) \cdot d\mathbf{x}/d\lambda$, and it is instructive to apply the Lagrangian covariance property to field lines.

So let us introduce an arbitrary curvilinear coordinate system in space, $x^\mu = x^\mu(\mathbf{x})$, $\mu = 1, 2, 3$. For example, x^μ could represent cylindrical, toroidal, helical, flux, Hamada, or field-line coordinates. Then the variational principle becomes

$$\delta \int A_\mu \frac{dx^\mu}{d\lambda} d\lambda = 0 \tag{3}$$

in which A_μ are the covariant components of the vector potential.

So far I have not said what the parameter λ is. It can actually be anything, but it is often convenient to let λ be one of the three coordinates x^μ. For example, if field lines generally wind around the z axis, as in a stellarator, then one might want to use cylindrical coordinates $x^\mu = (x^1, x^2, x^3) = (r, z, \phi)$, and take $\lambda = x^3 = \phi$. Then one would have $\mathbf{x} = \mathbf{x}(\phi)$ or $r = r(\phi)$, $z = z(\phi)$ for the equation of the field line. This usage does not assume azimuthal symmetry.

With $\lambda = x^3$, the variational principle can be written

$$\delta \int \left(A_1 \frac{dx^1}{dx^3} + A_2 \frac{dx^2}{dx^3} + A_3 \right) dx^3 = 0 \tag{4}$$

The connection between field lines and mechanics becomes clear when we write down the analogous variational principle for mechanics:

$$\delta \int \left(p \frac{dq}{dt} + 0 \frac{dp}{dt} - H \right) dt = 0 \tag{5}$$

Hamiltonian's equations of motion are implicitly contained in this variational principle.

Equations (4) and (5) are formally identical except for the fact that the middle term of Eq. (5) vanishes, as I have indicated by the coefficient of zero. This is an indication of the fact that the variables (q, p) are canonical, whereas (x^1, x^2) are not necessarily canonical. But already a number of analogies are clear. For example, the parameter $\lambda = x^3$ is analogous to the time; the coordinates (x^1, x^2) are coordinates (albeit noncanonical) on a phase space of one degree of freedom; and $-A_3$ is the Hamiltonian for the field-line flow.

To bring out the association between field-line flow and Hamiltonian mechanics more clearly, and to show that the two variational principles are indeed formally identical, it is convenient to introduce some new notation. First, we define a covariant vector γ on phase space, with components $\gamma_1 = p$, $\gamma_2 = 0$, and we write (z^1, z^2) for the two phase-space coordinates, with $z^1 = q$, $z^2 = p$. Then Eq. (5) becomes

$$\delta \int \left(\gamma_1 \frac{dz^1}{dt} + \gamma_2 \frac{dz^2}{dt} - H \right) dt = 0 \tag{6}$$

Of course, this is no different than Eq. (5), because we still have $\gamma_2 = 0$. But we may now perform an arbitrary change of coordinates, taking us from (z^1, z^2) to new coordinates $(\bar{z}^1, \bar{z}^2)$, and the covariance properties of Lagrangian variational principles guarantee that the new variational principle in the new coordinates will be valid. The new coordinates $(\bar{z}^1, \bar{z}^2)$ need not be canonical, and the transformation may be time dependent. The result is

$$\delta \int \left(\bar{\gamma}_1 \frac{d\bar{z}^1}{dt} + \bar{\gamma}_2 \frac{d\bar{z}^2}{dt} - \bar{H} \right) dt = 0 \tag{7}$$

where

$$\bar{\gamma}_i = \sum_j \frac{\partial z^j}{\partial \bar{z}^i} \gamma_j$$

$$\bar{H} = H - \sum_j \gamma_j \frac{\partial z^j}{\partial t}, \quad i, j = 1, 2 \tag{8}$$

The point is that the variational principle of Eqs. (6) or (7) is form invariant under arbitrary changes of coordinates, and that, in general, both γ_1 and γ_2 will be nonzero. The vanishing of γ_2 is a sign of a canonical coordinate system, and we can see from Eq. (4) that the usual coordinate systems of physical interest for field-line flow (e.g., rectangular, cylindrical) are actually noncanonical.

Henceforth we will recognize this form invariance and drop the overbars, so that (z^1, z^2) represent an arbitrary coordinate system on phase space. A further change of notation is even more suggestive. We extend the covariant vector γ to a third component by writing $\gamma_3 = -H$, and we also put $z^3 = t$. Then the generally covariant variational principle for mechanics is

$$\delta \int \gamma_\mu \frac{dz^\mu}{dt} dt = 0 \tag{9}$$

This should be compared to Eq. (3), to which it bears a perfect analogy.

In a Hamiltonian system of N degrees of freedom, the same formulas apply, but we must promote q and p into N vectors, and γ_μ and the coordinates z^μ into $(2N+1)$ vectors, with $z^{2N+1} = t$ and $\gamma_{2N+1} = -H$. We see that field-line flow represents a Hamiltonian system of one (or one and a half) degrees of freedom.

Now that we have established the formal identity of field-line flow and Hamiltonian mechanics, we may ask what can be done with it. One interesting and useful thing, as first pointed out by John Cary, is to apply Noether's theorem. The familiar forms of Noether's theorem apply to configuration-space Lagrangians and phase-space Hamiltonians in ordinary mechanics, and they state that if the Lagrangian or Hamiltonian has an ignorable coordinate q_i, then the corresponding momentum $p_i = \partial L/\partial \dot{q}_i$ is a constant of the motion. An ignorable coordinate is equivalent to a symmetry, because translations in the ignorable coordinate, $q_i \to q_i + \alpha$, leave the system invariant.

Similarly, a version of Noether's theorem is applicable to noncanonical Hamiltonian mechanics in the language of the γ vector, and it runs like this: If all of the quantities $\gamma_\mu = \gamma_\mu(z)$ are independent of one of the z's, say z^α, that is, if $\partial \gamma_\mu/\partial z^\alpha = 0$ for all μ, then γ_α is a constant of the motion, that is,

$$\frac{d\gamma_\alpha}{dt} = 0 \tag{10}$$

All of the conservation laws of mechanics are special cases of this theorem. For example, conservation of energy, which applies for a time-independent system, comes about by taking $\alpha = 2N+1$, $z^\alpha = t$, so that $dH/dt = 0$.

Noether's theorem is easily applied to field-line flow. Tokamaks are supposed to be azimuthally symmetric, so let us consider this case. In the cylindrical coordinates $x^\mu = (r, z, \phi)$, an azimuthally symmetric magnetic field can be represented by the vector potential,

$$\mathbf{A} \cdot d\mathbf{x} = A_r(r, z) dr + A_z(r, z) dz + A_\phi(r, z) d\phi$$

The covariant components A_μ are indicated. Since we have $\partial A_\mu / \partial \phi = 0$, Noether's theorem says

$$\frac{dA_\phi}{d\phi} = 0 \tag{11}$$

that is, A_ϕ is constant along field lines.

In more ordinary language, A_ϕ is the flux function ψ, and we see that its conservation in azimuthally symmetric systems is exactly like the conservation of energy in time-independent mechanical systems. In fact, A_ϕ or ψ (apart from sign) is the Hamiltonian for field-line flow. (This is true even when the system is not azimuthally symmetric, but then A_ϕ is not conserved.)

When A_ϕ is conserved, the surfaces $A_\phi = $ constant are "good flux surfaces." Most generally, good flux surfaces exist when the covariant components A_μ of the vector potential have an ignorable coordinate in some coordinate system x^μ. (Contrary to the impression one would get by reading much of the plasma-physics literature, this is an exceptional situation!)

We note that, in general, nothing can be said about the contravariant components A^μ, or the various components of B, even if good flux surfaces do exist, because an ignorable coordinate of A_μ may not coincide with an ignorable coordinate of the metric tensor $g_{\mu\nu}$, or the Jacobian $[\det(g_{\mu\nu})]^{1/2}$. The reason for this is that a symmetry of the metric tensor is an "isometry," that is, a rigid motion of space that preserves lengths and angles. These must be some combination of translations and rotations, which form a rather restricted class of symmetry operations. The possible symmetries of A_μ that give rise to ignorable coordinates and good flux surfaces form a much larger class than the isometries. These considerations are relevant to the existence of magnetohydrodynamic (MHD) equilibria in various configurations, because the MHD equations involve the metric tensor $g_{\mu\nu}$ quite explicitly. Thus, one should not expect the mere existence of good flux surfaces to mean that the MHD equations are going to be particularly simple.

However, to make the matter more confusing, several important symmetries in plasma physics actually are isometries, so that the existence of good flux surfaces does in fact coincide with simplified MHD equations. These are rotations about the z axis, applicable to azimuthally symmetric systems such as tokamaks, and screw motions about the z axis, applicable to helically symmetric systems such as straight stellarators. It is for this reason that nice Grad–Shafranov equations can be found for these systems.

For other types of symmetries, such as symmetries involving a simultaneous toroidal and poloidal rotation, the existence of good flux surfaces may not be easy to relate to other conditions one would like the magnetic field to satisfy. For example, even so simple a condition as $\nabla \times \mathbf{B} = 0$ may give considerable trouble. Since this is the subject of Cary's chapter in this volume and paper,[2] I will move on to other matters.

There is a considerable lore that has been built up over the years in the tokamak business concerning various kinds of flux coordinates, and it is interesting to bring this into a Hamiltonian perspective. In particular, for azimuthally symmetric systems people like to use flux coordinates (χ, θ, ϕ), where χ is the toroidal flux, θ is a particular poloidal angle, and ϕ is the toroidal angle. Hamada coordinates are very similar. In these coordinates, the vector potential can be written

$$\mathbf{A} = \chi \nabla \theta - \psi \nabla \phi \qquad (12)$$

where ψ is the poloidal flux. The Hamiltonian significance of this can be seen by making the comparison

$$\mathbf{A} \cdot d\mathbf{x} = \chi\, d\theta - \psi\, d\phi$$
$$\gamma_\mu dz^\mu = p\, dq - H\, dt \qquad (13)$$

From this we easily see that the coordinates (χ, θ) are actually canonical coordinates on the poloidal plane, which acts as the phase space for field-line flow. Furthermore, since the Hamiltonian ψ is a function only of the generalized momentum χ, we see that (χ, θ) are not just canonical, they are actually action/angle variables for the field-line system.

It is well known from mechanics that action/angle variables properly exist only when a system is integrable. And the tokamak field-line system is integrable because of azimuthal symmetry. Therefore, if we break the symmetry, we can expect to lose the relation $\psi = \psi(\chi)$. However, since canonical variables always exist,[6] we can expect to be able to retain relation (12), for some choice of variables (χ, θ), even without good flux surfaces. And indeed, a more-careful analysis bears out these expectations, and gives the Hamiltonian function $\psi = \psi(\chi, \theta)$.

Before leaving this subject, let me point out that canonical variables allow one to use Hamilton's equations in the usual sense. For the field-line system, this gives

$$\frac{d\theta}{d\phi} = \frac{\partial \psi}{\partial \chi}, \qquad \frac{d\chi}{d\phi} = -\frac{\partial \psi}{\partial \theta} \qquad (14)$$

which should be compared with

$$\frac{dq}{dt} = \frac{\partial H}{\partial p}, \quad \frac{dp}{dt} = -\frac{\partial H}{\partial p} \tag{15}$$

If the system is azimuthally symmetric, then $\partial \psi / \partial \theta = 0$, so that χ is constant along field lines, and $\partial \psi / \partial \chi = \iota(\psi)$, where "$\iota$" is the rotational transform. (This is analogous to the frequency $\omega = \partial H / \partial J$ in mechanics.)

Finally, it is possible to use the noncanonical γ mechanics to do perturbation theory. This is perhaps the most substantial and nontrivial result of the present theory. Suppose, for example, that one has

$$\mathbf{A} = \mathbf{A}_0 + \varepsilon \mathbf{A}_1 \tag{16}$$

where $\mathbf{A}_0$ represents an integrable magnetic field and $\mathbf{A}_1$ is small. Then one can carry out systematic perturbation calculations on this magnetic field, based on a perturbation theory developed by Cary and Littlejohn.[3] Traditionally, perturbation calculations of this sort have been effected by means of *ad hoc* methods, valid only to lowest order. The systematic theory is based on Lie transforms and may be carried to any order. It also has applications other than to magnetic-field-line flow. The theory gives what one would expect on the basis of mechanics, that is, adiabatic invariants, island structure, etc.

Throughout this chapter I have conveniently ignored the issue of gauge transformations, but these are quite essential to the transformation theory of the γ vector. For example, the usual generating function S of a canonical transformation in mechanics is actually a guage scalar. For more details I refer you to the paper by Cary and Littlejohn.[3]

In conclusion, let met say that noncanonical Hamiltonian mechanics is interesting and useful in its own right, and that magnetic-field-line flow is a simple but very illustrative example.

ACKNOWLEDGMENT

This work was supported by the U.S. Department of Energy under contract DOE-DE-AM03-SF7600010 PA26, VIA.

REFERENCES

1. John R. Cary, *Phys. Rep.* **79**, 131 (1981).
2. John R. Cary, *Phys. Rev. Lett.* **49**, 276 (1982).
3. J. R. Cary and R. G. Littlejohn, *Ann. Phys.* **151**, 1 (1983).
4. Daniel H. E. Dubin, John A. Krommes, C. Oberman, and W. W. Lee, *Phys. Fluids* **26**, 3524 (1983).

5. Robert H. G. Helleman and Tassos Bountis, "Periodic Solutions of Arbitrary Period, Variational Methods," in *Stochastic Behavior in Classical and Quantum Hamiltonian Systems*, Lecture Notes in Physics No. 93, G. Casati and J. Ford, eds. (Springer-Verlag, New York, 1979).
6. R. G. Littlejohn, *J. Math. Phys.* **20**, 2445 (1979).
7. R. G. Littlejohn, *Phys. Fluids* **24**, 1730 (1981).
8. R. G. Littlejohn, "Variational Principles of Guiding Center Motion," in *J. Plasma Phys.* **29**, 111 (1983).

SEARCHING FOR INTEGRABLE SYSTEMS

John R. Cary
Institute for Fusion Studies, University of Texas, Austin, Texas

Lack of integrability leads to undesirable consequences in a number of physical systems. The lack of integrability of the magnetic field leads to enhanced particle transport in stellarators[1] and tokamaks[2] with tearing-mode turbulence. Limitations of the luminosity of colliding beams may be due to the onset of stochasticity.[3] Enhanced radial transport in mirror machines caused by the lack of integrability and/or the presence of resonances may be a significant problem in future devices.[4]

To improve such systems one needs a systematic method for finding integrable systems. Of course, it is easy to find integrable systems if no restrictions are imposed; textbooks are full of such examples. The problem is to find integrable systems given a set of constraints.

An example of this type of problem is that of finding integrable vacuum magnetic fields with rotational transform. The solution to this problem is relevant to the magnetic-confinement program.[5-7]

In vacuum Ampere's law dictates that magnetic fields be curl free in addition to being divergence free. Thus,

$$\mathbf{B} = \nabla \Phi \tag{1}$$

where

$$\nabla^2 \Phi = 0. \tag{2}$$

In the coordinate system (ξ, η, ϕ), where

$$R = \frac{R_0(1-\xi^2)^{1/2}}{1-\xi \cos \eta} \tag{3a}$$

$$Z = \frac{-R_0 \xi \sin \eta}{1-\xi \cos \eta} \tag{3b}$$

and (R, ϕ, Z) are the usual cylindrical coordinates, the general solution to Eq. (2) allowing for a current on the z axis is

$$\Phi = I\phi + (\xi^{-1} - \cos \eta)^{1/2} \sum_{l,m} \alpha_{lm} Q^m_{l-1/2}(\xi^{-1}) e^{il\eta + im\phi} \qquad (4)$$

in which the coefficients I and α_{lm} are arbitrary, within the reality restriction $\alpha_{lm} = \alpha^*_{-l,m}$, and $Q^\mu_\nu(x)$ is the modified Legendre function of the second kind.

An integrable magnetic field is one for which the magnetic fields lie in nested toroidal "flux surfaces." The condition of nonzero rotational transform is dictated by physical requirements imposed by particle-orbit theory. This condition also eliminates the trivial solutions, in which all field lines are closed, such as the field due to a single current on the z axis.

A method for obtaining integrable vacuum magnetic fields has recently been published.[8,9] This method is iterative. The first iteration has been shown to significantly decrease stochasticity. A complete discussion of this method has been presented in Refs. 8 and 9. In this chapter a discussion of one of the ideas is presented.

A model stochastic system is that of the flow caused by the Hamiltonian

$$h = h_0(p) + \varepsilon[f_1(p) \cos(2\pi q) + f_2(p) \cos(2\pi q - 2\pi t)] \qquad (5)$$

in which h_0, f_1, and f_2 are particular functions to be specified shortly, and ε is a constant. For the choices

$$h_0 = \tfrac{1}{2} p^2 \qquad (6a)$$

$$f_1 = 1 \qquad (6b)$$

and

$$f_2 = 1 \qquad (6c)$$

this system has been studied in detail.[10,11] In general, this system is not

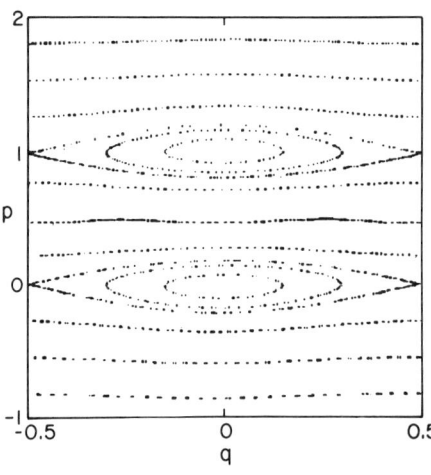

Figure 1. Surface of section for the Hamiltonian of Eqs. (5) and (6) with $\varepsilon = 0.01$.

integrable because two resonances are present. This system would be integrable if either f_1 or f_2 vanished.

The two resonances are illustrated in the surface of section of Fig. 1, which is for the case of Eqs. (5) and (6) with $\varepsilon = 0.01$. The calculation of the widths is standard.[12] The half-width for resonance i (=1 or 2, corresponding to f_i) is

$$\Delta p_i = 2\sqrt{\varepsilon f_i(v_i)} \qquad (7a)$$
$$= 2\sqrt{\varepsilon} \qquad (7b)$$

where

$$v_i = \begin{cases} 0, & i=1 \\ 1, & i=2 \end{cases} \qquad (8)$$

is the velocity of resonance i.

As ε is increased, the resonances overlap. The result is stochastic motion. This is illustrated in Fig. 2, a surface of section for the case $\varepsilon = 0.05$. One notes that the stochastic orbit covers a range in the momentum variable of $\Delta p = 1.9$.

Suppose, instead, one is given the Hamiltonian

$$H = h_0(p) + \varepsilon[f_1(p)\cos(2\pi q) + f_2(p)\cos(2\pi q - 2\pi t)$$
$$+ \alpha_1 g_1(p)\cos(2\pi q) + \alpha_2 g_2(p)\cos(2\pi q - 2\pi t)] \qquad (9)$$

with g_1 and f_1, and g_2 and f_2 being linearly independent functions. Suppose, furthermore, that ε is fixed as before, but one is allowed to choose the values of α_1 and α_2 with the goal being to have a system as integrable as possible. How are α_1 and α_2 to be chosen? Obviously, no choice of α_1 and α_2 will completely eliminate stochasticity, because there will always be two resonances.

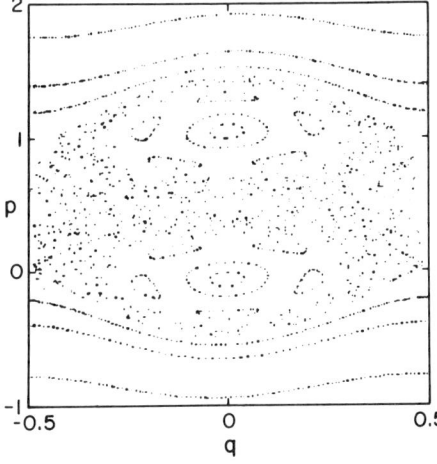

Figure 2. Surface of section for the Hamiltonian of Eqs. (5) and (6) with $\varepsilon = 0.05$.

Since stochasticity increases with island size, it is natural to use α_1 and α_2 to diminish the island size. According to Eq. (7), this prescription gives

$$\alpha_i = -\varepsilon f_i(v_i)/g_i(v_i) \tag{10}$$

Of course, in this case the formula (7a) is no longer accurate, since its derivation relied on the resonance amplitude being a slowly varying function of p.

To determine the island width for the case when Eq. (10) is imposed, we consider the island structure for the Hamiltonian,

$$H = \tfrac{1}{2}p^2 + \varepsilon p \cos(2\pi q) \tag{11}$$

which contains just one of the resonances. Being time independent, the flow due to the Hamiltonian (11) conserves the Hamiltonian. The level curves are shown in Fig. 3. One notes that there are two islands for this resonance. The separatrix is now defined by the two curves, $p=0$ and $p=-2\varepsilon \cos q$. Most important is the fact that the island half-width is now

$$\Delta p = 2\varepsilon \tag{12}$$

This implies that given a small value of ε, the island is much smaller than before [Eq. (7b)]. Figure 4 is a surface of section for the Hamiltonian

$$H = \tfrac{1}{2}p^2 + \varepsilon\{p\cos(2\pi q) + (p-1)\cos[2\pi(q-t)]\} \tag{13}$$

with $\varepsilon = 0.05$. This is a model Hamiltonian of the type (5), but for which the resonance amplitudes $f_i(p)$ vanish at the resonance, yet their derivatives are $\mathcal{O}(\varepsilon)$. As one can see, in comparison with Fig. 2, the islands are much smaller and the stochastic region is greatly diminished.

Another way of looking at the problem is to consider perturbation theory for a Hamiltonian of the form,

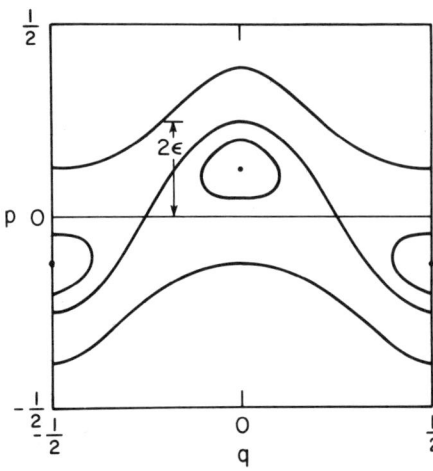

Figure 3. Structure of the resonance of the Hamiltonian of Eq. (11).

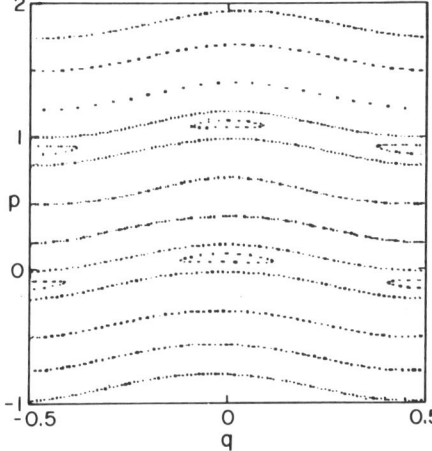

Figure 4. Surface of section for the Hamiltonian of Eq. (13) for $\varepsilon = 0.05$.

$$h = h_0(p) + \varepsilon \sum_{mn} f_{mn}(p) e^{i2\pi(mq-nt)} \qquad (14)$$

which includes all systems which are periodic in q and t. Suppose one now applies perturbation theory to attempt to determine a quantity P which is conserved to $\mathcal{O}(\varepsilon)$. Standard analysis[12] shows that the generating function

$$F(q, P, t) = qP + \varepsilon \sum_{mn} F_{mn}(P) e^{i2\pi(mq-nt)} \qquad (15)$$

must satisfy

$$F_{mn}(P) = \frac{i}{2\pi} f_{mn}(P) \Big/ \left(m \frac{\partial h_0}{\partial p}(P) - n\right) \qquad (16)$$

Thus, the analysis is invalid unless each resonance amplitude vanishes at its resonant surface:

$$f_{mn}(P_{mn}) = 0$$

where $\qquad (17)$

$$\frac{\partial h_0}{\partial p}(P_{mn}) = \frac{n}{m}$$

The perturbation theory indicates how one might proceed to higher order. One has to be able to choose the Hamiltonian to make the resonance amplitude vanish at the resonant surface to each order. Thus, higher-order calculations may slightly modify the values of coefficients like α_i as given by lower-order theory. Of course, like any perturbation technique this method likely relies on the smallness of the stochasticity.

Application of these ideas to the problem of improving the integrability of vacuum magnetic fields has been successful.[8,9] In that case it seems, intuitively, that sufficient freedom of choice is available. However, a mathematical proof has not been given. This is the subject of further research.

ACKNOWLEDGEMENTS

The author would like to thank John Greene, Robert Littlejohn, and Jim Hanson for useful discussions. This work was supported by the U.S. Department of Energy under grant DE-FG05-80ET-53088.

REFERENCES

1. K. Miyamoto, *Nuclear Fusion* **18**, 243 (1978).
2. R. B. White, in *Handbook of Plasma Physics*, M. N. Rosenbluth and R. Z. Sagdeev, eds. (North-Holland, Amsterdam, 1983), Vol. I, Chap. 3.5.
3. J. L. Tennyson, in *Physics of High Energy Particle Accelerators*, R. A. Carrigan, F. R. Huson, and M. Month, Eds. (American Institute of Physics, New York, 1982) p. 345.
4. D. D. Ryutov and G. V. Stupakov, *Pis'ma Zh. Eksp. Teor. Fiz.* **26**, 186 (1979) [*JETP Lett.* **26**, 174 (1977)].
5. L. Spitzer, *Phys. Fluids* **1**, 253 (1958).
6. M. D. Kruskal and R. M. Kulsrud, *Phys. Fluids* **1**, 265 (1958).
7. H. Grad, *Phys. Fluids* **10**, 137 (1967).
8. J. R. Cary, *Phys. Rev. Lett.* **49**, 276 (1982).
9. J. R. Cary, *Phys. Fluids* **27**, 119 (1984).
10. G. M. Zaslavskii and N. N. Filonenko, *Zh. Eksp. Teor. Fiz.* **54**, 1590 (1968) [*Sov. Phys. JETP* **27**, 851 (1968)].
11. P. K. Kaw and W. L. Kruer, *Phys. Fluids* **14**, 190 (1971).
12. B. V. Chirikov, *Phys. Reps.* **52**, 263 (1979).

HAMILTON'S PRINCIPLE AND THE SPLITTING OF PERIODIC ORBITS

George Schmidt
Stevens Institute of Technology, Hoboken, New Jersey

Abstract. *It is shown that bifurcation of periodic orbits, and orbit splitting in general, are closely related to the development of new saddle-point directions of the Lagrangian integral I_1. Critical parameter values for orbit splitting can be found by looking at zeros of the second variation of I_1 over a suitably chosen function space of periodic functions. I_1 is a minimum for unstable periodic orbits and KAM trajectories. The rapid convergence of this method for finding bifurcations is illustrated with the example of a particle in a standing wave field.*

INTRODUCTION

As is well known, Hamilton's principle states that a mechanical system characterized by the Lagrangian $L(q_i, \dot{q}_i, t)$ evolves in such a way that the integral

$$I = \int_{q_{i0}(t_0)}^{q_{if}(t_f)} L \, dt \tag{1}$$

is stationary along the physical trajectory, $\delta I = 0$. In performing the variation the two points are fixed and so is the time for which these points are reached.

It was already known by Jacobi that if q_{i0} and q_{if} are sufficiently close, I is actually a minimum at the actual trajectory, but it becomes a saddle point once they are further removed.[5,12] It is worth repeating here Jacobi's illuminating example.

A mass point moves with constant speed on the surface of a sphere. Then I is proportional to the distance traveled, $\int ds$. For the particle starting at the South Pole the actual trajectory is a meridian. Clearly, for trajectories that end before reaching the North Pole the meridian is actually the *shortest* path, but beyond the North Pole it is a saddle point; both shorter and longer adjacent trajectories can be constructed. The two poles are *conjugate* points in the sense that there is more than one actual adjacent trajectory (meridians) connecting them. It can be proven in general that I is a minimum if the end point is closer than the first point conjugate to the starting point.

We are mainly interested in Hamiltonian chaos and therefore in periodic orbits. Hamilton's principle can easily be generalized to this case. In evaluating $\delta I = 0$ one obtains a term

$$R = \left[\frac{\partial L}{\partial \dot{q}_i} \delta q_i\right]_{q_{i0}}^{q_{if}}$$

If this term is zero, the Euler–Lagrange equations for the actual trajectory follow. If the end points are fixed, R obviously vanishes, but it also vanishes if the orbit as well as the perturbed orbits are periodic and the integral in Eq. (1) is carried out over the period.

We will study integrals

$$I_1 = \oint_{mP} L(q_i, \dot{q}_i, t, K)\, dt$$

where P is the period of the actual orbit, m is an integer, and K is a parameter which describes the nonintegrability of the system. When $K = 0$, the Hamiltonian is integrable.

Hamilton's principle for nonintegrable systems was first discussed by Helleman[5] and Helleman and Bountis.[6] They noted that I is in general a saddle point, and devised a method of calculating orbits. Percival[8] conjectured that KAM tori can be calculated by *minimizing* the Lagrangian integral. Some of his conjectures have recently been rigorously proven by Mather.[7]

This chapter is concerned with the question of orbit splitting, the periodic analog of conjugate points described earlier. In a nonintegrable problem a periodic orbit is surrounded by orbits of longer periods, which are "emitted" from the "mother" orbit as K is increased.[1,9] The most important is the bifurcation into a stable orbit of double period leading to destabilization of the original orbit. When an orbit of period P gives birth to an orbit of period mP, there are two actual closeby orbits, and for each of them $\delta I_1 = 0$. This implies that at this value of K, $\delta^2 I_1 = 0$ in some direction in function space for the set of functions of period mP. It is reasonable to conjecture that when $\delta^2 I_1$ first becomes zero, a new saddle point is borne simultaneously with the new orbit. The equivalent of reaching the conjugate point in the fixed-end-point problem is reaching multifurcation for the period problem. Several interesting conjectures follow. Since a periodic hyperbolic orbit, which is hyperbolic in the $K \to 0$ limit, never produces new adjacent orbits, it never develops a saddle point, I_1 is a minimum for all periodic function spaces. Similarly, a KAM torus, consisting of an orbit with an irrational rotation number, stands alone. There are no adjacent tori with everywhere-adjacent trajectories. Thus, I_1 is a minimum for such tori supporting Percival's conjecture.

THE INTEGRABLE PROBLEM

Consider the set of nonintegrable equations

$$\dot{\mathbf{P}} = \boldsymbol{\phi}(\mathbf{x}, \mathbf{P}, \mathbf{K}, t) \qquad (3)$$

$$\dot{\mathbf{X}} = \mathbf{f}(\mathbf{P}) \qquad (4)$$

where $\boldsymbol{\phi} \to 0$ as $\mathbf{K} \to 0$, and $\mathbf{f}$ is a monotonously increasing function of $\mathbf{P}$ so as to give a twist mapping on the surfaces of sections. One may consider $\mathbf{P}$ and $\mathbf{x}$ as action-angle variables. Let $\boldsymbol{\phi}$ be periodic in t, as well as the components of $\mathbf{x}$ with period one.

The problem becomes integrable once $\mathbf{K} \to 0$, to yield $\mathbf{P} = $ (const) $\mathbf{x} = \mathbf{f}(\mathbf{P})t + \mathbf{x}_0$. The Hamiltonian $H(\mathbf{P})$ can be reduced by a contact transformation to the free-particle Hamiltonian $H = \mathbf{P}^2/2$. For periodic orbits of period T

$$\mathbf{P}T = \mathbf{N} \qquad (5)$$

where all components of $\mathbf{N}$ are integers. The Lagrangian integral over the period is

$$I_1 = \int_0^T \tfrac{1}{2}\dot{\mathbf{x}}^2 \, dt = \tfrac{1}{2}\mathbf{P}^2 T \qquad (6)$$

Consider now other adjacent orbits with the same period or one of its multiples

$$\mathbf{X} + \delta\mathbf{x} = \mathbf{P}t + \mathbf{x}_0 + \sum_{n=0}^{\infty} \left(\delta\mathbf{a}_n \sin \frac{2\pi nt}{mT} + \delta\mathbf{b}_n \cos \frac{2\pi nt}{mT} \right) \qquad (7)$$

where m is an integer. The Lagrangian integral becomes

$$I_1 + \delta I_1 = \tfrac{1}{2}\mathbf{P}^2 mT + \tfrac{1}{2}\sum \left(\frac{2\pi n}{mT}\right)^2 (\delta a_n^2 + \delta b_n^2) \qquad (8)$$

with δI_1 definite positive. So I_1 for the actual orbit is indeed a minimum for all periodic orbits of period mT.

Note that the integrable case is degenerate. For each periodic orbit starting at $\mathbf{P}$ and $\mathbf{x}_0$ there are other actual orbits with the same period starting at $\mathbf{P} + \mathbf{x}_0 + \delta\mathbf{x}_0$ with the same value of I_1. We know, however, that this degeneracy disappears once the problem becomes nonintegrable for

$\mathbf{K} \neq 0$. The continuum of periodic orbits breaks up into a finite set of elliptic and hyperbolic orbits.[1]

SADDLE POINTS

We know that for the nonintegrable case I_1 has saddle points for the actual orbits. We also know that a stable periodic orbit is surrounded by orbits of higher period and these orbits are emitted as K is increased. For small K, nearby orbits of very large period are present only, and, as K is increased, lower and lower periodic orbits—all multiples of T—are emitted.

Suppose K has just reached the value where an orbit of period T emits an orbit of period mT. For both of these adjacent orbits $\delta I_1 = 0$ for functions of period mT. Consequently, $\delta^2 I = 0$ at the emitting orbit with a saddle point developing in function space in the direction that connects the original orbit with the emitted one. With increasing K more saddles develop in function space for lower and lower periods.

To find the K value where this happens one evaluates

$$\delta^2 I_1(K) = 0 = \int_0^{mT} \left(\frac{\partial^2 L}{\partial x_i^2} \xi_i^2 + 2 \frac{\partial^2 L}{\partial x_i \partial \dot{x}_i} \xi_i \dot{\xi}_i + \frac{\partial^2 L}{\partial \dot{x}_i^2} \dot{\xi}_i^2 \right) dt \qquad (9)$$

where L is evaluated at the unperturbed orbit and $\xi(t)$ is a small perturbation of period mT. The integral is a function of K and a functional of ξ_i, for a given orbit and period. The functions ξ_i can be conveniently given in Fourier representation. There is a well-defined threshold value of K where Eq. (9) can be satisfied for some set of the Fourier coefficients. These coefficients characterize the function describing the new orbit.

What about unstable (hyperbolic) orbits? These orbits are not emitters of new ones, in fact no orbits adjacent to hyperbolic orbits can exist. Hence, saddle points cannot develop anywhere in the space of periodic functions, and I_1 stays a minimum.

A similar argument can be made for KAM tori, which represent nonperiodic orbits. They can be viewed as limits of orbits of very long periods, by approaching them via a continued-fraction approximation of their rotation number.[4] Since they cannot emit orbits adjacent to themselves everywhere, one concludes that I_1 is a minimum for KAM tori.

PARTICLE IN THE FIELD OF A STANDING WAVE

Consider the motion of a particle in the field of a standing wave

$$\ddot{x} = K \cos 2\pi x \sin 2\pi t \qquad (10)$$

This problem has an obvious importance in plasma physics. The ponderomotive force, usually employed to treat this problem, arises from a second-order perturbation expansion of this nonintegrable equation. It was shown[10] that this expansion breaks down for large amplitudes, and the particle motion as well as the motion of the oscillation center become chaotic. The last vestiges of the ponderomotive force disappear once the fixed points at $x = 1/4$, $3/4$ destabilize. We will now determine this critical amplitude using our variational method:

$$L = \tfrac{1}{2}\dot{x}^2 + \frac{K}{2\pi} \sin 2\pi x \sin 2\pi t \tag{11}$$

Consider the orbit $x = 1/4$, $\dot{x} = 0$, and the value of $K = K_c$ where this orbit bifurcates via period doubling. Before employing the variational method we note that the tangent mapping around this orbit results in the Mathieu equation,[3] for $x = 1/4 + \delta x$

$$\delta\ddot{x} + (2\pi K \sin 2\pi t)\delta x = 0 \tag{12}$$

The point K_c where this orbit destabilizes can be obtained from tabulated values to yield[10] $K_c = 2.853$.

Let us see now how one obtains K_c using the variational method. Equation (9) yields the equation

$$\int_0^2 (\dot{\xi}^2 - 2\pi K \sin 2\pi t\, \xi^2)\, dt = 0 \tag{13}$$

Expanding

$$\xi(t) = \sum_n (a_n \sin n\pi t + b_n \cos n\pi t) \tag{14}$$

and substituting into Eq. (13) gives

$$K = \frac{1}{2\pi} \frac{S_1}{S_2} \tag{15}$$

where

$$S_1 = \int_0^2 \sum_n (n\pi)^2 (a_n \sin n\pi t + b_n \cos n\pi t)^2\, dt$$

$$= \pi^2 \sum n^2 (a_n^2 + b_n^2) \tag{16}$$

and

$$S_2 = \int_0^2 \sin 2\pi t \sum (a_n \sin n\pi t + b_n \cos n\pi t)^2\, dt$$

$$= a_1 b_1 + \sum (a_{n+2} b_n - a_n b_{n+2}) \tag{17}$$

so the period-doubling bifurcation, leading to destabilization, occurs for

$$K_c = \frac{1}{2\pi} \min\left[\frac{\pi^2 \sum(a_n^2 + b_n^2) n^2}{a_1 b_1 + \sum(a_{n+2} b_n - a_n b_{n+2})}\right] \quad (18)$$

Upper limits of K_c can be obtained by truncating the Fourier series. We will see that this process converges very rapidly. Keeping only the $n = 1$ Fourier mode ($a_n = b_n = 0$ for $n \geq 2$) gives

$$K_c < \min \frac{\pi}{2} \frac{a_1^2 + b_1^2}{a_1 b_1} = \pi \quad (19)$$

surprisingly close to the exact value for such a crude approximation. Next we keep the third harmonic as well, to find

$$K_c < \min \frac{\pi}{2} \frac{a_1^2 + b_1^2 + (3a_3)^2 + (3b_3)^2}{a_1 b_1 - a_1 b_3 + a_3 b_1}$$

$$= \frac{\pi}{2} \min \frac{1 + \beta_1^2 + (3\alpha_3)^2 + (3\beta_3)^2}{\beta_1 - \beta_3 + \alpha_3 \beta_1} \quad (20)$$

where

$$\beta_1 = \frac{b_1}{a_1}, \quad \alpha_3 = \frac{a_3}{a_1}, \quad \text{etc.}$$

Minimizing with respect to these variables yields, after trivial algebra,

$$\beta_1^2 = 9\alpha_3(1 + \alpha_3) \quad (21)$$

$$\beta_3 = \frac{\alpha_3}{\beta_1} \quad (22)$$

Therefore, Eq. (20) is a function of α_3 alone, and minimizing numerically one obtains

$$K_c < 2.8536 \quad (23)$$

an excellent approximation. One may include additional terms, but the corrections are very small. The coefficients minimizing Eq. (20) are $\alpha_3 = 0.101$, $\beta_1 = 1.0004$, and $\beta_3 = -0.10096$. These numbers are the normalized Fourier coefficients for the bifurcated orbits.

A similar calculation can be carried out to find the values K_m where the period m orbit is born, to yield

$$K_m = \frac{2\pi}{m^2} \min \frac{\sum_n n^2 (a_n^2 + b_n^2)}{\sum_n (a_{n+m} b_n - a_n b_{n+m}) + \sum_{n=1}^{m} a_n b_{m-n}} \quad (24)$$

The first crude approximation can be obtained by keeping only the lowest Fourier coefficients a_1 and b_{m-1}, to yield

$$K_m < 4\pi \frac{m-1}{m^2} \quad (25)$$

As expected $K_m \to 0$ as $m \to \infty$, the point is always a saddle point for sufficiently long periods.

THE STANDARD MAPPING

As a second example consider the standard mapping[2,4,9,11] arising from the equation of motion

$$\ddot{x} = \frac{K}{2\pi} \sin 2\pi x \delta_1(t) \qquad (26)$$

where

$$\delta_1(t) = \sum_{n=-\infty}^{\infty} \delta(t - \varepsilon - n), \qquad \varepsilon \to 0 \qquad (27)$$

The corresponding Lagrangian is

$$L = \tfrac{1}{2}\dot{x}^2 - \frac{K}{(2\pi)^2} \cos 2\pi x \delta_1(t) \qquad (28)$$

The second variation of I_1 about the periodic orbit $x_0(t)$ is

$$\delta^2 I_1 = \oint [K \cos 2\pi x_0 \delta_1(t) \xi^2 + \dot{\xi}^2] \, dt \qquad (29)$$

integrated over some period. The period-one fixed points (periodic orbits) are $x_0(t) = 0$ and $x_0(t) = 1/2$; the former hyperbolic, the latter elliptic. Evidently for $x_0 = 0$, $\delta^2 I_1 = 0$ can never be satisfied for $K > 0$, as expected. There is no saddle point for a hyperbolic orbit.

Although all orbits of the standard mapping consist of straight-line segments, it is useful to see how our variational method, using truncated-Fourier-series orbits, approximates $K_c = 4$.

Consider again the period-doubling bifurcation. Substituting Eq. (14) into Eq. (29) yields

$$K_c = \min \frac{\pi^2 \sum n^2 b_n^2}{(\sum b_n)^2 + (\sum b_n \cos n\pi)^2} \qquad (30)$$

The $n = 1$ term alone gives

$$K_c < \frac{\pi^2}{2} \approx 4.93$$

One can easily see that only odd terms contribute. Considering $n = 1$ and $n = 3$, one finds $K_c < (9/10)\pi^2 = 4.44$ giving an 11% error. Higher approximations seem to approach $K_c = 4$. Similar calculations have been carried out to find the K_m values for the emissions of period-m orbits giving again, upon truncation, very good approximations to the exact values.

ACKNOWLEDGMENTS

It is a pleasure to acknowledge stimulating discussions with A. Weinstein, R. Mackay, and J. Bialek. Thanks are also due to the Aspen Center for Physics, where part of this work was carried out.

REFERENCES

1. M. V. Berry, in *Topics in Nonlinear Dynamics*, S. Jorna, ed. (AIP, New York, 1978).
2. B. V. Chirikov, *Phys. Reps.* **52**, 265 (1979).
3. D. F. Escande and F. Doveil, *Phys. Lett.* **83A**, 307 (1981); **84A**, 399 (1981).
4. J. M. Greene, *J. Math. Phys.* **20**, 1183 (1979).
5. R. H. G. Helleman, *Topics in Nonlinear Dynamics*, S. Jorna, ed. (AIP, New York, 1978).
6. R. H. G. Helleman and T. Bountis, in *Volta Memorial Conference*, G. Casati and J. Ford, Eds. (Springer, New York, 1978), pp. 353–375.
7. J. N. Mather, "Existence of Quasi-Periodic Orbits for Twist Homeomorphisms of the Annulus," *Topology* **21**, 457 (1982).
8. I. C. Percival, "Variational Principles for Invariant Tori and Cantori," in *Nonlinear Dynamics and Beam-Beam Interactions*, M. Month and J. C. Herrara, eds. (AIP, New York, 1980), pp. 302–310.
9. G. Schmidt, *Phys. Rev. A* **22**, 2849 (1980).
10. G. Schmidt, *Comments Plasma Phys. Controlled Fusion* **7**, 87 (1982).
11. G. Schmidt and J. Bialek, *Physica* **5D**, 397 (1982).
12. E. T. Whittaker, *Analytical Dynamics*, 4th ed. (Dover, New York, 1944), Chap. IX.

RESONANCE BEHAVIOR OF THE PERTURBED TODA LATTICE

R. de Fainchtein and L. E. Reichl

Center for Studies in Statistical Mechanics, University of Texas, Austin, Texas

Abstract. Numerical data are presented to supplement previous theoretical work on dynamic external-field-induced resonance behavior and chaos in the two-particle Toda lattice.

1. INTRODUCTION

Conservative anharmonic-oscillator systems with two or more degrees of freedom have been the focus of much interest in recent years for two quite different reasons. On the one hand, it has been shown that internal resonances can induce chaos in such systems and render them nonintegrable. On the other hand, one particular anharmonic system has been found by Toda[3] that does not exhibit chaos regardless of the number of degrees of freedom. This system is completely integrable[4,5] and is one of the simplest of a class of lattice systems that can sustain the propagation of solitons. A problem of growing concern is to what extent an anharmonic lattice can sustain soliton propagation in the presence of internal chaos or dynamic external perturbations.

In order to gain insight into this problem, Reichl et al.[6] have studied the effect of a sinusoidal external field on the two-particle Toda lattice. This system can be studied analytically and therefore provides a basis for understanding the effect of external fields on Toda lattices with more than two particles. In this chapter, we wish to supplement the theory and numerical results presented in Ref. 6 with additional numerical results on external-field-induced resonance behavior in the two-particle Toda lattice. We will begin in Sections 2 and 3 with a brief review of theoretical results obtained in Ref. 6, and in Section 4 we will present a systematic study of the appearance of resonance zones at a given field amplitude.

2. UNPERTURBED TWO-PARTICLE TODA LATTICE

The Hamiltonian for the two-particle Toda lattice with cyclic boundary conditions can be written

$$H = \frac{P_1^2}{2m} + \frac{P_2^2}{2m} + \frac{A}{B}e^{B(Q_2-Q_1)} + \frac{A}{B}e^{B(Q_1-Q_2)} - \frac{2A}{B} \quad (1)$$

where P_i and Q_i are the positions and coordinates, respectively, of the ith particle, m is the mass of *each* particle, and A and B are constants which fix the scale of the potential. If the center of mass is assumed to be at rest, the Hamiltonian takes the form

$$H = \frac{p^2}{m} + \frac{A}{B}(e^{-Bq} + e^{Bq} - 2) \quad (2)$$

where the relative momentum p, and relative position, q, are defined as $p = (P_1 - P_2)/2$ and $q = (Q_1 - Q_2)$, respectively. We can make a canonical transformation to new coordinates (J, Φ) defined by

$$p = \sqrt{\frac{mA}{B}} \frac{2k}{\sqrt{1-k^2}} \operatorname{sn}(\Phi, k) \quad (3)$$

$$q = \frac{2}{B} \ln\left(\frac{\operatorname{dn}(\Phi, k) - k \operatorname{cn}(\Phi, k)}{\sqrt{1-k^2}}\right) \quad (4)$$

Here

$$k = \frac{J^2 - 4m_0 A/B}{J^2} \quad (5)$$

with $m_0 = 4m/B^2$, and $\operatorname{sn}(\Phi, k)$, $\operatorname{dn}(\Phi, k)$, and $\operatorname{cn}(\Phi, k)$ are Jacobi elliptic functions with modulus k. The Hamiltonian, in terms of these canonical coordinates takes the form

$$H = \frac{J^2}{m_0} - \frac{4A}{B} \quad (6)$$

Since $k^2 \geq 0$, the canonical variables J can only take on values

$$|J| \geq 2\sqrt{\frac{m_0 A}{B}} = J_{\min}$$

Thus, the (J, Φ) phase space is divided into two disconnected parts. In subsequent sections we will consider this system for values $A = 0.15$, $B = 3.1$, and $m = 11.68 \times 10^5$. Then $J_{\min} = 306$.

3. PERTURBED TODA LATTICE

We will apply an external force to this system in such a way that the center of mass remains fixed. The perturbed Hamiltonian then takes the form

$$H = \frac{p^2}{m} + \frac{A}{B}(e^{Bq} + e^{-Bq} - 2) + \varepsilon q \cos(\Omega_t) \tag{7}$$

where ε is the amplitude and $\Omega_t = \omega_0 t + \delta$ (ω_0 is the angular frequency, and δ is the phase of the applied field; in our strobe plots, $\delta = \pi/2$). In terms of the canonical variables (J, Φ) the Hamiltonian becomes

$$H = \frac{J^2}{m_0} - \frac{4A}{B} + \frac{2\varepsilon}{B} \cos(\Omega_t) \ln\left(\frac{\mathrm{dn}(\Phi, k) - k\,\mathrm{cn}(\Phi, k)}{1 - k^2}\right) \tag{8}$$

and can be written in the form

$$H = \frac{J^2}{m_0} - \frac{4A}{B} + \varepsilon \sum_{\substack{n=-\infty \\ (\text{odd})}}^{\infty} g_n(J) \cos\left(\frac{n\Phi\pi}{2K(J)} - \Omega_t\right), \tag{9}$$

where K is the complete elliptic integral of the first kind.
The sum over n involves only odd integers. The function $g_n(J)$ is defined as

$$g_n(J) = -\left[\frac{Bn}{2} \sinh\left(\frac{n\pi K'}{2K}\right)\right]^{-1} \tag{10}$$

The magnitude of $g_n(J)$ decreases roughly exponentially with increasing n.

It is useful to note that this one-degree-of-freedom system, which is perturbed by a time-dependent external field, is equivalent to a conservative two-degrees-of-freedom system with Hamiltonian

$$H(J_1, J_2, \theta_1, \theta_2) = \frac{J_1^2}{m_0} - \frac{4A}{B} + J_2 + \varepsilon \sum_{n=-\infty}^{\infty} g_n(J_1) \cos\left(\frac{n\pi\theta_1}{2K(J_1)} - \omega_0 \theta_2 - \delta\right) \tag{11}$$

From the form of the Hamiltonian in Eq. (9) we see immediately that a number of resonance zones (called primary resonances) will be created in the phase space of the perturbed system. The hyperbolic and elliptic fixed points associated with these zones are located by the condition

$$\dot{\Phi} = \frac{2\omega_0 K(J)}{n\pi} = \frac{\partial H}{\partial J} \tag{12}$$

For small ε, this reduces to

$$J^c = \frac{m_0 \omega_0}{n\pi} K(J^c) \tag{13}$$

A plot of the J^c vs $f_0 = \omega_0/2\pi$ for $n = 1, 3, 5, 7, 9$ is given in Fig. 1. We see

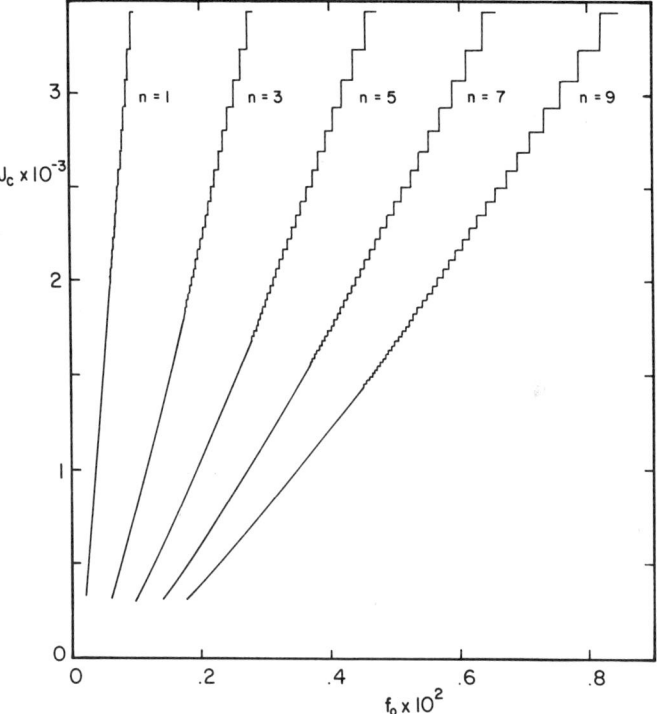

Figure 1. Plot of J_n^c vs f_0 for primary zones $n = 1, 3, 5, 7,$ and 9 for $A = 0.15$, $B = 3.1$, and $m = 11.68 \times 10^5$.

that no primary zones are expected below $f_0 \approx 2 \times 10^{-4}$. The primary zones emerge at $J^c \approx 306$ and move to higher energies as the frequency of the applied field is increased.

The resonance condition for the primary resonance zones may be obtained in another way and in so doing we find that additional resonance zones may also appear. Let us attempt to solve the equations of motion obtained from the full perturbed Hamiltonian (9) by means of a series of canonical transformations. For example, let us change from canonical variables (J, Φ) to canonical variables (J_1, Φ_1) via the generating function

$$F_0(J_1, \Phi, t) = J_1\Phi + \varepsilon G_0(J_1, \Phi, t) \tag{14}$$

where

$$G_0(J_1, \Phi, t) = \sum_{\substack{n \\ (\text{odd})}} \frac{m_0 g_n(J_1) \sin\left(\frac{n\pi}{2K(J_1)}\Phi - \Omega_t\right)}{m_0\omega_0 - n\pi J_1/K(J_1)} \tag{15}$$

The new canonical variables are given in terms of the old variables via the conditions

$$\Phi_1 \equiv \frac{\partial F_0}{\partial J_1} = \Phi + \varepsilon \frac{\partial G_0}{\partial J_1} \tag{16}$$

and

$$J = \frac{\partial F_0}{\partial \Phi} = J_1 + \varepsilon \frac{\partial G}{\partial \Phi} = J_1 + \varepsilon \sum_n \frac{m_0 g_n(J_1) n\pi \cos\left(\frac{n\pi}{2K(J_1)}\Phi - \Omega_t\right)}{2K(J_1)(m_0\omega_0 - n\pi J_1/K(J_1))} \tag{17}$$

The new Hamiltonian may be written

$$H_1(J_1, \Phi_1, t) = H(J, \Phi, t) + \frac{\partial F_0}{\partial t} = \frac{J_1^2}{m_0} + O(\varepsilon^2) \tag{18}$$

If we neglect terms of order ε^2 and higher in the Hamiltonian $H_1(J_1, \Phi_1, t)$, then J_1 is constant and $\Phi_1 = (2J_1/m_0)t$. The variables $(J$ and $\Phi)$ then take the form

$$\Phi = \frac{2J_1 t}{m_0} - \varepsilon \frac{\partial}{\partial J_1}\left[\sum_n \frac{m_0 g_n(J_1) \sin[(n\pi J_1 t/m_0 K(J_1)) - \Omega_t]}{m_0\omega_0 - n\pi J_1/K(J_1)}\right] \tag{19}$$

$$J = J_1 + \varepsilon \sum_n \frac{n\pi}{2K(J)} \frac{m_0 g_n(J_1) n\pi \cos[(n\pi J_1 t/m_0 K(J_1)) - \Omega_t]}{m_0\omega_0 - n\pi J_1/K(J_1)} \tag{20}$$

The expression for J and the canonical transformations (14) and (15) diverge when the resonance condition (13) is satisfied, thus indicating a strong distortion of the phase space.

We can write expressions for J and Φ to order ε^l by a series of canonical transformations similar to that given above. After l transformations we obtain a Hamiltonian of the form

$$H_l(J_l, \Phi_l, t) = \frac{J_l^2}{m_0} + O(\varepsilon^{l+1}) \tag{21}$$

and we can write expressions for J and Φ in terms of J_l and Φ_l. However, this perturbation expansion contains resonance denominators that cause it to diverge when conditions of the type

$$J_l^c = \frac{m_0 \omega_0 K(J_l^c)}{\eta \pi} \tag{22}$$

are satisfied. Here η can take on rational fractional values. These new resonances, which appear in higher-order perturbation theory, are called secondary or fractional resonances. When resonance conditions (22) are translated back into conditions in terms of (J, Φ) space, the structure of the secondary zones can be somewhat more complicated than that of the primary zones.

The primary resonance zones in (J, Φ) space can be described approximately by the pendulum Hamiltonian when ε is small. The value J^c

given by Eq. (13) locates the position of the hyperbolic and elliptic fixed points. The half-width of the nth primary resonance zone is

$$\Delta J_{sx}^{(n)} \approx \sqrt{2 m_0 \varepsilon |g_n(J_n^c)|} \tag{23}$$

for small ε.

As long as $g_n(J)$ falls off fairly rapidly with increasing n, these resonance zones or highly distorted regions occupy a small fraction of the phase space for small ε. This was shown by KAM[7-9] and this is what we find in numerical studies of this system.[6]

4. NUMERICAL RESULTS

We wish to supplement some of the numerical results of Ref. 6, with additional numerical results on the behavior of the resonance zones in the Toda lattice. As in Ref. 6, we probe the flow of points in a particular region of phase space. We always start at $p = 600$ and $q = 0$ or, equivalently, $J = 491$ and $\Phi = K$. When $\varepsilon = 0$, the flow in (p, q) space is given in Fig. 2. When $\varepsilon \neq 0$, the system has a four-dimensional phase space, and we can only obtain strobe plots of p and q. That is, we plot points p and q at time intervals $t = 1/f_0$, where $f_0 = \omega_0/2\pi$ is the frequency of the external field. As we can see from Fig. 1 (the resonance conditions for secondary zones will have somewhat similar behavior, at least for small ε) as the frequency of the applied field increases, more and more resonance zones emerge in the phase space at $J \approx 306$ and then move to larger values of J as the frequency increases. Thus, if we focus on the point $J = 491$ we will see resonance

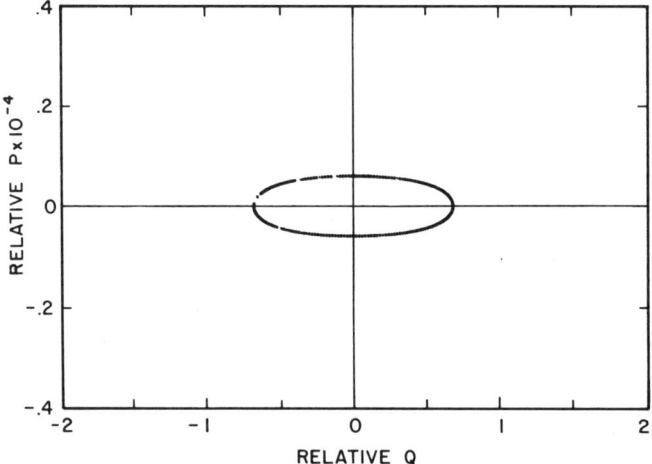

Figure 2. The (p, q) phase-space trajectory of the unperturbed Toda lattice for energy for $A = 0.15$, $B = 3.1$, and $m = 11.68 \times 10^5$ and energy $E = 0.302$ or $J = 491$.

zones moving past from lower values of J to higher values as frequency increases.

In all of the subsequent figures, we will use $\varepsilon = 1.3$. This is a fairly large amplitude and we expect deviations from the resonance condition (13), although Fig. 1 will still give qualitatively correct behavior for the primary resonance zones.

In Fig. 3, we show the approach and passage of the primary $n = 3$ resonance zone. For ε small we expect its hyperbolic fixed point to enter the phase space at $f_0 \approx 6.0 \times 10^{-4}$ and reach $J = 491$ at about $f_0 = 7.89 \times 10^{-4}$. From Eq. (23) we estimate its half-width to be $\Delta J_{sx}^{(3)} \approx 87$. As we see from the figure we enter the primary $n = 3$ resonance zone at one of its hyperbolic fixed points at a frequency 9.1×10^{-4}. This is slightly higher than that predicted by the small-ε resonance condition. It is possible that the $n = 3$ resonance zone is being "repelled" somewhat by the very large $n = 1$ zone, which is higher in energy. From the figure we see that its width is

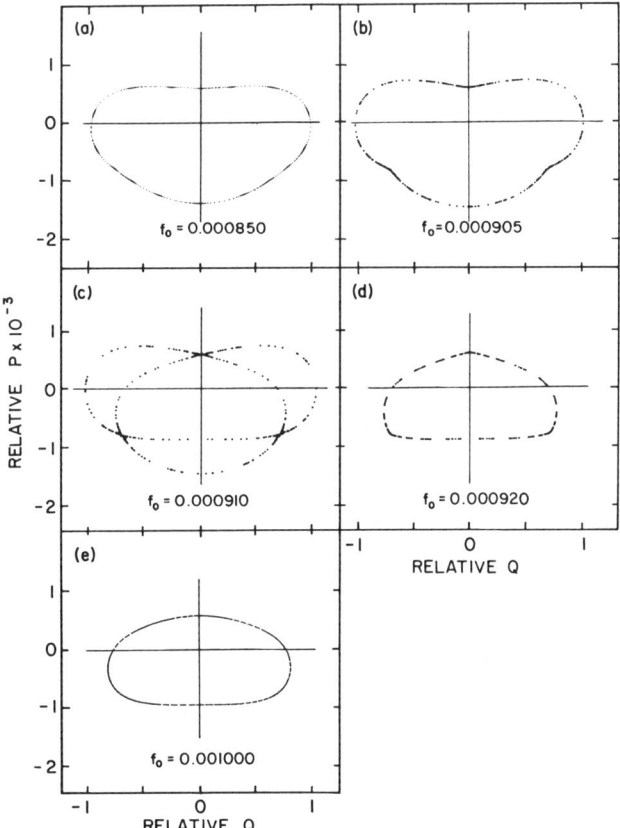

Figure 3. The primary $n = 3$ resonance zone.

approximately $\Delta p = 320$. Using Eq. (3), this translates to a half-width $\Delta J = 95$, in good agreement with our estimate. In Fig. 3b we are just above the resonance zone. In Fig. 3d we are just below it. In Fig. 3e we see that the phase space is still somewhat perturbed by the resonance zone, but it has settled down almost to its unperturbed behavior. It is interesting to note that the stochastic layer at the separatrix is just barely observable in Fig. 3c.

In Fig. 4, we show the approach and passage of the $n = 5$ primary resonance zone. For ε small we expect its hyperbolic and elliptic fixed points to enter the phase space at $f_0 = 10.0 \times 10^{-4}$ and arrive at $J = 491$ at $f_0 = 13.15 \times 10^{-4}$. From Fig. 4 we see that it arrives at $f_0 = 14 \times 10^{-4}$. From Eq. (23) we estimate its half-width to be $\Delta J_{sx}^{(5)} \approx 16$. From the figure we find $\Delta J \approx 23$. Again we have good agreement. This figure differs from Fig. 3 in that we enter the resonance zone at its widest part. Thus in Fig. 4 we hit the separatrix from above and are pulled to lower energy. In Fig. 4c–4e the resonance zone is moving to higher energy and we are sampling the inner

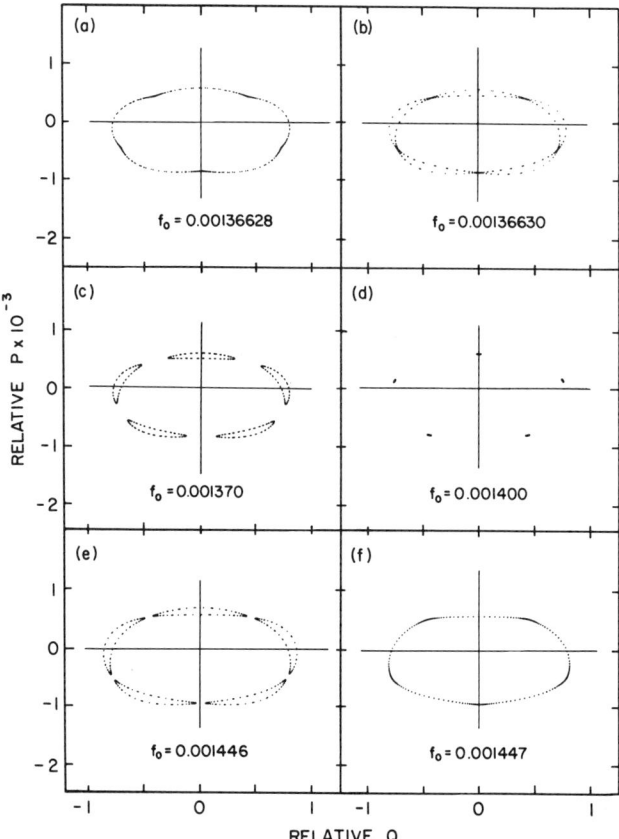

Figure 4. The primary $n = 5$ resonance zone.

islands. In Fig. 4d we are very close to the elliptic fixed point. In Fig. 4e we are again approaching the separatrix and trajectories are carried to higher energy. Finally, in Fig. 4f the resonance zone has passed but we still see the effects of the separatrix. For this zone, the stochastic layer on the separatrix is unobservable. The well-defined curves in Figs. 3 and 4 indicate that there is an approximate conserved quantity governing the motion of the system. For both cases this conserved quantity is the quasienergy

$$H = J^2 + aJ + V \cos\left(\frac{n\pi\Phi}{2K(J)} - \frac{\pi}{2}\right)$$

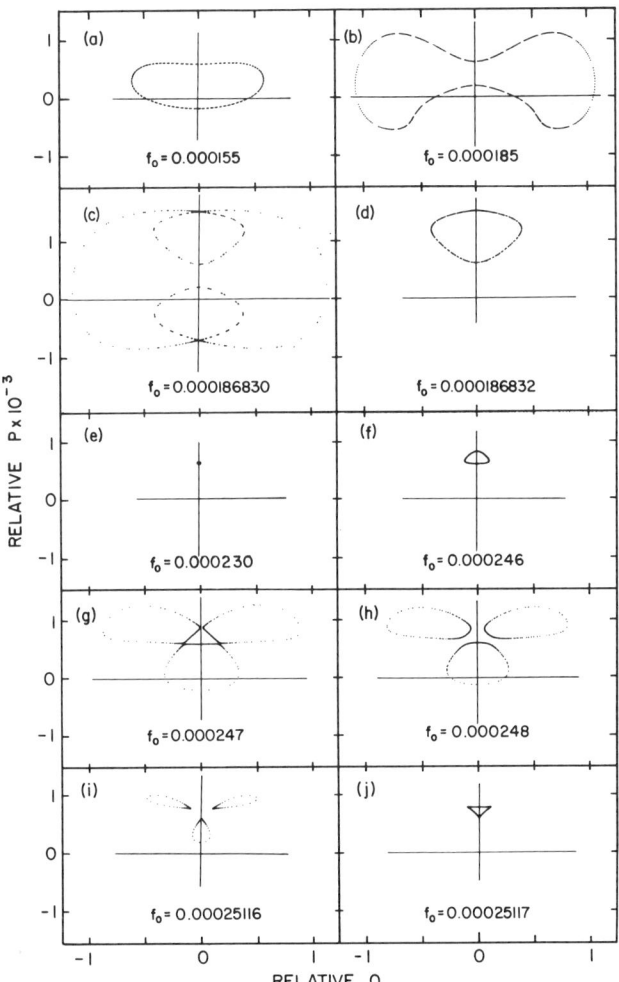

Figure 5. (a–h) A sequence of resonances, which contains the primary $n = 1$ zone. (Note the change in scale in going from Fig. 5o to Fig. 5p.) (Fig. 5a to Fig. 5r continued on pages 74 and 75.)

where $n = 3$ for Fig. 3 and $n = 5$ for Fig. 4. The values of a and V can be found from the figures.

We now wish to study the primary $n = 1$ resonance zone. However, as we shall see, the region of phase space about the $n = 1$ zone is so strongly affected by secondary resonance zones that we cannot clearly locate the entrance point into the primary $n = 1$ zone. Let us first note that hyperbolic or elliptic fixed points of the primary $n = 1$ zone should enter the phase space at frequency $f_0 = 2.0 \times 10^{-4}$ and should reach $J = 490$ at about $f_0 = 2.63 \times 10^{-4}$. In Fig. 5 we give the sequence that leads us into the primary $n = 1$ zone. If we look at Fig. 5k at frequency $f = 2.60 \times 10^{-4}$ we

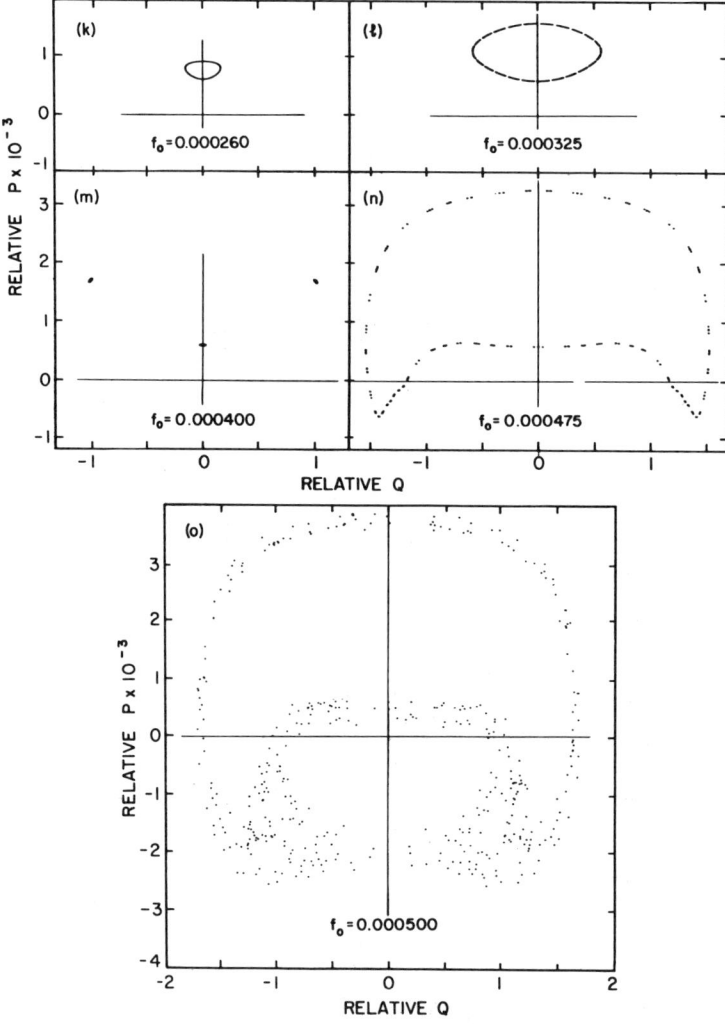

Figure 5. $(k-o)$.

see we are on a small island, and this island begins to grow with increasing frequency until we pass out of this resonance zone on a clearly defined although chaotic $n = 1$ separatrix (see Ref. 6 for more discussion of this separatrix). Thus, this whole sequence appears to end by passing out of the primary $n = 1$ resonance zone, and the theory serves to locate this zone fairly accurately.

However, the sequence that ends with the primary $n = 1$ zone appears to begin in Fig. 5c at frequency $f_0 = 1.8683 \times 10^{-4}$ when we pass into what appears to be an $n = 2$ secondary zone. (We have given plots at lower frequency in Figs. 5a and 5b to show how the phase space becomes distorted as we approach this separatrix.) However, there appears to be no simple description of this zone in terms of a pendulum Hamiltonian in (J, Φ) space as there is for the primary zones. In Fig. 5d we are on an inner island of the $n = 2$ zone and in Fig. 5e we reach its elliptic fixed point. In Fig. 5f, we are again on a island, which is now inverted in shape relation to that in Fig. 5d, and in Fig. 5g we suddenly find ourselves on the separatrix secondary $n = 3$ zone. We then proceed through the main islands of this zone and leave through a separatrix (not shown), which is inverted in relation to the first. The island in Fig. 5j then proceeds to grow until we reach Fig. 5k, which appears to be part of the $n = 1$ primary zone. In Fig. 5p the $n = 1$ zone appears to have passed although the strobe plot is still somewhat chaotic. In Fig. 5q a secondary $n = 4$ zone appears to be passing. Note that the hyperbolic fixed points do not occur at regular angles in Fig. 5q. Finally, in Fig. 5r this whole sequence has passed through and we appear to be back to a relatively unperturbed phase space, although the system oscillates somewhat in energy.

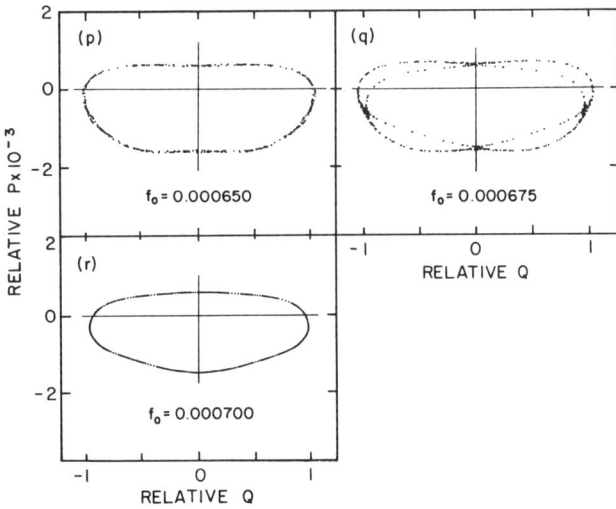

Figure 5. (p–r).

It is interesting that this same sequence occurs at lower frequency ($f_0 = 0.86019 \times 10^{-4}$ to $f_0 = 1.232675 \times 10^{-4}$) as is shown in Fig. 6. We again enter via a secondary $n = 2$ zone. We then enter a $n = 3$ zone (but in reverse sequence to that in Fig. 5) and then pass onto an $n = 1$ zone before leaving this region. Finally, in Figs. 7a–7f we show an $n = 3$ secondary zone that passes between $f_0 = 1.4674 \times 10^{-4}$ and $f_0 \approx 1.5303 \times 10^{-4}$. This zone is fairly normal in behavior but is inverted relative to the primary $n = 3$ zone in Fig. 3.

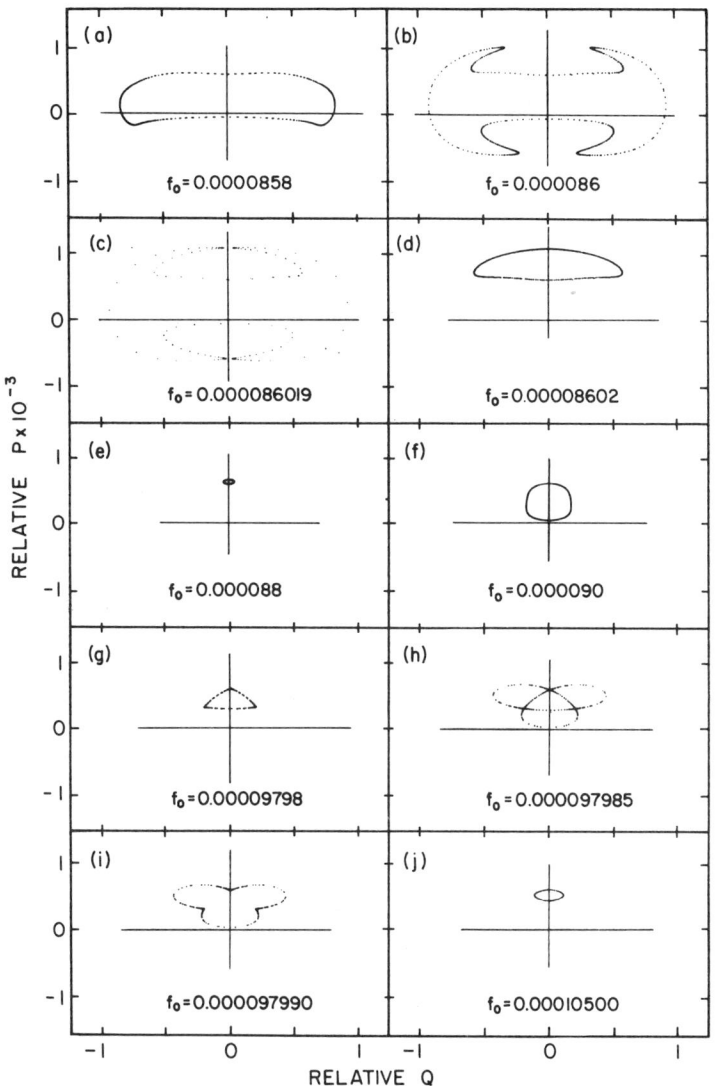

Figure 6. (a–j) A sequence of secondary resonances similar to that in Fig. 5. (Fig. 6k to Fig. 6m continued on next page.)

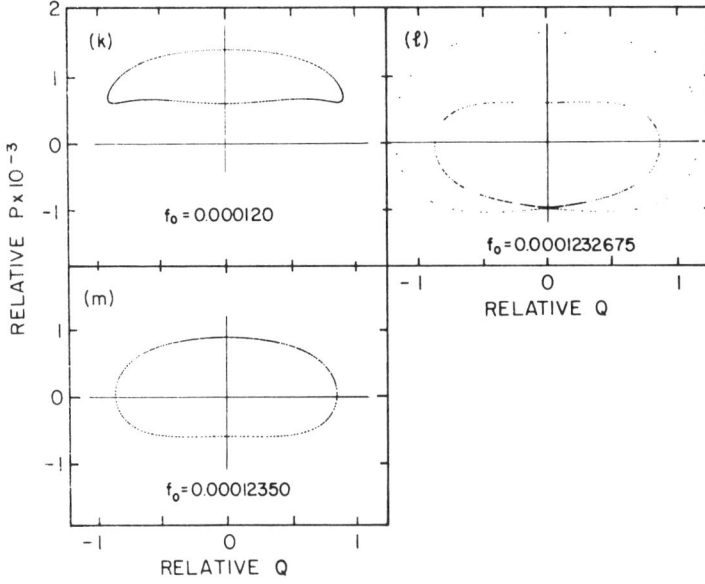

Figure 6. $(k-m)$.

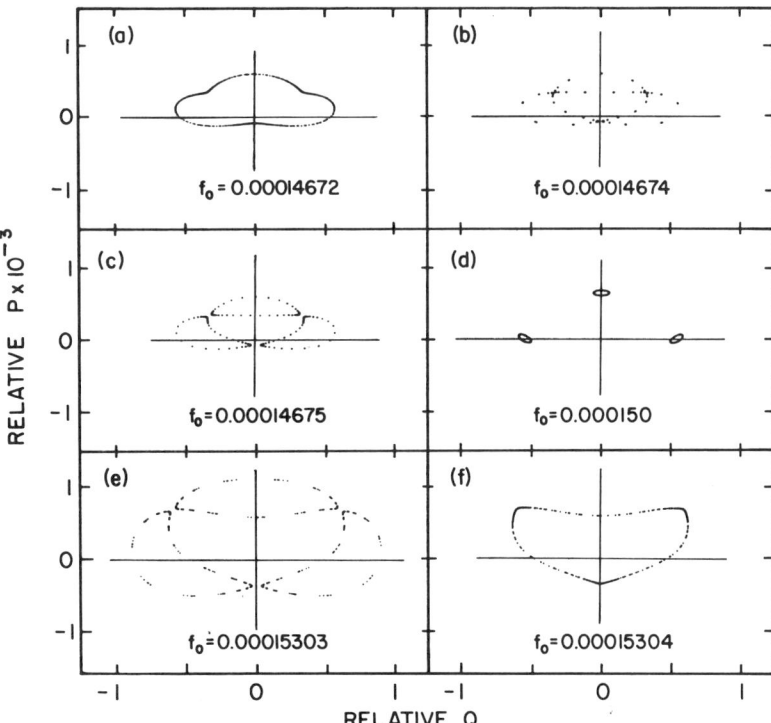

Figure 7. An isolated secondary $n = 3$ resonance zone.

In Figs. 3–7 we have shown all the large-scale resonance zones we observed at $\varepsilon = 1.3$ starting at $f_0 = 0.25 \times 10^{-4}$ and ending at 15.00×10^{-4}. Our initial step size in scanning this interval was 0.25×10^{-4}. When we observed disturbance or resonance we focused in with smaller step size. It is interesting to note that there appears to be some order in the way zones appear as frequency is increased. The large scale zones that we see appear in the sequence (2–3–1) (3) (2–3–$\underline{1}$) (4) ($\underline{3}$) ($\underline{5}$) (we underline primary zones) in order of increasing frequency. To summarize, it appears that the primary zones dominate the region of space in which they are located, but their size decreases rapidly with increasing n. The primary $n = 1$ zone is by far the most important. Since the width of the nth primary zone is proportional to $\sqrt{g_n}$, and g_n falls off exponentially; this is what we expect. The secondary zones near the primary $n = 1$ zone have a strong effect on the phase space. However, secondary zones appear to have little effect in the neighborhood of higher-order primary resonance zones.

ACKNOWLEDGMENTS

The authors wish to thank W. M. Zheng for valuable discussions, and the U.S. Air Force for partial support of this work through contract No. F33615-78-D-0617.

REFERENCES

1. C. H. Walker and J. Ford, *Phys. Rev.* **188**, 416 (1969).
2. B. V. Chirikov, *Phys. Reps.* **52**, 263 (1979).
3. M. Toda, *J. Phys. Soc. Japan* **22**, 431 (1967); **23**, 501 (1967); *J. Phys. Soc. Japan Suppl.* **26**, 235 (1969); *Prog. Theor. Phys. Suppl.* **45**, 179 (1970).
4. H. Flaschka, *Phys. Rev. B* **9**, 1929 (1979).
5. M. Kac and P. Van Moerbeke, *Proc. Natl. Acad. Sci. (USA)* **72**, 2879 (1975).
6. L. E. Reichl, R. de Fainchtein, T. Petrosky, and W. M. Zheng, "Field Induced Chaos in the Toda Lattice", *Phys. Rev. A* **28**, 3051 (1983).
7. A. N. Kolmogorov, in *Foundations of Mechanics* R. Abraham, Ed. (W. A. Benjamin, New York, 1967), App. D.
8. V. I. Arnol'd, *Russian Math. J.* **18**, 9 (1963); **18**, 85 (1963).
9. J. Moser, *Nachr. Akgd. Wiss. Gottingen II, Math. Physik Kl.* **1**, (1962).

RECENT DEVELOPMENTS OF SOLITON RESEARCH IN PLASMA PHYSICS

Y. H. Ichikawa

Institute of Plasma Physics, Nagoya University, Nagoya, Japan

N. Yajima

Research Institute of Applied Mechanics, Kyushu University, Fukuoka, Japan

ABSTRACT. Summary report is presented of recent advancements in soliton research in plasma physics. In particular, the two-loop soliton solution, soliton-resonance phenomema, and studies of solitons and chaos have been discussed by emphasizing the importance of experimental research on nonlinear plasma properties.

1. INTRODUCTION

Soliton phenomena have been the focus of ever growing interest in many fields of natural science, such as plasma physics, fluid dynamics, solid-state physics, particle physics, and electric circuit engineering. In the studies of these vast branches of sciences, solitons are becoming the key concept used to disentangle complex features of nonlinear phenomena.

In plasmas, magnetosonic waves,[1] the Trivelpiece–Gould mode of electron plasma waves,[15] and ion acoustic waves[16] are known to be able to propagate solitons. Solitons associated with self-modulation of the Langmuir wave[25] are also well known. Many theoretical and experimental studies have explored the characterstics of solitons, that is, the dependence of shape and velocity of solitons on their amplitude, preservation of identities of solitons through their collisions, and formation of solitons from initial disturbances. As a result, the solitons are now regarded as nonlinear normal modes that play an essential role in nonlinear processes of plasma waves; for example, several attempts have been made to study plasma turbulence by using the concept of soliton gas.[9,14,24]

In this chapter, we would like to survey the scope of soliton research carried out in the Japanese plasma-physics community and to report some recent developments. Since Washimi and Taniuti[20] showed that soliton phenomena of the ion acoustic wave are governed by the Kortweg–de Vries equation, many studies on solitons in plasmas have been undertaken, from the mathematical foundation of solving the nonlinear evolution equation to numerical simulation of vortex collisions and various experiments in a double plasma device. We may classify them as follows;

1. Generalization of the inverse-scattering-transformation method.[6,18,19]
2. Solitonlike behavior of vortex motion.[8,21]
3. Resonant interaction of ion acoustic solitons.[4,5,11,17,22]
4. Solitons in an unstable plasma with ion beam.[29]
5. Negative potential solitary wave and weak double layer.[2,12]
6. Solitons and chaos.[10,13,23]

Studies of mathematical methods of solving the nonlinear evolution equation have been carried out very actively as the research collaboration program of Institute of Mathematical Sciences, Kyoto University, while Institute of Plasma Physics has been conducting theoretical and experimental studies of solitons in plasmas as one of its research collaboration programs during the past 10 years. In the following sections, we discuss some of the recent progress.

2. THE LOOP SOLITONS

Through generalization of the inverse scattering transformation, we have shown that the following nonlinear evolution equation

$$\frac{\partial}{\partial t}y_x + \text{sgn}\left(\frac{ds}{dx}\right)\frac{\partial^2}{\partial x^2}\left\{\frac{y_{xx}}{[1+y_x^2]^{3/2}}\right\} = 0 \tag{1}$$

is integrable by means of the Wadati–Konno–Ichikawa scheme of the inverse scattering transformation.[6] In Eq. (1), ds stands for an increment of the arc length, and the subscript x denotes differentiation with respect to x.

Although the subject is a little different than the principle of topics of the present volume, one might be interested to hear about further developments in the study of the loop soliton.[3] Recently, Konno and Jeffrey have constructed the N-loop soliton of Eq. (1).[27] Here, let us present explicitly the two-loop soliton solution for the two eigenvalues, $\lambda_k = i\eta_k$, with $k = 1$ and 2, given by the following set of equations:

$$y(x, t) = 2\frac{S(x, t; \varepsilon_+)}{R(x, t; \varepsilon_+)} \tag{2a}$$

$$\varepsilon_+(x, t) = 2\left\{\frac{1}{\eta_1} + \frac{1}{\eta_2} - \frac{T(x, t; \varepsilon_+)}{R(x, t; \varepsilon_+)}\right\} \tag{2b}$$

where the functions R, S, and T are defined as

$$R = 1 + C_1^2 e^{2\delta_1} + C_2^2 e^{2\delta_2} + \frac{8\eta_1\eta_2}{(\eta_1 + \eta_2)^2}C_1 C_2 e^{\delta_1+\delta_2} + \left(\frac{\eta_1 - \eta_2}{\eta_1 + \eta_2}\right)^2 C_1^2 C_2^2 e^{2\delta_1+2\delta_2} \tag{3a}$$

$$S = \frac{C_1}{\eta_1}e^{\delta_1} + \frac{C_2}{\eta_2}e^{\delta_2} + \frac{(\eta_1 - \eta_2)^2}{\eta_2(\eta_1 + \eta_2)^2}C_1^2 C_2 e^{2\delta_1+\delta_2} + \frac{(\eta_1 - \eta_2)^2}{\eta_1(\eta_1 + \eta_2)^2}C_1 C_2^2 e^{\delta_1+2\delta_2} \tag{3b}$$

$$T = \frac{C_1^2}{\eta_1}e^{2\delta_1} + \frac{C_2^2}{\eta_2}e^{2\delta_2} + \frac{4}{\eta_1 + \eta_2}C_1 C_2 e^{\delta_1+\delta_2} + \frac{(\eta_1 - \eta_2)^2}{\eta_1\eta_2(\eta_1 + \eta_2)^2}C_1^2 C_2^2 e^{2\delta_1+2\delta_2} \tag{3c}$$

with the abbreviation

$$\delta_i = -2\eta_i(x - 4\eta_i^2 t + \varepsilon_+) - \ln 2\eta_i \quad (i = 1, 2) \tag{4}$$

C_1 and C_2 represent initial values of the normalization factors of the bound-state eigenfunctions of the associated eigenvalue problem. The negative value of C_2 represents the negative y displacement, so that we call this as an antiloop soliton.

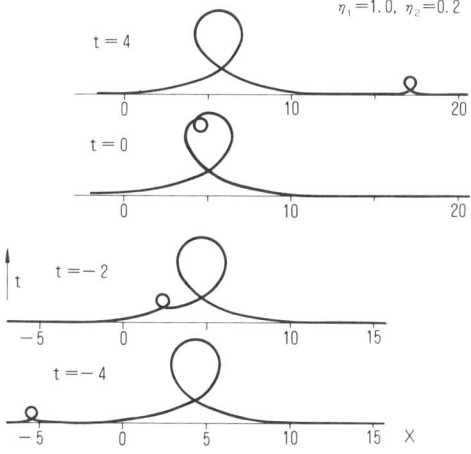

Figure 1. Collision process of two-loop solitons with dissimilar amplitudes.

The collision processes of two-loop solitons are classified into the three cases:

1. Two-loop solitons with dissimilar amplitudes.
2. Two-loop solitons with similar amplitudes.
3. A loop soliton and an antiloop soliton.

We show these three collision processes in Figs. 1, 2, and 3, respectively.

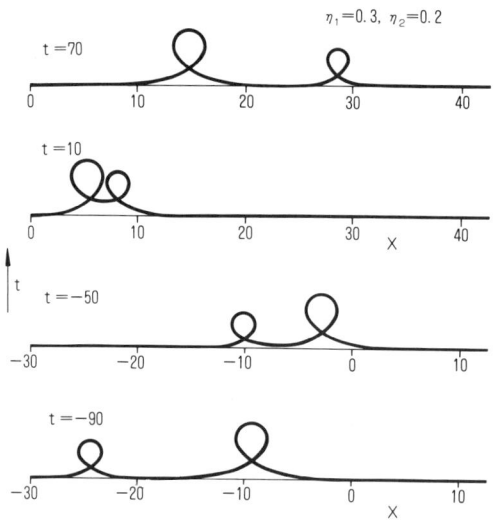

Figure 2. Collision process of two-loop solitons with similar amplitudes.

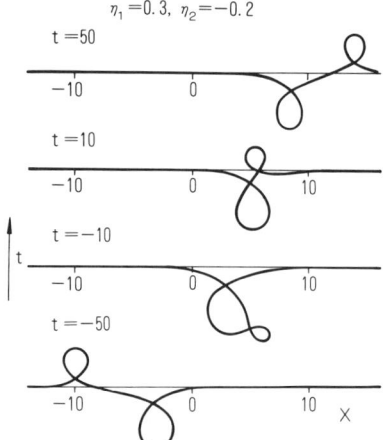

Figure 3. Collision process of a loop soliton and an antiloop soliton.

3. RESONANT INTERACTION OF SOLITONS

It is well known that the one-dimensional motion of ion acoustic waves of small amplitude and long wavelength is governed by the Korteweg–de Vries equation,[20] which is expressed in a nondimensional form as

$$\frac{\partial}{\partial t}u + \frac{\partial}{\partial x}u + u\frac{\partial}{\partial x}u + \frac{1}{2}\frac{\partial^3}{\partial x^3}u = 0 \tag{5}$$

where u is the dimensionless electrostatic potential normalized by Te/e (Te is electron temperature, $-e$ is the electric charge of an electron), x is normalized by λ_D (the Debye length), and t is multiplied by the ion plasma frequency. Equation (1) has a soliton solution

$$u = 6K^2 \operatorname{sech}^2(Kx - \Omega t) \tag{6a}$$

with

$$\Omega = K(1 + 2K^2) \tag{6b}$$

In isotropic media, solitons can propagate in every directions, and they interact with each other with a finite angle of intersection. Such interactions of solitons were first investigated by Miles in the study of oblique reflection of shallow water waves on a rigid wall.[28] A typical example of oblique interaction between two shallow water solitons is illustrated by a beautiful photograph of Toedtemier.[26] Two solitons interact with a peculiar spatial structure such that two wedges made by interacting soliton crests are connected with an intermediate wave crest. Miles has shown that three solitons interact strongly when the resonance conditions

$$\Omega_3 = \Omega_1 \pm \Omega_2 \quad \text{and} \quad \mathbf{K}_3 = \mathbf{K}_1 \pm \mathbf{K}_2 \tag{7}$$

are fulfilled, where each wave satisfies the soliton dispersion relation, Eq. (6b); that is, $\Omega_i = |K_i|(1 + 2K_i^2)$. If the resonance condition is satisfied, the intermediate wave crest extends to infinity to become a real soliton.

In plasmas, the resonant interaction of solitons is also possible. Yajima et al.[22] have studied this phenomenon by solving a multidimensionally extended form of the Kortweg–de Vries equation,

$$\frac{\partial^2}{\partial t^2}f - \nabla^2 f + \frac{\partial}{\partial t}(\nabla f)^2 - \frac{\partial^2}{\partial t^2}\nabla^2 f = 0 \tag{8}$$

where f is the dimensionless velocity potential of the ion fluid. The soliton solution for Eq. (8) is the same as Eq. (6a), but Kx should be replaced by $\mathbf{K}\cdot\mathbf{x}$ and Ω by

$$\Omega = \pm|\mathbf{K}|/(1 - 4K^2)^{1/2} \tag{9}$$

which agrees with Eq. (6b) in the case of small dispersion, $|\mathbf{K}| \ll 1$.

Two soliton solutions are illustrated in Fig. 4. The wave patterns are quite similar to the above-mentioned photograph.[26] When the resonance

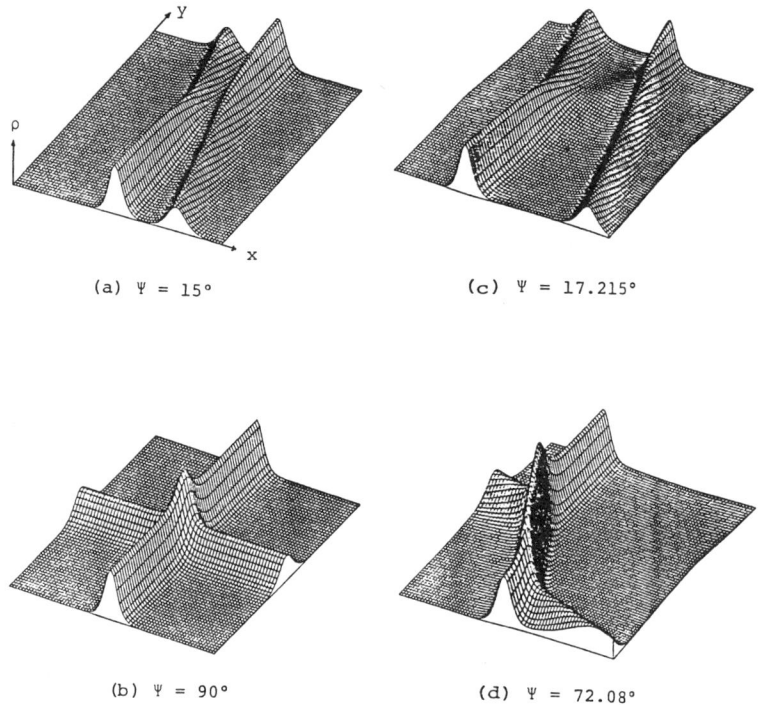

Figure 4. Two-soliton solutions for various interaction angles of solitons. Ψ = interaction angle of solitons.

condition [Eq. (7)] is satisfied, the internal wave crest elongates unrestricted and triple wave structure appears (see Fig. 5). If the interaction angle of solitons is in the range of two resonance angles determined by the upper and the lower sign of the resonance condition [Eq. (7)], the two-soliton solution no longer has the simple structure of Fig. 4, but a trapezoidal structure composed of internal (virtual) solitons appears to join the two colloiding solitons (see Fig. 6).

Experimental studies on oblique interactions of two ion acoustic solitons were made by Nishida and Nagasawa,[11] Tsukabayashi and Nakamura,[17] and Lonngren et al.[7] They have confirmed experimentally the above theoretical predictions. Typical wave patterns observed in their experiments are shown in Figs. 7 and 8.[30] These results point out the importance of the concept of solitons in the multidimensional space. Solitons interact strongly with each other by exchanging internal (virtual) solitons, which satisfy the resonance condition (7) approximately. Thus, the soliton resonance produces a complicated evolution of the system so that the system hardly exhibits recurrence.

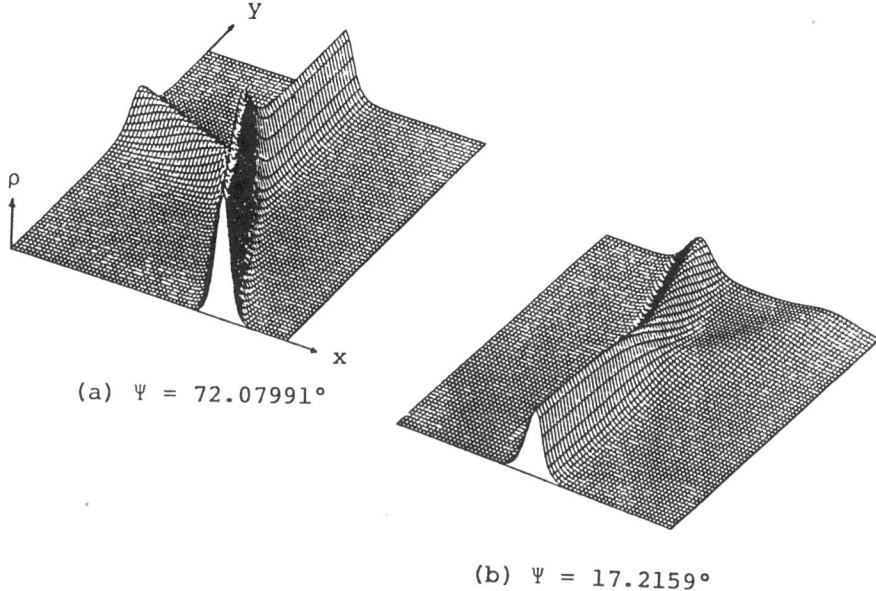

(a) Ψ = 72.07991°

(b) Ψ = 17.2159°

Figure 5. Resonant interaction of two solitons. Ψ = interaction angle of solitons.

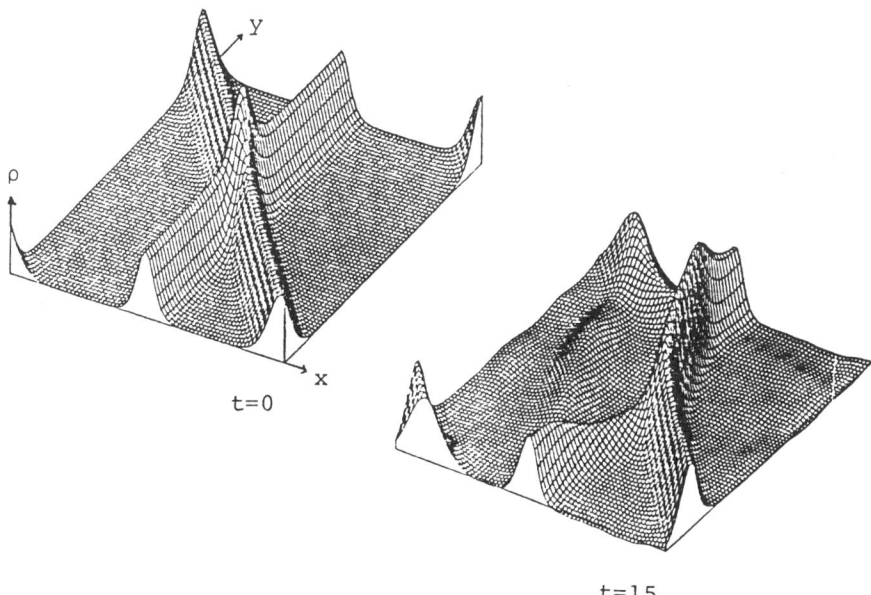

Figure 6. Temporal evolution of two-soliton solutions, in which the interaction angle of solitons is in the range of two resonance angles.

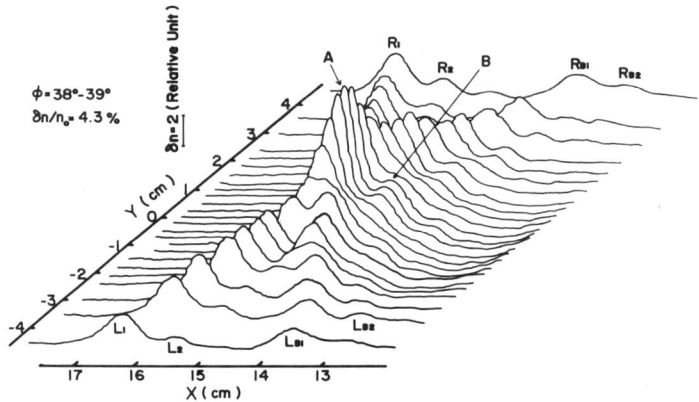

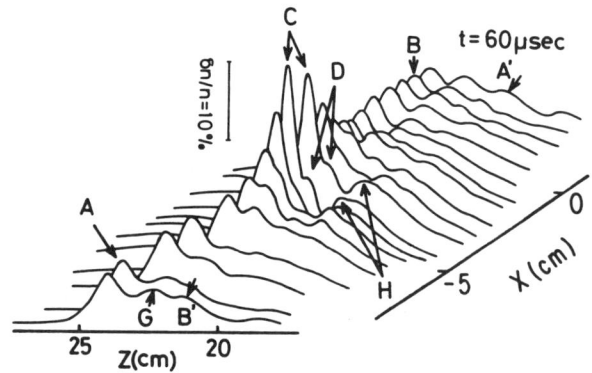

Figure 7. Experimental observations of oblique interactions of two ion acoustic solitons, by Nishida–Nagasawa (above) and Tsukabayashi–Nakamura (below).

4. SOLITON AND CHAOS

Since the soliton equations are completely integrable, the solitons do not interact with the chaos. However, the completely integrable soliton equations are idealized equations as limiting cases of the fundamental set of equations; they cover a wide variety of complex phenomena occurring in real systems, and they are valid in a limited range. For example, the Korteweg–de Vries equation is derived in the lowest order of the reductive perturbation expansion, which accounts for the balance of the nonlinear steepening effect and the dispersion effect as compared with the Korteweg–de Vries system; this might be a range where the soliton picture is valid for small amplitudes, but chaotic behavior prevails when the amplitude is increased. Nagashima and Kuwahara[10] and Yoshimura and Watanabe[23]

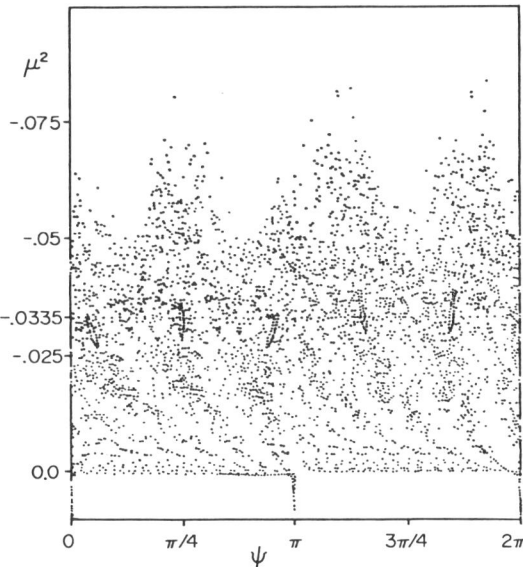

Figure 8. Trajectories of the breather mode under the action of the oscillatory external force.

have undertaken studies of such problems by examining the following equation:

$$\frac{\partial}{\partial t}u + u\frac{\partial}{\partial x}u - \frac{\partial^5}{\partial x^5}u = 0 \qquad (10)$$

which has three polynomial conserved quantities given by

$$I_1 = \int_{-\infty}^{+\infty} u\,dx \qquad (11a)$$

$$I_2 = \int_{-\infty}^{+\infty} u^2\,dx \qquad (11b)$$

$$I_3 = \int_{-\infty}^{+\infty} \left[u^3 - 3\left(\frac{\partial^2 u}{\partial x^2}\right)^2\right] dx \qquad (11c)$$

They have shown that two solitary waves, having oscillating tails, remain approximately stable through the collision process, but the three-soliton interaction does not remain stable. Occasionally, they observed formation of bounded solitons out of the three-soliton interaction. Upon increasing the amplitude of the initial perturbation, the number of solitons increases, and randomization takes place owing to the frequent collisions of the three solitons.

Another feature of the interaction between solitons and chaos could be found in the soliton systems that are subjected to the action of external force. Nozaki[13] has studied stochastic instability of the sine-Gordon solitons by analyzing the solution of

$$\frac{\partial^2}{\partial t^2}u - \frac{\partial^2}{\partial x^2}u + \sin u = \varepsilon \cos \omega t \tag{12}$$

Under action of the oscillating force, the breather mode will gain its energy stochastically, and it may break up eventually into a kink–antikink pair. Though production of kink–antikink pairs takes place randomly, once the pair is created it exhibits coherent motion driven by the oscillatory external force.

Nozaki[13] starts with a breather solution of the sine-Gordon equation

$$u = 4 \arctan[T(t) \operatorname{sech} 2Kx] \tag{13a}$$

$$T(t) = \sin(\tfrac{1}{16}\alpha) \tan \theta \tag{13b}$$

$$K = \tfrac{1}{2}\theta \tag{13c}$$

where θ and α form the canonical conjugate variables. Their time dependence is given by

$$\theta(t) = \theta(0) \tag{14a}$$

$$\alpha(t) = (16 t \sin \theta) + \alpha(0) \tag{14b}$$

as far as $\varepsilon \to 0$. When $|\tan \theta| \ll 1$, the energy of the breather could be very small. Hence, the breather mode could be excited at a thermal level. Since the breather mode is a genuine nonlinear mode, it contains higher harmonic components that can resonate with the external oscillating force. When these resonance domains overlap, the Zaslavsky–Chirikov stochastic instability occurs. Under the action of the oscillating force, θ increases, stochastically and reaches a value of $\theta = \pi/2$, where the breather modes turn into the kink–antikink pair. The resonance overlap condition determines the stochastic region of θ as

$$\frac{\pi}{2} > \theta > \frac{\pi}{2} - \left(\frac{\pi \varepsilon \omega}{2}\right)^{1/2} \tag{15}$$

Changing the variables θ and α into μ^2 and ψ through the following relations.

$$\mu^2 = -\cot^2 \theta \quad \text{and} \quad \psi = -\tfrac{1}{16}\alpha \tag{16}$$

Nozaki performed a numerical integration of the nonlinear dynamical equation for the parameters $\varepsilon = 0.01$ and $\omega = 0.9$. According to Eq. (15), the stochastic region is estimated to be

$$0 > \mu^2 > -0.06 \tag{17}$$

Now, for the initial value of $\mu^2 = -0.0355$, the trajectories of 52 points with various values of ψ has been traced up to 240 times steps in the unit of $2\pi/\omega$. Figure 8 shows all of the trajectories at each time step, in which 31 points reach to the value of $\mu^2 = 0$. That is to say, the breather mode turns into a kink–antikink pair. This observation illustrates the generation of the coherent structure out of the chaotic state.

5. CONCLUDING REMARKS

We have briefly described recent developments of studies of solitons in plasmas. The soliton is an idealized concept for specific integrable systems; it is the opposite of chaos. Since plasmas are genuine nonlinear systems, we believe that the interaction of solitons and chaos would be very fundamental process to disentangle the complex phenomena occurring in plasmas.

Because of space limitations, we are not able to discuss such problems as solitons in an unstable plasma and the formation of electric double layer. We would like to conclude this chapter by emphasizing that continued experimental studies of nonlinear phenomena in plasmas are crucial to stimulate fresh mathematical approaches. At the same time, as shown in the studies of soliton resonance experiments, theoretical analysis provides experimentalists with new ways to grasp the essence of their observations.

REFERENCES

1. C. S. Gardner and G. K. Morikawa, *Courant Inst. Math. Sci. Rep.*-9082, NYO (1960).
2. A. Hasegawa and T. Sato, *Phys. Fluids* **25**, 632 (1982).
3. Y. H. Ichikawa, K. Konno, and M. Wadati, *Long-Time Prediction in Dynamics*, C. W. Horton, Jr., L. E. Reichl, and A. G. Szebehny, eds. (Wiley, New York, 1983), p. 345.
4. F. Kako and N. Yajima, *J. Phys. Soc. Japan* **49**, 2063 (1980).
5. F. Kako and N. Yajima, *J. Phys. Soc. Japan* **51**, 311 (1982).
6. K. Konno, Y. H. Ichikawa, and M. Wadati, *J. Phys. Soc. Japan* **50**, 1025 (1981).
7. M. Khazei, J. Bulson and K. E. Lonngren, *Phys. Fluids* **25**, 759 (1982).
8. M. Makino, T. Kamimura, and T. Taniuti, *J. Phys. Soc. Japan* **50**, 980 (1981).
9. J. D. Meiss and W. Horton, Jr., *Phys. Rev. Lett.* **48**, 1362 (1982).
10. H. Nagashima and M. Kuwahara, *J. Phys. Soc. Japan* **50**, 3792 (1981).
11. Y. Nishida and T. Nagasawa, *Phys. Rev. Lett.* **45**, 1626 (1980).
12. K. Nishihara, H. Sakagami, T. Taniuti, and A. Hasegawa, *Rev. Rep. Inst. Laser Eng.*, ILE-8213p (1982).
13. K. Nozaki, *Phys. Rev. Lett.* **49**, 1883 (1982).
14. L. I. Rudakov, A. S. Kingsep, and R. N. Sudan, *Phys. Rev. Lett.* **31**, 1482 (1973).
15. K. Saeki, P. Michelesen, H. L. Peseli, and J. Rasmussen, *Phys. Rev. Lett.* **42**, 501 (1979).

16. R. Z. Sagdeev, *Reviews of Plasma Physics*, M. A. Leontovich, ed. (Consultants Bureau, New York, 1966), Vol. 4, p. 23.
17. I. Tsukabayashi and Y. N. Nakamura, *Phys. Lett.* **85A**, 152 (1981).
18. M. Wadati, K. Konno, and Y. H. Ichikawa, *J. Phys. Soc. Japan* **46**, 1965 (1979).
19. M. Wadati, K. Konno, and Y. H. Ichikawa, *J. Phys. Soc. Japan* **47**, 1698 (1979).
20. H. Washimi and T. Taniuti, *Phys. Rev. Lett.* **17**. 996 (1966).
21. K. Watanabe and T. Taniuti (private communication).
22. N. Yajima, M. Oikawa, and J. Satsuma, *J. Phys. Soc. Japan* **44**, 1711 (1978).
23. K. Yoshimura and S. Watanabe, *J. Phys. Soc. Japan* **51**, 3028 (1982).
24. V. E. Zakharov, *Soviet Phys. JETP* **33**, 538 (1971).
25. V. E. Zakharov, *Soviet Phys. JETP* **35**, 908 (1972).
26. M. J. Ablowitz and H. Segur, *Solitons and the Inverse Scattering Transform* (SIAM, Philadelphia, 1981) p. 291.
27. K. Konno and A. Jeffrey, *J. Phys. Soc. Japan* **52**, 1 (1983).
28. J. W. Miles, *J. Fluid Mech.* **79**, 171 (1979).
29. N. Yajima, M. Kono and S. Ueda, *J. Phys. Soc. Japan* **52**, 3414 (1983).
30. I. Tsukabayashi, Y. Nakamura, F. Kako and K. E. Lonngren, *Phys. Fluids* **26**, 790 (1983).

Part 2

DISSIPATIVE DYNAMICS

ONSET OF CHAOS IN CONTINUOUS MEDIA: CASE OF REACTION–DIFFUSION SYSTEMS

Y. Kuramoto
Research Institute for Fundamental Physics, Kyoto University, Kyoto, Japan

1. COMPLEX GINZBURG–LANDAU EQUATION

The present study concentrates on a single specific class of nonlinear diffusion equations, namely, the complex Ginzburg–Landau (GL) equation

$$\dot{W} = (\lambda - g|W|^2)W + D\nabla^2 W \tag{1}$$

where λ, g, and D are complex parameters, and we discuss a variety of turbulent behaviors exhibited or possibly exhibited by it. To begin with, some historical background of Eq. (1) is touched upon. The complex GL in one space dimension was first derived by Stewartson and Stuart[15] in connection with the stability problem in the plane Poiseuille flow. Equation (1) (with $\nabla^2 W = W_{xx}$) then describes the evolution of slightly unstable velocity fluctuations about the laminar flow of the well-known parabolic profile. The same equation (in arbitrary dimension) was later derived for reaction-diffusion systems[5] for the case when a spatially uniform steady state loses stability and small-amplitude oscillations of concentration fluctuations set in. The complex GL would never be restricted to a few fluid-mechanical problems or chemical reactions, but there are good reasons for expecting its appearance in many other physical problems. In fact, Newell[14] showed that a rather general class of nonlinear partial differential equations with quadratic nonlinearity reduces universally to the form of Eq. (1) near a certain stability threshold.

A few words would be needed as to why the complex GL represents such a universal class of partial differential equations. This seems to be related to the universal way in which oscillatory instabilities or Hopf bifurcations take place in a variety of physical situations. Take a general system of partial differential equations (for simplicity, in one space dimension) for an n-component vector $\mathbf{X}$ as

$$\dot{\mathbf{X}} = \mathbf{F}_\mu(\partial_x, \mathbf{X}), \quad \mu \text{ is a control parameter} \tag{2}$$

and consider how to describe the bifurcation of oscillatory motion from an equilibrium solution $\mathbf{X}_0(x, \mu)$. Let the equation be linearized in the deviation $\mathbf{u}(x, t) \equiv \mathbf{X}(x) - \mathbf{X}_0(x, \mu)$, and get the corresponding eigenvalue problem

$$\mathscr{L}(\partial_x, \mu)\mathbf{u} = \lambda \mathbf{u} \tag{3}$$

Let the system length L be finite and not too long so that the spectrum of λ may well be discrete. Then the Hopf bifurcation occurs when a pair of complex-conjugate eigenvalues cross the imaginary axis, while the rest of the eigenvalues are in the left-half plane and remain at a finite distance from the imaginary axis. Corresponding to the pair of eigenvalues that are becoming critical we have in an infinite-dimensional state space (e.g., the mode-amplitude space) an eigenplane P. In the crudest picture, the system dynamics near criticality ($\mu \simeq \mu_c$) is confined within P. To be a little more precise, there exists a special two-dimensional surface M (called the center manifold) tangent to P at the equilibrium point $\mathbf{X}_0(x, \mu)$ such that the dynamics is confined within M except for relatively short initial transients. In any case, the original infinite-dimensional system reduces to a two-dimensional system. As a consequence, if expressed in terms of a suitably defined complex amplitude W, the system reduces universally to a very simple equation, that is, $\dot{W} = (\lambda - g|W|^2)W$, which is called the Stuart–Landau (SL) equation. The above picture should be modified when the system length L becomes much longer and hence the eigenvalue spectrum becomes almost continuous. The previous picture (and hence the SL equation, too) still holds in the limit $\mu \to \mu_c$, but its validity would be limited to an infinitesimal range of μ about μ_c as L goes to infinity; for practical purposes, it seems therefore more desirable to take account not only the critical pair of eigenmodes, but also a large number of eigenmodes whose eigenvalues are close to the critical pair. One may then imagine the corresponding infinite-dimensional eigenspace $\tilde{P}$ or center manifold $\tilde{M}$ tangent to it (by no means a mathematically well-defined object, however), and expect that the dynamics is practically confined within $\tilde{M}$. The original n-component system is then expected to reduce to a two-component system but with almost continuous spatial degrees of freedom. In fact, a certain asymptotic method applied to the present situation leads to the complex GL equation, which is simply the modification of the SL equation by adding a diffusion term.

The complex GL involves some interesting extreme cases. First, let λ, g, and D all be imaginary numbers. Then, Eq. (1) reduces to a nonlinear Schrödinger equation, which is known to be an integrable system at least for one space dimension, and, hence, shows no turbulent behavior. Second, let λ, g, and D all be real numbers. Then we get a simple GL equation of phase transitions, and, hence, no turbulence again. Therefore, turbulence is possible only for some intermediate range between the above two. In the following, we restrict our attention to the case that λ', g', and D' (i.e., real parts of λ, g, and D) are all positive. Physically, this is equivalent to saying that one is considering situations above criticality, assuming the bifurcation type to be supercritical, and also assuming no diffusive instabilities to occur about X_0 up to criticality.

Note that the diffusionless complex GL (i.e., the SL equation) describes a perfectly smooth periodic motion along a perfect circle in the complex W plane. If we write the equation as $\dot{W} = (\lambda - gR^2)W$ $(R \equiv |W|)$, it may be clear that g'' ($\equiv \text{Im } g$) gives rise to the amplitude dependence of the oscillation frequency; this fact will become important later. The complex GL thus represents the field of such ideal limit cycle oscillators coupled through diffusion. For one space dimension, the profile of W for given t may be visualized by a state line threading through the three-dimensional space composed of two-dimensional W and space axis x. Such a representation will frequently be used in later discussions.

2. THREE TYPES OF PERIODIC SOLUTIONS

There exist at least three types of time-periodic solutions to the complex GL equation. The first two refer to systems of arbitrary space dimension, while the third is peculiar to two-dimensional systems. As to the former types of solutions, we restrict ourselves for simplicity to one-dimensional systems in the present chapter.

1. *Uniform Oscillations* $W_0 = R \exp(i\omega t)$. With the above-mentioned representation of the system state in the $W - x$ space, the solution W_0 may be visualized as a straight line circulating round a cylinder of radius R at a constant angular velocity ω (Fig. 1).

2. *Plane Waves* $W_Q = R(Q) \exp[i\omega(Q)t - iQx]$. In this case, we have a one-parameter family; both the amplitude and frequency are functions of wavenumber Q. In particular, R is a decreasing function of Q, and there exists a maximum value of Q for which R vanishes (Fig. 2). In the $W - x$ space, these plane waves are represented by "wires" coiling with pitch $2\pi/Q$ round a cylinder of radius $R(Q)$ and circulating with angular frequency $\omega(Q)$. For vanishing Q, these solutions reduce to type I. However, solution I deserves separate consideration.

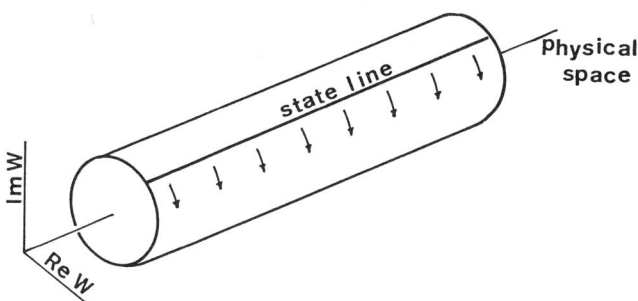

Figure 1. Geometrical representation of a uniformly oscillating state.

3. *Rotating Waves* $W_{rot}(r, \theta, t) = R(r) \exp\{i[\theta + S(r) + \omega t]\}$ (Ref. 18). Analytical forms of $R(r)$ or $S(r)$ are unknown; but numerical calculations imply that they behave as in Fig. 4. Specifically, $R(r) \to \text{const} (\neq)$ and $S(r) \to -Qr$ as $r \to \infty$. Thus, the contour of constant θ (phase of W) is given by a steadily rotating Archimedean spiral $\theta - Qr - wt = \text{const}$ as $r \to \infty$. A peculiar feature about this solution is the existence of a phase-singular point (i.e., $R = 0$ at $r = 0$). The contours Re $W = 0$ and Im $W = 0$ intersect vertically at this point. Let us define a topological number n by

$$n = \frac{1}{2\pi} \oint \nabla \theta \cdot d\mathbf{r} \qquad (4)$$

for a given contour C. If C represents a closed loop round the phase singularity, then $N = +1$ or -1 depending on the direction of wave rotation. This kind of rotating wave is known as the Belousov–Zhabotinsky reaction[16]. Recent experimental studies revealed the existence of rotating waves with $|n| > 1$[1]. However, such "multiarmed spiral waves" will not be considered here.

We now pose some questions on the above periodic solutions:

1. Are these periodic solutions stable?
2. If unstable, what happens?

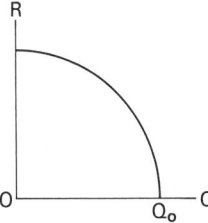

Figure 2. Wave amplitude R versus wavenumber Q for type-II solution.

It turns out that each of the solutions I–III may or may not be stable depending on parameter values. For every known case, their instability leads to turbulence (possibly after some bifurcations). Then:

3. Can different types of turbulence (corresponding to different types of periodic solutions) coexist?
4. If so, what is their interrelationship?

3. TURBULENCE ARISING FROM TYPE-I SOLUTION

Formal stability analysis of the uniform oscillation is very easy, but what is more important is the physical interpretation of the destabilizing mechanism, which we describe briefly. We employ the previous visualization of system states. Suppose, first, a straight state line is circulating around a cylinder as in Fig. 1. Let this line be given a small wavy deformation as in Fig. 3a, where the deformation is confined to the cylinder surface, that is,

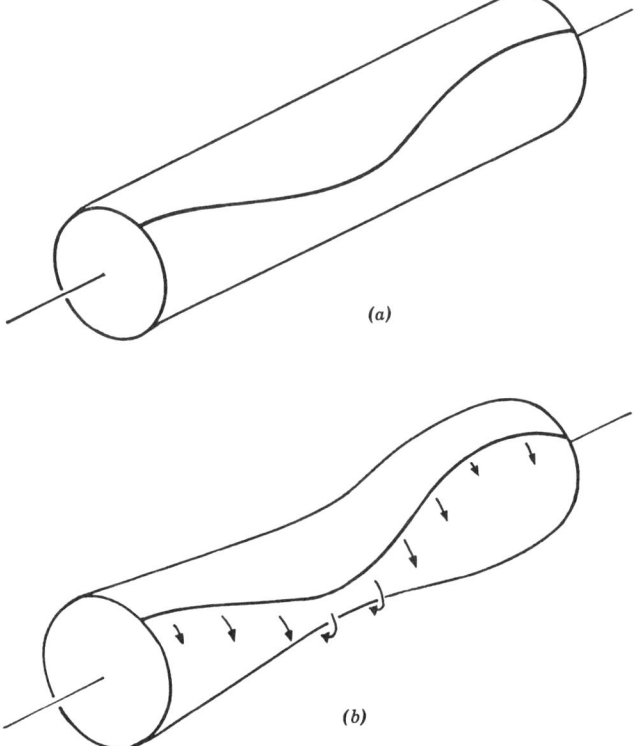

Figure 3. Destabilizing mechanism of a slowly phase-modulated oscillating state.

what is perturbed is only phase and not amplitude. If D is real, such a phase-deformed state profile will simply be relaxed back to the uniform state. In contrast, if D has the imaginary part D'', it has the effect, crudely speaking, of rotating this wavy line around itself. Then, some parts of the line will be pushed inward of the cylinder while the other parts will be pushed outward. Thus, the wavy phase disturbance has partly been transformed into a wavy amplitude disturbance. But we have already seen that the oscillation frequency of the local oscillators generally depends on their amplitude through the term g'', and, hence, we also get a wavy distribution in the local frequencies. It may occur (if D'' and g'' have right relative signs) that the phase-advanced parts acquire higher frequencies while the phase-retarded parts acquire lower frequencies (Fig. 3b). Obviously, this has the effect of amplifying the phase distortion. Of course, this effect should at least be partly cancelled by the ordinary stabilizing effect coming from the real part of the diffusion. These two conflicting effects are sharply reflected on the instability condition[14]

$$\nu \equiv 1 + \frac{D''}{D'} \frac{g''}{g'} < 0 \tag{5}$$

It may be realized from the preceding argument that instability of the uniform oscillation is essentially the "phase instability." In fact, the linear dispersion curves about W_0 show (Fig. 4) that the instability occurs in the *phaselike branch* and by no means in the *amplitudelike branch*.

If such a phase instability is sufficiently weak (i.e., if $\nu \lesssim 0$), then only long-wavelength phase modes are unstably excited, and this leads to the notion of *phase turbulence*.[8] In order to describe phase turbulence, it is crucial to make use of the fact that we have a good separation between time scales associated with the unstable phase fluctuations, on the one hand, and rapidly decaying amplitude fluctuations, on the other hand. Thus, we are allowed to eliminate the amplitude modes adiabatically, which enables the complex GL to be contracted to the *phase turbulence equation*[6,8]

$$\dot{\theta} = \nu \theta_{xx} + \theta_x^2 - \theta_{xxxx} \tag{6}$$

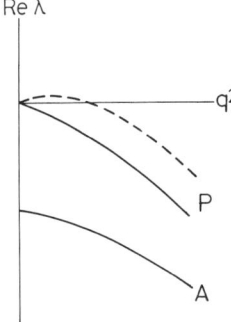

Figure 4. Linear dispersion curves about the uniform oscillation W_0 for the complex Ginzburg–Landau equation (1). P and A denote, respectively, the phase- and amplitudelike branches. The phaselike branch in a broken line corresponds to instability.

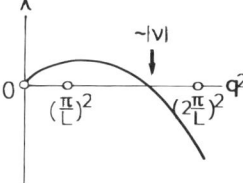

Figure 5. For the system size L of order $|\nu|^{-1/2}$, the number of the unstable modes is order 1.

where θ is the phase of W. [Extension to higher space dimensions is trivial, and leads to $\dot{\theta} = \nu\nabla^2\theta + (\nabla\theta)^2 - \nabla^4\theta$.] Equation (6) represents the simplest partial differential equation showing turbulence. The effective phase diffusion constant ν should of course be small negative, and the solution of Eq. (5) has the scaling form

$$\theta(x, t) = |\nu|\tilde{\theta}(|\nu|^3 t, |\nu|^{1/2} x) \qquad (7)$$

Let the periodic or no-flux boundary condition be imposed. Even if $|\nu|$ is very small, one may expect an arbitrarily large number of unstable modes by letting the system length L be sufficiently long. Specifically, if $L \gg |\nu|^{-1/2}$, the number of unstable modes is much larger than 1, while if $L \sim |\nu|^{-1/2}$, we have a few or no unstable modes (Fig. 5). Considering the latter case is suited to investigate the onset of phase turbulence. Put $L = L_0 |\nu|^{-1/2}$ and increase L_0 continuously, and numerically integrate the complex GL. In this way, one may investigate the structure of bifurcation sequence and transitions to chaos. It was revealed[9] that the transition to chaos in the present

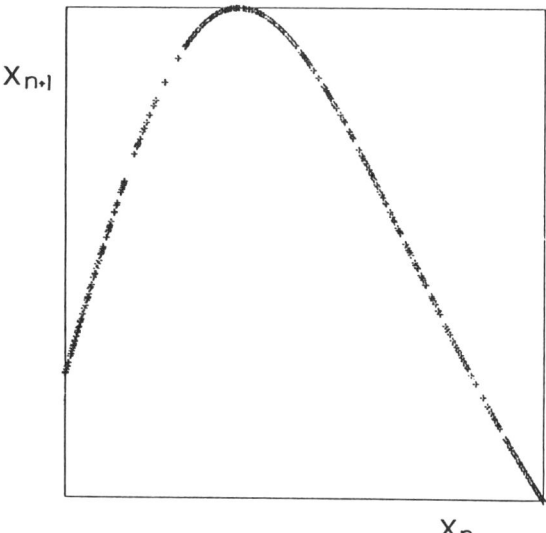

Figure 6. Approximate one-dimensional map obtained from Eq. (6).

case is characterized by the accumulation of period-doubling bifurcations, which is implied from a quasi-one-dimensional map (Fig. 6) constructed from the phase portrait in a suitably defined mode amplitude space. Note that no other routes to chaos can exist. This is because there is only one parameter involved as can be seen from the elimination of spurious parameter(s) by rescaling.

When L is made larger, the turbulent behavior becomes more and more complicated, which does not seem to allow for any reasonable description. In the limit of large L, however, the dynamics seems to recover simplicity but only in a statistical sense. In fact, the spectrum of the phase fluctuations, that is, $S_k = \overline{|\theta_k|^2}$ (average taken over a sufficiently long time interval), as a function of the wavenumber k seems to obey a certain simple law as suggested from Fig. 7.[19] In particular, the long-wavelength fluctuations seem to behave like $S_k \sim k^{-2}$, which is analogous to the frequency spectrum of the displacement in one-dimensional Brownian motion. This fact implies that the large-scale behavior of phase turbulence is statistically characterized by *spatial phase diffusion* (on a cylindrical surface). For further numerical analysis and statistical theories on the phase turbulence equation, see Fujisaka and Yamada,[3] Maneville,[12] and Yakhot.[17]

When ν decreases further, the turbulence becomes stronger, until it deviates from the picture of phase turbulence; we would rather call it

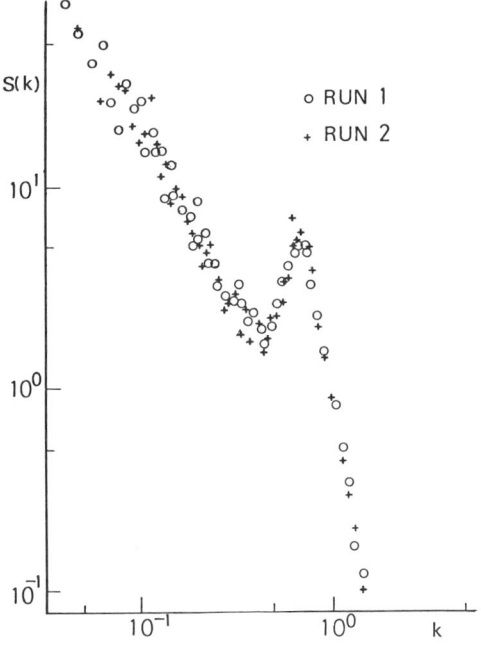

Figure 7. Phase fluctuation spectrum for the system of Eq. (6).

amplitude turbulence, for which the amplitudelike branch becomes important in addition to the phaselike branch. Then, we have to work with the full complex GL equation. Again, we should consider separately the cases of smaller L and larger L.

For amplitude turbulence, the route to chaos may differ from that of phase turbulence. Moreover, there may be many possible routes because the system now involves a number of parameters. In order to study bifurcations and transitions to chaos, we should first note a certain symmetry property of the complex GL, namely, its invariance under rotation $W \to We^{i\phi}$. This implies, and it may in fact be proved, that the dimension of attractors is lowered by one if we work with the new variable $\tilde{W}$ defined by

$$\tilde{W}(x, t) = W(x, t)e^{-i\theta_0(t)}, \tag{8}$$

where $\theta_0(t)$ is the phase of the spatially uniform component of $W(x, t)$. In this new picture, possible routes to chaos discovered so far are listed as follows:

Route 1: limit cycle $\to$ 2^n bifurcations $\to$ chaos (Ref. 13).

Route 2: limit cycle $\to$ 2-torus $\to$ limit cycle (phase-locking) $\to$ chaos (Ref. 13).

Route 3: limit cycle $\to$ anomalous 2^n bifurcations $\to$ chaos (Ref. 11).

Of course, there may be other routes. Here we only make a brief comment

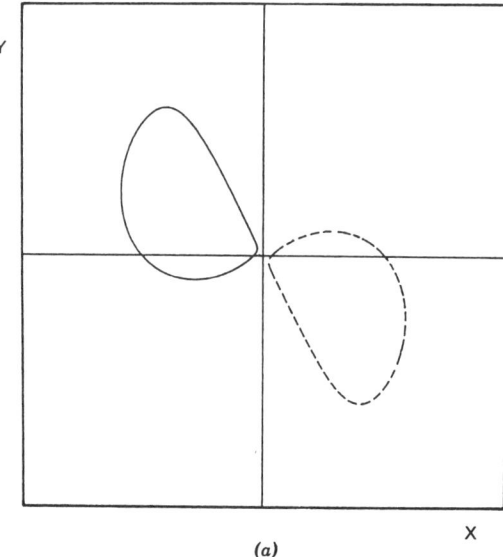

Figure 8. $(a-f)$ Anomalous 2^n bifurcations shown by the complex Gl equation. (Continued on pages 102–104.)

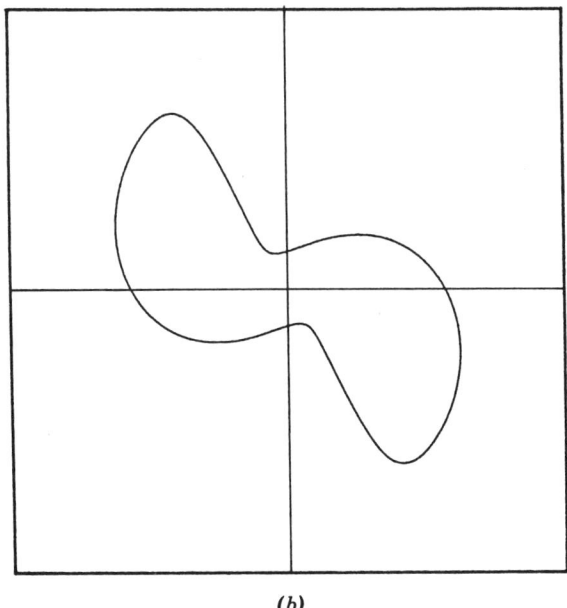

(b)

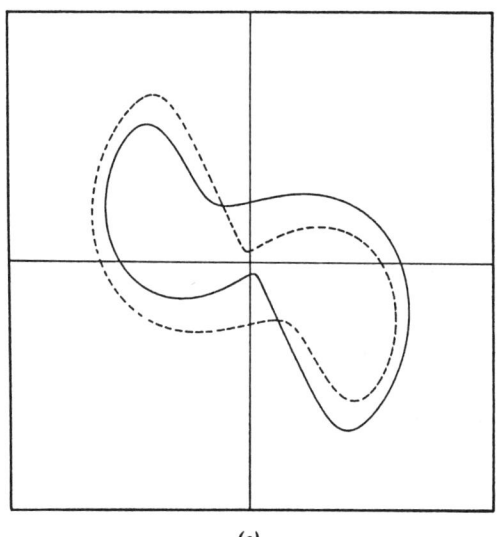

(c)

Figure 8. (b–c).

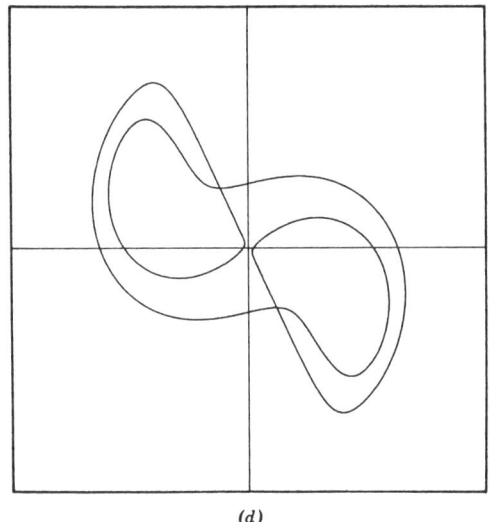

(d)

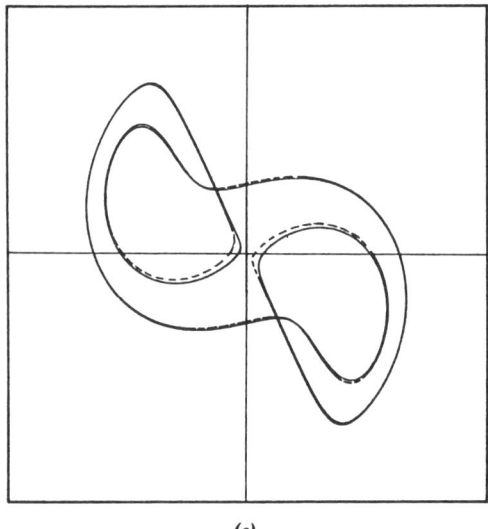

(e)

Figure 8. (d–e).

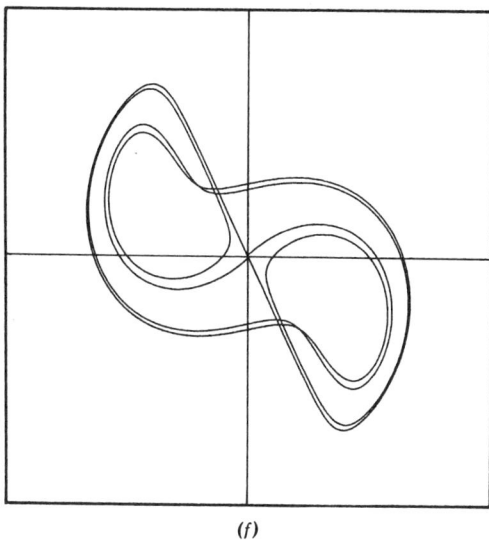

(f)

Figure 8. (f).

on what the anomalous 2^n bifurcations mean. A numerical simulation of Eq. (1) over the interval $[-L/2, L/2]$ under the no-flux boundary conditions was carried out. We changed g'' with all the other parameters kept constant. In a suitably defined phase space [specifically, the space spanned by the amplitudes of some Fourier modes of $\tilde{W}(x, t)$], we first have a limit cycle C_1 as indicated by a solid line in Fig. 9a. Owing to the invariance property of the system under spatial inversion, there exists another closed orbit $\bar{C}_1$, which is indicated by a broken line. (The spatial inversion transforms C_1 and $\bar{C}_1$ into each other.) As g'' is changed, C_1 and $\bar{C}_1$ come close to each other and also to the unstable fixed point O (situated at the center of symmetry), which corresponds to the uniform oscillations of W. For some critical value of g'', C_1 and $\bar{C}_1$ form a pair of homoclinic orbits, after which they are joined to form a single closed orbit D_1 as in Fig. 8b. D_1 is symmetric, namely, invariant under the spatial inversion. As g'' increases further, D_1 splits into a pair of symmetry-broken orbits C_2 and $\bar{C}_2$, and the situation becomes topologically the same as in Fig. 9a. As before, C_2 and $\bar{C}_2$ are connected into a single closed orbit D_2, which splits into C_3 and $\bar{C}_3$, and so forth. Such a process seems to repeat itself an infinite number of times, and an accumulation of bifurcations seems to occur at some finite value of g''. The possibility of such an exotic bifurcation cascade was first pointed out by Arneodo et al.[2] Slightly above the accumulation point, a suitably constructed Lorenzian plot gives a smooth unimodal curve as in Fig. 9. However, the curve is not analytic near its maximum (empirically

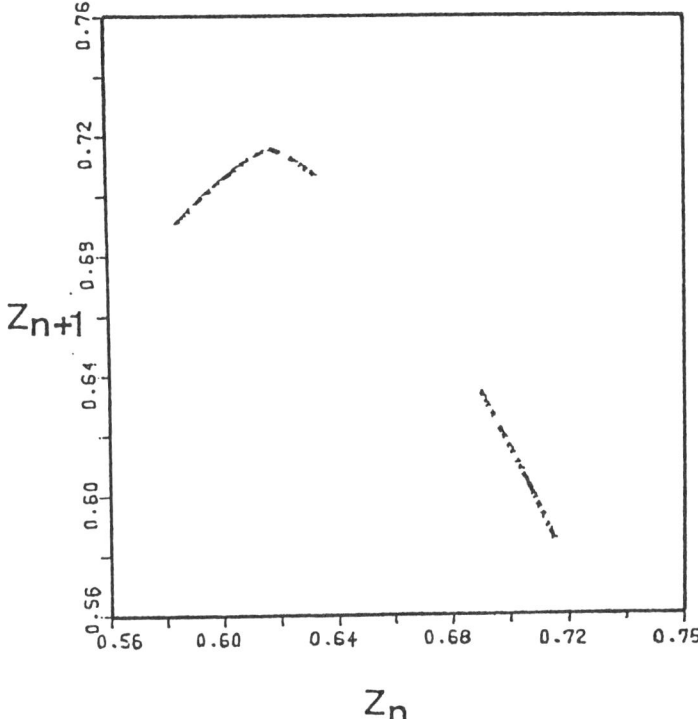

Figure 9. Approximate one-dimensional map obtained from the complex GL equation.

and also theoretically). As a consequence, the Feigenbaum constant δ would differ from its standard value $4.669\cdots$.

In the limit of large L, some simple *statistical* properties may be expected, but there are virtually no numerical studies in this direction as yet. We only show the spatial correlation of Re W as a function of distance, which in fact shows a smooth behavior.[7] While observing the behavior of $W(x, t)$, it was noted that the state line spontaneously intersects the phase-singular line $W = 0$ (Fig. 10). If we define a rotation number n by $n \equiv \Delta\theta/2\pi$, where $\Delta\theta$ is the phase difference between the endpoints of the system, then the above feature of the state line means the spontaneous change of n. In other words, the system may jump from one plane-wave state to another in a stochastic manner because various plane waves were seen to be distinguished by n. Such stochastic transitions characterize amplitude turbulence, while this was absolutely impossible for phase turbulence since the state line is almost confined to the cylinder surface. The difference between the two types of turbulence is symbolically represented in Fig. 11.

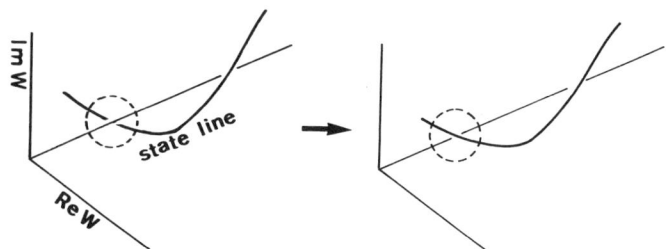

Figure 10. State line crossing the phase-singular line $W = 0$.

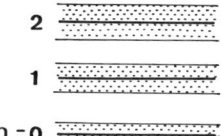

Figure 11. Plane-wave states (thick lines) are specified by the rotation number n ($=\Delta\theta/2\pi$). Each plane-wave state, if weakly unstable, is accompanied by a narrow turbulent regime about it, which also has a definite rotation number.

4. TURBULENCE ARISING FROM TYPE-II SOLUTION

In the nonlinear-Schrödinger limit, the plane waves are totally unstable or totally stable (depending on the sign of $d''g''$). This is not the case for the general complex GL. The stability of W_Q may be known by first putting

$$W(x, t) = W_Q(x, t)\{1 + \rho(x, t)\} \exp[i\phi(x, t)]$$

and then linearizing the complex GL in ϕ and ρ. Analysis shows that the degree of stability generally decreases as $|Q|$ is increased. Since the sign of ν determines the stability of W_Q, which is the most stable "plane wave," it follows that if $\nu > 0$, there exists some critical value Q_c below which W_Q is stable and beyond which it is unstable; if $\nu < 0$, all plane waves are unstable (Fig. 12). We concentrate attention upon the former case. If Q is only slightly larger than Q_c, then we expect that something like phase turbulence will happen. A perturbation theory shows that the phase-turbulence equation appororiate to this case includes some dispersion effects like

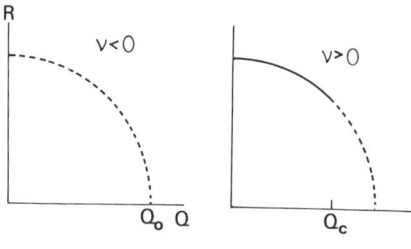

Figure 12. For $\nu > 0$, there exists a critical wavenumber Q_c above which the plane waves are unstable, while for $\nu < 0$, all plane waves are unstable.

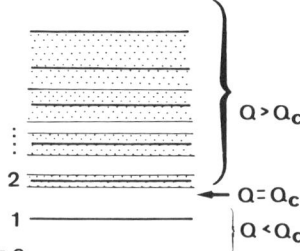

Figure 13. For the region of large rotation number n, the turbulence is so strong that n itself cannot be conserved. Compare with Fig. 11.

$$\dot{\phi} = \mu\phi_{xx} + \phi_x^2 + \phi_{xxx} - \phi_{xxxx} \qquad (9)$$

where μ is small negative corresponding to the weak instability of W_O. In Eq. (9), not all terms can have comparable magnitude. We conjecture the scaling form of the solution as

$$\phi = |\mu|^{1/2}\tilde{\phi}(|\mu|^{3/2}t, |\nu|^{1/2}x) \qquad (10)$$

Then terms ϕ_x^2 and ϕ_{xxx} are of $O(|\mu|^2)$ and dominant, while $\mu\phi_{xx}$ and ϕ_{xxxx} are of $O(|\mu|^{5/2})$. Thus, Eq. (10) represents a weakly dissipative K–dV equation (via the transformation $\theta_x = -v$). Note that the weak dissipation terms include both instability and damping, and hence as $t \to \infty$, some statistically definite state would be realized. In any case, numerical studies of Eq. (6) would be valuable, and there are no such works except for Kawahara's very recent study (private communication, 1983).

Also, no studies exist about the amplitude turbulence expected for $Q \gg Q_c$. We may only say at present that at least the qualitative picture as indicated in Fig. 13 would be valid. That is, we would have stochastic transitions among various plane-wave states in the region $Q \gg Q_c$ where the mean Q or the rotation number n would possibly drift toward lower values, while there would be no such transitions in the region near Q_c.

5. TURBULENCE ARISING FROM TYPE-III SOLUTION

No analytical theories exist toward the stability of rotating waves of the two-dimensional complex GL. Still some qualitative predictions seem to be possible. For simplicity, let D'' vanish, which stabilizes solutions I and II except for too large Q. Let g'' be very large, which means strong dependence of frequency on amplitude. Then, the amplitude profile $R(r)$ as shown in Fig. 2 implies that the distribution of the local frequencies shows a strong dependence on r especially near the core region. In contrast, if g'' were not so large, the frequency profile of the distributed oscillators would show a weak dependence on r, so that radially coupled such oscillators could show a synchronized motion that would appear as a steady rotation of waves about the phase-singular point. For too large g'', however, such a synchronized behavior would be impossible. Roughly speaking, the outer region would tend to rotate very fast, while in the inner region its slow rotation would persist, which would necessarily cause a strong "shearing force" leading to the breakdown of the coherent motion. Figure 14 shows the turbulization of a rotating spiral pattern possibly due to the previously mentioned destabilization mechanism.[10] Careful observations of this process show that there occur spontaneous pairwise formations of phase singularities. Some of these phase singularities may be annihilated quickly, while others may enjoy a rather long lifetime. Since the phase singularity itself turns out to be the very source of instability, the above implies a kind of instability cascade; the turbulent region will spread indefinitely. Note that such a violent turbulence is caused only by a single initial phase singularity; without it, the system would only repeat perfectly uniform monotone oscillations. It may also be said that the domain of attraction of a rotating-wave state, which may be turbulent or nonturbulent, is topologically

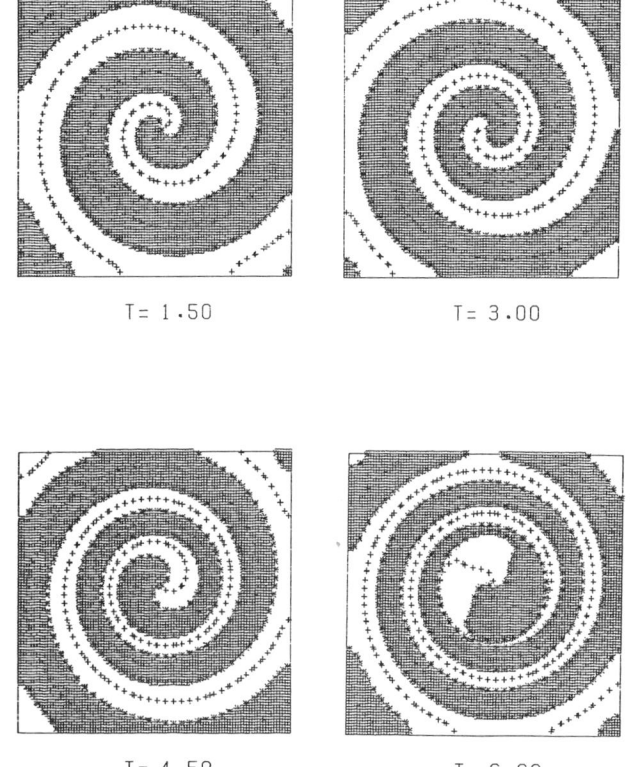

Figure 14. Turbulization process of a rotating spiral pattern after suddenly increasing the value of g'' across the stability threshold. (*Continued on next page*)

T= 7.50   T= 9.00

T= 10.50   T= 12.00

Figure 14.

separated from that of uniform oscillation by different values of the topological number defined in Eq. (4) (C being chosen as enclosing the entire system). This makes a clear contrast to one-dimensional cases for which sufficiently strong turbulence can unite various domains of attraction of plane-wave states.

It may be questioned why a destabilizing mechanism similar to the above should not cause turbulence in one space dimension. In fact, this kind of instability in one-dimensional systems turns out possible if one fixes the W value at one endpoint to zero. Then we have an amplitude profile similar to that of Fig. 4a, and, hence, the foregoing reasoning may hold. Specifically, a too steep gradient in the local frequency distribution would be sufficient for causing destabilization and turbulence. Preliminary computer simulation in fact confirms this view.[4] Actually, even the fixed boundary condition would not be necessary; some local dip in R seems to be sufficient. There are some preliminary computer calculations of this problem, but it is too early to state their results.

REFERENCES

1. K. I. Agladze and V. I. Krinsky, "Multi-Armed Vortices in an Excitable Chemical Medium," *Nature* **296**, 424 (1982).
2. A. Arneodo, P. Coullet, and C. Tresser, "A Possible New Mechanism for the Onset of Turbulence," *Phys. Lett.* **81A**, 197 (1981).
3. H. Fujisaka and T. Yamada, "Theoretical Study of Chemical Turbulence," *Prog. Theor. Phys.* **57**, 734 (1977).
4. S. Koga and Y. Kuramoto (unpublished).
5. Y. Kuramoto and T. Tsuzuki, "Reductive Perturbation Approach to Chemical Instabilities," *Prog. Theor. Phys.* **52**, 1399 (1974).
6. Y. Kuramoto and T. Tsuzuki, "Persistent Propagation of Concentration Waves in Dissipative Media Far from Thermal Equilibrium," *Prog. Theor. Phys.* **55**, 356 (1976).
7. Y. Kuramoto and T. Yamada, "Turbulent State in Chemical Reaction," *Prog. Theor. Phys.* **55**, 679 (1976).
8. Y. Kuramoto, "Diffusion-Induced Chaos in Reaction Systems," *Prog. Theor. Phys. Suppl.* **64**, 346 (1978).
9. Y. Kuramoto, "Diffusion-Induced Chemical Turbulence," In *Dynamics of Synergetic Systems*. H. Haken, ed. (Springer, Heidelberg, 1980), p. 134.
10. Y. Kuramoto and S. Koga, "Turbulized Rotating Chemical Waves," *Prog. Theor. Phys.* **66**, 1081 (1981).
11. Y. Kuramoto and S. Koga, "Anomalous Period-Doubling Bifurcations Leading to Chemical Turbulence," *Phys. Lett.* **92A**, 1 (1982).
12. P. Manneville, "Statistical Properties of Chaotic Solutions of a Unidimensional Model for Phase Turbulence," *Phys. Lett.* **84A**, 129 (1982).
13. H. T. Moon, P. Huerre, and L. G. Redekopp, "Three-Frequency Motion and Chaos in the Ginzburg–Landau Equation," *Phys. Rev. Lett.* **49**, 458 (1982).
14. A. C. Newell, "Envelope Equations," *Lectures in Applied Mathematics* **15**, 157 (1974).
15. K. Stewartson and J. T. Stuart, "A Non-Linear Instability Theory for a Wave System in Plane Poiseuille Flow." *J. Fluid Mech.* **48**, 529 (1971).
16. A. T. Winfree, "Spiral Waves of Chemical Activity," *Science* **175**, 634 (1972).
17. V. Yakhot, "Large-Scale Properties of Unstable Systems Governed by the Kuramoto–Sivashinsky Equation," *Phys. Rev. A* **24**, 642 (1981).
18. T. Yamada and Y. Kuramoto, "Spiral Waves in a Nonlinear Dissipative System," *Prog. Theor. Phys.* **55**, 2035 (1976).
19. T. Yamada and Y. Kuramoto, "A Reduced Model Showing Chemical Turbulence," *Prog. Theor. Phys.* **56**, 681 (1976).

PROGRESS IN COMPUTING LYAPUNOV EXPONENTS FROM EXPERIMENTAL DATA

Alan Wolf and Jack Swift
Department of Physics, University of Texas, Austin, Texas

Abstract. *Of the many diagnostics proposed for deterministic chaotic systems, the spectrum of Lyapunov exponents appears to be the most physically relevant. A well-known technique for computing these exponents for systems defined by sets of differential equations may be reformulated for use with experimental data. Here we discuss the calculation of the largest positive exponent with this new technique and compare it to the use of an underlying one-dimensional map to compute the same exponent.*

1. INTRODUCTION

Irregular temporal behavior of a physical system, evidenced by a continuous rather than a discrete power spectrum, can arise from at least two different sources: (1) the system may be driven by stochastic forces such as thermal noise, which represents Avogadro's number of degrees of freedom, or (2) the system contains a "strange attractor," that is, it can be described by a small number of completely deterministic differential equations in which there is a "sensitive dependence on initial conditions."[1]

There has been much recent effort toward determining which of the two mechanisms best describes the irregular behavior in a particular experiment and characterizing the chaotic motion associated with the latter mechanism. Methods that have been proposed include computation of the fractional dimension* of the attractor[2,3], D_f, directly from its definition; computing

---

*By fractional dimension we mean the capacity, fractal dimension, information dimension, dimension of the natural measure, and anything else one finds in the recent literature. We do not distinguish between these because it is not clear that any one of them is more physically relevant than the others and often they are nearly the same (or identical) in value. For a lucid discussion of the many definitions of dimension, see Ref. 3.

the dimension by studying the behavior of a spatial correlation integral[4]; computing entropylike quantities by applying ideas from symbolic dynamics[5]; computing the fractional dimension of power spectra[6]; and computing the spectrum of Lyapunov exponents.[7,8] Symbolic dynamics appears promising but has not yet provided useful tools for characterizing chaos. Based on our experience with computations of fractional dimension of attractors[9] we do not believe such techniques can be applied in general to experimental data which are obtained in limited amounts and always contain noise†. We believe, however, that Lyapunov exponents can be determined in many cases from experimental data and are more useful in describing the dynamics than the other diagnostics listed (including providing an approximate value for D_f[10,11]). In this work we discuss some practical considerations for computing these exponents from imperfect data.

The Lyapunov exponents provide a *qualitative* description of the chaotic dynamics in that the number of positive exponents gives the number of directions in phase space in which there is an exponentially fast separation of nearby‡ orbits, while the number of negative exponents gives the number of directions in which there is an exponentially fast approach to the attractor. The positive exponents *quantify* the "sensitive dependence," giving an estimate of how far into the future one can predict the state of a deterministic system given some uncertainty in the knowledge of its present state.[12] The negative Lyapunov exponents *quantify* the stability of the system with respect to perturbations and are directly related to the average time required for a transient to decay.

In addition, the conjecture of Kaplan and Yorke[10,11] provides an estimate of the fractional dimension of an attractor if a certain subset of the Lyapunov exponents (the positive, zero, and sufficiently weak negative ones) are known. As Takens has discussed,[13] deterministic systems should be characterized by motion on finite (but not necessarily integer) dimensional sets in phase space, while stochastic systems move on infinite-dimensional sets. As a practical matter calculations of fractional dimension will never predict that a system is infinite dimensional, but the dimension may be taken as a relative measure of dynamical complexity.

In Section 2 we review the definition of Lyapunov exponents and the standard technique for calculating them exactly (that is, to any specified precision) from the differential equations defining a system. In Section 3 we discuss the calculation of the single positive Lyapunov exponent of an

---

†Despite the improvements[3,4] that have been made over the simple box-counting algorithms we considered in Ref. 9, we still believe that fractional dimension cannot in general be computed for strange attractors of dimension $\gtrsim 2$ by "static" techniques. It is also unclear to us whether the utility of fractional dimension justifies the cost of its computation (which for static techniques is exponential in the dimension).

‡Throughout this chapter terms such as "nearby" and "adjacent" mean either "separated by an infinitesimal distance" or, in the case of experimental data, "closest available."

approximately two-dimensional strange attractor through the construction of an underlying one-dimensional (1-D) map. Although this technique is well known[14-16] we find that its application to *experimental* data does not provide "robust" (numerically stable) answers. In Section 4 we present a new method for calculating the positive exponent of such attractors that should prove more effective. In a future work we will discuss the extension of this method to the determination of other exponents, for a large class of attractors. We conclude in Section 5 with a warning about the estimation of Lyapunov exponents in systems containing extrinsic noise.

2. DEFINITION AND PROPERTIES OF LYAPUNOV EXPONENTS

If $e_i(0)$ is the length of the ith principal axis of an n-dimensional ellipsoid of initial conditions in an n-dimensional phase space, then the ith one-dimensional Lyapunov exponent is defined by[17]

$$\lambda_i = \lim_{t \to \infty} \frac{1}{t} \ln \left[\frac{e_i(t)}{e_i(0)} \right] \tag{1}$$

where $e_i(0)$ and $e_i(t)$ are infinitesimal and the time evolution of each state is governed by the equations of motion defining the system. The labels of the principal axes are chosen so that $\lambda_i \geq \lambda_{i+1}$. We illustrate the interpretation of the Lyapunov spectrum for the Rössler strange attractor,[18] by discussing the spectral calculation using the method of Shimada and Nagashima.[8]

The attractor is defined by the equations

$$\frac{dx}{dt} = -y - z$$

$$\frac{dy}{dt} = x + 0.15y$$

$$\frac{dz}{dt} = 0.2 + z(x - 10.0) \tag{2}$$

After transients decay, phase-space motion takes place on an attractor of dimension $D_f \approx 2.01$. The attractor is very nearly locally planar with fractal structure not visible to the "naked eye" (that is, on the largest length scales).

Shimada and Nagashima linearize the equations of motion so they can study the long-time evolution of small perturbations. In particular they take an orthonormal frame of (three) perturbation vectors and use the linearized system (sometimes called the first variational equations) and the equations of motion to mimic the evolution of an ellipsoid of initial conditions. Long-time integration of the perturbation vectors results in a divergence of their magnitudes and an inability to distinguish between their directions owing to the existence of a most rapidly growing direction in phase space. To avoid these problems they periodically reorthonormalize the vectors by a Gram–Schmidt procedure which in essence allows one vector to seek the direction which is most rapidly growing locally, a second vector to seek the second most rapidly growing direction, and so on. When the average growth rate of these vectors converges, all of the Lyapunov exponents are determined. Implementing this technique for the Rössler attractor we find a Lyapunov spectrum of (+0.09, 0.00, −9.80) bits/second. Convergence to this precision for the positive and zero exponent requires an integration time corresponding to a few hundred orbits. The positive exponent governs the exponentially fast separation (on the average) of points on nearby orbits *within* the almost locally planar attractor. The zero exponent corresponds to the fact that nearby points on the *same* orbit do not show any exponentially fast relative motion. Finally, the negative exponent governs the exponential approach (on the average) to the attractor of points near to it.

The Lyapunov exponents have a property that follows clearly from their definition, Eq. (1). If two adjacent points are picked at random near a strange attractor as initial conditions, the long-time evolution of the length they define goes as $D(t) = D_0 e^{\lambda_1 t}$, where λ_1 is the largest Lyapunov exponent. (Initially there are additional contributions from the other exponents.) If three nearby points are chosen at random near the attractor, the long-time evolution of the area they define goes as $A(t) = A_0 e^{(\lambda_1 + \lambda_2)t}$, where λ_1 and λ_2 are the two largest Lyapunov exponents. Similarly, four points define a three-dimensional volume whose growth rate is governed by the sum of the three largest exponents. For systems in higher-dimensional phase spaces this property generalizes in an obvious manner.

Although this property was well known,[7,8] apparently it was not realized that it provides a practical means for calculating Lyapunov spectra from experimental data where the equations of motion are not available. We describe this technique in Section 4.

3. A LYAPUNOV EXPONENT FROM A 1-D MAP

We now consider systems with (+, 0, −, −, −, ...) Lyapunov spectra where all of the negative exponents are much larger in magnitude than the positive one. Many of the standard model systems show (+, 0, −) spectra[18,19] and some experimental systems seem to indicate (+, 0, −, −, −, ...) spectra,[20,21] so this is not a terribly limiting restriction.

Given samples of a single dynamical variable, $x(t_i)$ at times t_i, from such a system we may obtain the underlying 1-D map responsible for its chaotic behavior. Labeling the value of the *i*th local maxima (or minima) of this time series as $x(i)$ we have a sequence of numbers that could be considered the output of the one-dimensional process, $x(i+1) = f(x(i))$.[19] We may define a Lyapunov exponent for this map which will be positive. This exponent, λ (with units of bits/iteration), is simply related to the exponent for the associated continuous system, λ_1 (bits/"second"), by the equation $\lambda = \lambda_1 \tau$ if the distribution of times between maxima is sharply peaked about the value τ ("seconds"/iteration).

Given the functional form of a 1-D map the Lyapunov exponent can be computed from[22]

$$\lambda = \lim_{n\to\infty} \frac{1}{n} \sum_{i=0}^{n-1} \ln|f'(x(i))| \qquad (3)$$

Alternately, if we define

$$p(x) = \lim_{n\to\infty} \frac{1}{n} \sum_{i=0}^{n-1} \delta(x - x(i)) \qquad (4)$$

as the probability density for our 1-D map, Eq. (3) is (assuming ergodicity) equivalent to

$$\lambda = \int p(x) \ln|f'(x)| \, dx \qquad (5)$$

where the integral is taken over the domain of the map. We briefly review the origin of these formulas. A nonzero Lyapunov exponent indicates that a pair of infinitesimally separated points, initially a distance $D(0)$ apart, diverge (converge) exponentially fast on the average according to $D(n) = D(0)e^{\lambda n}$. We may write

$$\frac{D(n)}{D(0)} = \frac{D(n)}{D(n-1)} \cdots \frac{D(1)}{D(0)} \qquad (6)$$

$$= |f'(x(n-1))| \, |f'(x(n-2))| \cdots |f'(x(0))|$$

$$= \prod_{i=0}^{n-1} |f'(x(i))|$$

Putting this into the desired exponential form we have

$$\lambda = \frac{1}{n} \ln \prod_{i=0}^{n-1} |f'(x(i))|$$

which is equivalent to Eq. (3) when we take the limit of large n (which simultaneously averages over all initial conditions and tells us about long-time behavior). Thus, λ reflects nothing more than the mean stretching or contracting of the uncertainty in the state of the system by the slope of the map at the points sampled by a trajectory. The implicit probability density built into this sum by evaluating the derivatives at the points $x(i)$ is made explicit in Eq. (5) by the use of Eqs. (3) and (4). We now discuss various numerical problems involved in calculating λ from these definitions.

Given only the sequence of maxima (or minima) $x(i)$, we may attempt to find the slope of the map at points $x = x(i)$ by finding neighboring points $x(j)$, $x(k)$ and evaluating $[x(j+1) - x(k+1)]/[x(j) - x(k)]$. This might suffice if $|x(j) - x(k)| = \delta$ were small enough (but bigger than the noise). The problem is that regions of the map where accurate slope determination is most important (say where the slope is near zero) may be too infrequently visited by the time series to provide such pairs of points. In the presence of noise it is clear that any single pair $x(j)$, $x(k)$ could provide a very inaccurate slope estimate (perhaps even incorrect in sign!), and so we would need to average the angle of inclination of several nearby pairs to get a reasonable estimate for the local slope (implying an average over the arctangent of the slopes and not the slopes themselves). Our numerical work on a model 1-D map with noise {in particular $x(i+1) = \sin[\pi x(i)] + \xi(i)$, where $\xi(i)$ was a uniformly distributed random variable} indicated that the conflicting requirements on δ (small enough to resolve structure in the map, but large enough to bring the slope out of the noise) make local slope estimation unreliable for experimental data.*

Since accurate slope determination for noisy data cannot be done locally, we consider instead fitting the data to an appropriate functional form, $f(x)$. A common choice is a set of cubic splines,[24] cubic polynomials whose coefficients are chosen to ensure continuity of the function and its first and second derivatives and minimization of the sum squared error

$$S = \sum_{i=0}^{n-1} [f(x(i)) - x(i+1)]^2$$

For a given number of splines the IMSL routine ICSVKU[25] can vary their placement ("knot" locations) so as to find at least a local minimum for S. The routine is likely to place most of the splines in regions of large functional variation to take advantage of the extra "wiggles" they can provide. There are several problems associated with the use of cubic splines: (1) varying knot locations to minimize S depends on good initial placement, and one cannot be certain that a global minimum will be located; (2) it is not clear that S is the quantity we wish to minimize, especially in maps with regions of very large slope (not uncommon in experiments or model systems); (3) we must choose the number of splines, m, in the fit. If we use a very large number of splines, each will tend to follow the noisy local behavior and log(slope(splines)) will act increasingly wild (Fig. 1). We also need to be sure that enough splines have been used to allow a proper fit to the data. It is not useful to look at the sum squared error S, as a function of m, as this will certainly decrease for large m, which we have seen is no indication that our

---

*Crutchfield et al.[23] studied the effects of very small amounts of noise on Lyapunov exponents for 1-D maps. They defined the exponent by Eq. (3) where slopes are computed from the *known deterministic* map and used many more iterations (10,000–100,000) in their calculation than one could hope to recover from most experimental systems. Their results are not pertinent to our discussion for these reasons.

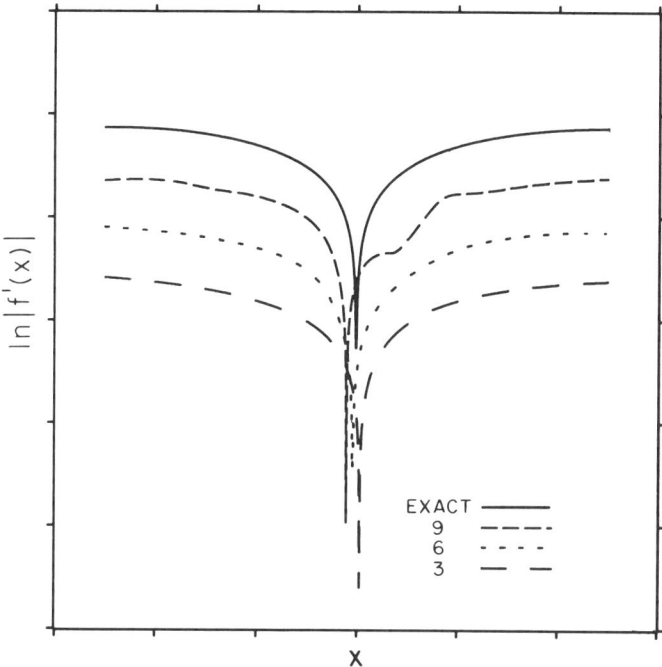

Figure 1. $\ln|f'(x)|$ vs x as determined by fitting 200 iterates of the map $f(x) = (1 + \xi)\sin(\pi x)$ with splines [ξ uniformly distributed in (−0.04, 0.04)]. The curves for three, six, and nine splines, and the exact result (with no noise) have been displaced by equal amounts vertically for clarity. The resulting Lyapunov exponents are, respectively, 0.58, 0.53, 0.59, 0.69.

slope profile will be well behaved. Once the data have been fit with splines, we may trivially compute either $f'(x(i))$ for use in Eq. (3) or $f'(x)$ for Eq. (5).

We now consider the determination of the map's probability density and its influence on the calculation of an exponent. We ignore the important question of the difference between the density for the underlying deterministic system and the system with extrinsic noise (however see the warning in Section 5) and examine a simpler problem. The generation of $p(x)$ for a 1-D map by accumulating a histogram, essentially the application of Eq. (4), is likely to be the only means available to an experimentalist. Typically the convergence of $p(x)$ with the number of points accumulated is not very rapid. For some experiments, for example, the extensively studied Belousov–Zhabotinsky chemical reaction,[16,26,27] the total number of points in a map (total number of orbits) is limited to a few hundred because of a long orbital period (on the order of minutes) and long-term drift. The influence of $p(x)$ (and the amount of noise in the system) on λ is illustrated in Table 1. There have been attempts to generate more accurate densities by iterating points through a least squares approximation to a map.[14] This is a very dangerous procedure. Except for a system well inside the stability interval of a

Table 1. The Effect of Noise and Finite Amount of Data on the Lyapunov Exponent for a 1-D Map.

N	$\varepsilon = 0.0$	$\varepsilon = 0.01$	$\varepsilon = 0.02$
25	0.386 ± 35%	0.373 ± 42%	0.355 ± 50%
50	0.381 ± 23%	0.371 ± 27%	0.349 ± 36%
75	0.380 ± 18%	0.369 ± 21%	0.347 ± 29%
100	0.379 ± 15%	0.369 ± 17%	0.348 ± 25%
200	0.379 ± 11%	0.368 ± 12%	0.345 ± 18%
500	0.378 ± 6%	0.366 ± 8%	0.345 ± 10%
1000	0.378 ± 4%	0.366 ± 6%	0.344 ± 7%

The mean and standard deviation (in 1000 trials) of the Lyapunov exponent for the map $f(x) = 22(1+\xi)xe^{-x}$ for different noise levels, ξ [uniformly distributed in $(-\varepsilon, \varepsilon)$] and different number of map iterates n. The standard deviations show approximately a $1/\sqrt{n}$ dependence. The exponent was computed as in Ref. 23 using the slope of the deterministic map to highlight the dependence on the degree of convergence of the probability density.

very-low-period periodic state, a very small change in $f(x)$ is likely to take a periodic orbit to a chaotic one, or vice versa, with distinctly different $p(x)$.

The largest contributions to λ [depending on the weighting by $p(x)$] will come from regions of large slope or near-zero slope where $\|\ln|f'(x)|\|$ is large. Many 1-D maps exhibit regions of large slope with large $p(x)$ [where an accurate determination of $p(x)$ is possible, but slope estimation is most sensitive to noise] and tails or interior regions of low slope and very small $p(x)$ (where the converse holds).[20,28] This suggests two ways of estimating the error in a determination of λ, by looking at the effect of a small change in slope for the regions where it is close to zero or large and by studying the dependence on $p(x)$ by removing a small set of points from the calculation.

Our experiences with the Zhabotinsky reaction[16,20,29] illustrate many of these problems. The system consists of a continuously stirred tank reactor with approximately 30 chemical species. The concentration of a single reaction product, bromide ion, is sampled about 100 times per orbit. The 1-D map we recover from this time series has the appearance of the function $x(i+1) = rx(i)e^{-x(i)}$ where the bifurcation parameter r appears to contain about 5–10% multiplicative noise. No more than 100 map points could be used in our calculations because of the problem of long-term drift. The technique of local slope determination proved unstable (often in the *sign* of the exponent) to such minor changes as the removal of a single data point or changing the bounds (minimum and maximum width allowed) on δ. Similarly, several attempts to compute the integral in Eq. (5) failed to provide stable answers. For the *single* data set shown in Fig. 2, corresponding to nearly three-cycle behavior, we found that an exponent obtained from Eq. (3) with a histogram evaluation of $p(x)$ and a splined map was stable to the

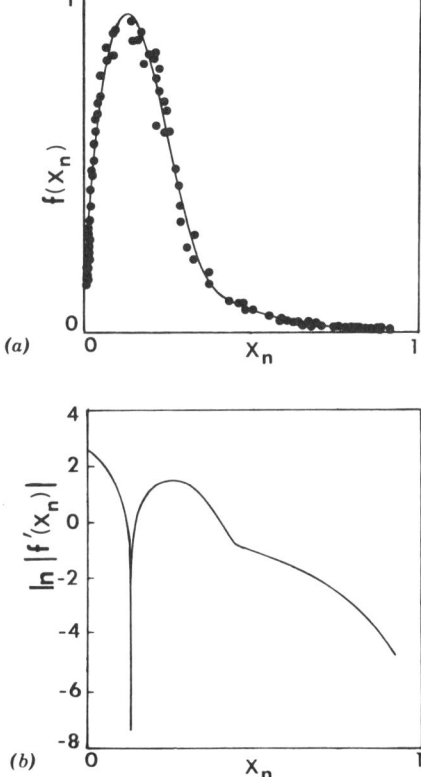

Figure 2. (a) Experimental data for the Belousov–Zhabotinsky reaction, fit with four splines. (b) $\ln|f'(x)|$ for the data in Fig. 2a as determined from the spline fit. Removal of a few data points or varying the number of splines from three to six had no significant effect on the Lyapunov exponent, $\lambda = 0.3 \pm 0.1$ bits/iteration. See Ref. 29.

removal of most single data points and, over a small range, to the number of splines. The result, $\lambda = 0.3 \pm 0.1$ bits/iteration was reported in Ref. 29. Our success for this particular set of data was most likely due to a stabilizing influence from the nearby three-cycle window.

To summarize, we find that the problems in accurately determining the slope of a 1-D map and its probability density from experimental data severely limit the usefulness of the standard approach for computing a Lyapunov exponent.

4. HYPERCUBE EVOLUTION

We now turn to a method for determining a positive Lyapunov exponent that we believe to be more stable than methods requiring the extraction of a 1-D map and that may be generalized to the computation of additional exponents.

Thus far we have ignored the question of recreating the full phase-space behavior of a system from experimental measurements. Generally, it is impossible to monitor (or even identify) all of the phase-space variables in a complex system. Fortunately the work of Takens[2] and others[30] justifies the reconstruction of a strange attractor from a single dynamical observable through the use of "staggered" time series, provided the system and choice of observable are "generic." Given the (post-transient) observable $x(t_i)$, the time evolution of the vector $(x(t_i), x(t_i + T), x(t_i + 2T), \ldots, x(t_i + NT))$ for suitably large N and almost any time delay T, defines an attractor whose Lyapunov spectrum and fractional dimension are identical to those of the original system.[31] In what follows we assume that all phase-space variables are available either from direct measurement or from a proper staggered-time-series reconstruction.

We consider attractors with a $(+, 0, -)$ spectrum with a nearly planar local structure (equivalent to the requirement that $|\lambda_3| \gg \lambda_1$*). In the (possibly reconstructed) attractor we locate three nearby points, $P(0)$, $Q'(0)$, and $R'(0)$. Point $P(0)$ is any point in the attractor, perhaps the first one in our time series. Points $Q'(0)$ and $R'(0)$ are its nearest (in the Euclidean sense) neighbors in the attractor, found by an exhaustive search or other means.† We then look ahead in the time series to find the new locations of these points, call them $P(1)$, $Q(1)$, and $R(1)$, at time Δt_1. The two triples of points define areas A_1 and $A_1(\Delta t_1)$, which tell us about the locally expanding or contracting nature of the attractor through the quantity $L_1 = \ln(A_1(\Delta t_1)/A_1)$. We would like to look at the long-term evolution of the original triple $P(0)$, $Q'(0)$, $R'(0)$ but the points may already have spread so far apart that we are not measuring local properties of the attractor. (In an extreme case the points will "bounce off the walls" of the attractor.) Keeping point $P(1)$ as one element of our triple, we look for its new nearest neighbors $Q'(1)$ and $R'(1)$. We evolve the new triple for a time Δt_2 and compute the quantity $L_2 = \ln(A_2(\Delta t_2)/A_2)$. This procedure is repeated until we run out of data. By the argument in Section 2

$$\lambda_1 + \lambda_2 \approx \sum_{i=1}^{n} L_i \bigg/ \sum_{i=1}^{n} \Delta t_i \qquad (7)$$

Since λ_2 is zero for this attractor, a direct estimate of λ_1 has been obtained.

When computing Lyapunov spectra for sets of differential equations, we did not worry about the exponentially growing magnitude of perturbation vectors (until they had reached computer limitations) because they were in tangent space rather than phase space. It was therefore impossible for the perturbations to experience any nonlinear effects. The key point to our technique is to minimize nonlinear effects in *phase space* by redefining the "perturbations" every time they grow to a nonnegligible size. This is justified

---

*For systems with $(+, 0, -)$ Lyapunov spectra the Kaplan–Yorke conjecture states that the noninteger part of the fractional dimension is $-\lambda_1/\lambda_3$.

†Exhaustive search through the data for neighbors is expensive. Algorithms that can do coarse-grained probability distributions of n data points in an m-dimensional phase space with order (nm) operations reduce this cost substantially.

because (1) we have retained the evolution of one point, $P(0)$, in the triple through the entire calculation, to weight contributions to the exponent by the probability density for the underlying 1-D map and (2) the attractor is locally planar so that there is no correction to the evolution of areas from other exponents. The sole effect of choosing new triples is to replace each term $L_i(\Delta t_i) = \ln A_i(\Delta t_i)/A_i$ in Eq. (7) with $\ln[A_i(\Delta t_i)K_i]/(A_iK_i)$, where K_i is the ratio of the area A_1 would have become by the ith step, to the essentially random value of A_i. The K_i's cancel and leave the exponent unaffected.

Note that we cannot take an identical approach to the direct calculation of λ_1 by length evolution. When a line segment has grown to substantial size, we will not find a new (shorter) replacement with an identical orientation. We lose the sense of the most expanding direction in the attractor and mix contributions from the positive and zero exponent in a system-dependent and complicated manner. It was therefore crucial for our determination of λ_1 to know that λ_2 was zero, as only the sum $\lambda_1 + \lambda_2$ can be obtained from area evolution. In a future work we will present an algorithm which, in principle, allows us to estimate all of the non-negative Lyapunov exponents from hypercube evolution. This is possible if the density of points defining the attractor is high enough so that replacement points defining hypercubes differing in orientation from the original hypercubes by a small amount can be found. We will be able to relax the condition that the system is so strongly dissipative that it is nearly locally hyperplanar.

For this calculation to succeed it is vitally important to choose the evolution times Δt_i in an "intelligent" manner. Ideally each triple starts out very small (but substantially larger than the noise level) and defines an approximately square area in a region of low phase-space velocity (so that area determination can be done accurately). We follow each triple until we can both safely compute its final area and find an adequate replacement for it. In addition to a general set of constraints there may also be system-dependent criteria that rule out certain portions of the attractor as good places to compute areas. Our implementation of these constraints is fairly simple, but we have designed an algorithm that seeks optimal results by re-doing earlier stages of the calculation when trouble arises.

Our method was tested on the Rössler and Lorenz strange attractors, whose positive exponents are 0.09 and 1.50 bits/second, respectively, for standard choices of the parameters in the defining differential equations.* Each calculation was done two ways, first making use of all three phase-space variables and second, recreating the attractor from staggered copies of the x-coordinate time series. (The latter calculations are identical to the procedure we would follow for experimental data.) In each case the equations of motion were integrated to produce one long trajectory, containing between 100 and 1000 orbits with 10–100 samples per orbit (usually totaling about

---

*Parameter values for the Rössler attractor are as in Eq. (2). For the Lorenz attractor, defined in Ref. 19, the parameter values are $b = 4.0$, $\sigma = 16.0$, $r = 45.92$.

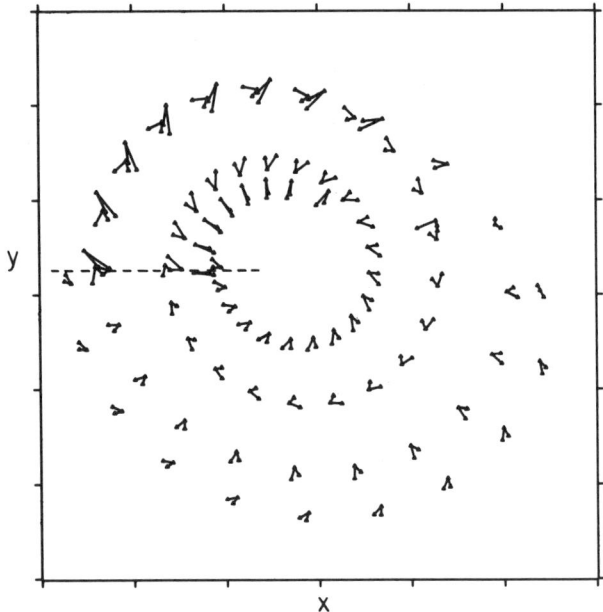

Figure 3. Triples of points propagated through the Rössler attractor (shown in $x - y$ projection). A new triple is initiated every time an orbit crosses the dotted line (Δt_i = one orbit).

32,000 samples). For the Rössler attractor we found a positive exponent of 0.089 ± 0.002 bits/second for both the original and reconstructed attractors where the error estimate reflects the range of values we obtained for several different runs (corresponding to different sampling rates, total number of orbits, delay times, and values of heuristically chosen constants determining evolution times). In Fig. 3 we show the evolution of triples for a few orbits around the Rössler attractor where the Δt_i's have been chosen to be exactly one orbit. For the Lorenz attractor we computed an exponent of 1.3 ± 0.2 bits/second. The significantly higher error for this attractor is due to its more singular structure† (the effect of which can be partially alleviated by filling out the attractor with many more points) and the lack of a sufficiently intelligent routine for monitoring the evolution of triples.

We would like to present data for these attractors supporting our claim that area evolution is a stable means of computing a positive exponent in the presence of noise, but this is difficult to do. The errors produced by noise are of two classes, the catastrophic and the statistical. An example of the former

---

†The Lorenz attractor consists of two "lobes", which orbits switch between at apparently random times, with an unstable fixed point located close to their intersection. A triple of finite size passing through this region may be split between the two lobes or suffer strong "tidal" forces from the nearby fixed point, both of which reflect undesirable sampling of nonlinear behavior.

would be the "accidental" inclusion of a point in a new triple whose short-time evolution is drastically different from the other points. Even if this error occurs very infrequently, the enormity of its contributions to an exponent can dominate the result.* An example of the latter is the error in computing an area due to the "jitter" of points in a triple. We believe the former class of problems can be completely eliminated by the intelligent monitoring of evolution, as discussed before. The problem of jitter can be reduced by improving the statistics in several ways, for example, finding n nearest neighbors (rather than two) to each point P and measuring the typical (not mean) growth of areas within this cluster of points. If we added a moderate amount of noise to the Lorenz and Rössler attractors, we would find increased errors in our exponents, but we could then make our algorithm correspondingly more complex to reduce these errors. We did perform a simple test of the effect of limited precision in experimental data for a fixed algorithm. After generating a time series we reduced its resolution from the specified integration tolerance, about 30 bits of precision, to k bits (for $k \le 16$, as it is in most experiments). For the Rössler attractor we found no significant dependence of λ_1 on k until k was decreased to about 9 bits, at which point λ_1 was very poorly determined. For the Lorenz attractor this transition took place at $k \approx 12$ bits. This suggests that the precision required to compute positive exponents from experimental data is system dependent and rather high.

5. CONCLUSIONS AND A WARNING

We have presented a method for calculating a positive Lyapunov exponent from experimental data that has several advantages over the usual technique of examining an underlying 1-D map. These include a reduction in the minimum data collection rate (since no Poincaré section is formed),† the ability to minimize the effects of noise on the calculation through the continuous "intelligent" monitoring of the behavior of trajectories, and the ease of generalizing this calculation to other Lyapunov exponents.

We must end on a somewhat pessimistic note. Strictly speaking, quantities such as Lyapunov exponents and fractional dimension are not well defined

---

*Catastrophic errors were discovered when calculations on slightly noisy two-tori yielded two large positive exponents rather than zero exponents. They have also been observed in data for the Zhabotinsky system.

†For our purpose the formation of a 1-D map by the detection of local maxima in a time series is equivalent to looking at the successive intersections of the reconstructed attractor with a specified plane (a Poincaré section). This typically requires a sampling rate of 100 points or more per orbit in order to determine the intersections to sufficient accuracy. For systems with very short orbital period[21] this poses a serious problem in data acquisition. Area evolution might allow data collection rates of 10 points/orbit or less, depending on the nature of the attractor.

(or do not have their usual interpretation) in the presence of *any* amount of noise. Both quantities require taking the limit of a length scale going to zero where noise is certain to dominate calculations. In the case of fractional dimension one can hope that scaling behavior will be visible on several length scales before space-filling noise takes over. One can then obtain the dimension that "would have been" in the absence of noise. For Lyapunov exponents it is not clear that one can find an equally simple argument. As an example of the problems in interpretation that might arise we consider the recent work of Matsumoto and Tsuda.[32] For a certain class of 1-D maps they found that the addition of a small amount of additive noise to a (deterministically) chaotic state reduces somewhat the erratic behavior of trajectories (though, of course, the motion is still aperiodic) but changes the positive Lyapunov exponent to a *negative* value. This suggests that the current diagnostics of chaotic systems may not be appropriate when extrinsic noise is present.[33]

ACKNOWLEDGMENTS

We thank Harry Swinney and Jean-Claude Roux for providing us with their experimental data. Discussions with Doyne Farmer were most helpful. This research was supported by National Science Foundation Grant No. CHE79-23627 and Robert A. Welch Foundation Grants No. F-805 and No. F-767.

REFERENCES

1. D. Ruelle, "Sensitive Dependence on Initial Conditions and Turbulent Behavior of Dynamical Systems," *Ann. N.Y. Acad. Sci.* **317**, 408 (1979).
2. F. Takens, "Detecting Strange Attractors in Turbulence," In *Lecture Notes in Mathematics*, D. A. Rand and L.-S. Young, eds. (Springer, New York, 1981), p. 366.
3. J. D. Farmer, E. Ott, and J. A. Yorke, "Dimension of Chaotic Attractors," *Physica* **7D**, 153 (1983).
4. P. Grassberger and I. Procaccia, "Characterization of Strange Attractors," *Phys. Rev. Lett.* **50**, 346 (1983).
5. J. P. Crutchfield and N. H. Packard, "Symbolic Dynamics of One-Dimensional Maps: Entropies, Finite Precision and Noise," *Int. J. Theor. Phys.* **21**, 433 (1982).
6. S. Blacher and J. Perdang, "Power of Chaos," *Physica* **3D**, 512 (1981).
7. G. Bennetin, L. Galgani, A. Giorgilli, and J.-M. Strelcyn, "Lyapunov Characteristic Exponents for Smooth Dynamical Systems and for Hamiltonian Systems; A Method for Computing All of Them," *Meccanica* **15**, 9 (1980).
8. I. Shimada and T. Nagashima, "A Numerical Approach to Ergodic Problem of Dissipative Dynamical Systems," *Prog. Theor. Phys.* **61**, 1605 (1979).
9. H. S. Greenside, A. Wolf, J. Swift, and T. Pignaturo, "The Impracticality of a Box-Counting Algorithm for Calculating the Dimensionality of Strange Attractors," *Phys. Rev. A* **25**, 3453 (1982).

10. P. Frederickson, J. Kaplan, E. Yorke, and J. Yorke, "The Lyapunov Dimension of Strange Attractors," *J. Diff. Eqs.* **49**, 185 (1983).
11. D. A. Russell, J. D. Hanson, and E. Ott, "The Dimension of Strange Attractors," *Phys. Rev. Lett.* **45**, 1175 (1980).
12. R. Shaw, "Strange Attractors, Chaotic Behavior, and Information Flow," *Z. Naturforsch.* **36a**, 80 (1981).
13. F. Takens, "Distinguishing Deterministic and Random Systems," in *Non-linear Dynamics and Turbulence* (to be published).
14. J. L. Hudson and J. C. Mankin, "Chaos in the Belousov–Zhabotinsky Reaction," *J. Chem. Phys.* **74**, 6171 (1981).
15. H. Nagashima, "Experiment on Chaotic Response of Forced Belousov–Zhabotinsky Reaction," *J. Phys. Soc. Japan* **51**, 21 (1982).
16. J.-C. Roux, J. S. Turner, W. D. McCormick, and H. L. Swinney, "Experimental Observations of Complex Dynamics in a Chemical Reaction," in *Nonlinear Problems: Present and Future*, A. R. Bishop, D. K. Campbell, and B. Nicolaenko, eds. (North-Holland, Amsterdam, 1982), p. 409.
17. J. D. Farmer, "Chaotic Attractors of an Infinite Dimensional Dynamical System," *Physica* **4D**, 366 (1982).
18. O. E. Rössler, "An Equation for Continuous Chaos," *Phys. Lett.* **57A**, 397 (1976).
19. E. N. Lorenz, "Deterministic Nonperiodic Flow," *J. Atmos. Sci.* **20**, 130 (1963).
20. R. H. Simoyi, A. Wolf, and H. L. Swinney, "One-Dimensional Dynamics in a Multi-component Chemical Reaction," *Phys. Rev. Lett.* **49**, 245 (1982).
21. J. Testa, J. Perez, and C. Jeffries, "Evidence for Universal Chaotic Behavior of a Driven Nonlinear Oscillator," *Phys. Rev. Lett.* **48**, 714 (1982).
22. V. I. Oseledec, "A Multiplicative Ergodic Theorem. Lyapunov Characteristic Numbers for Dynamical Systems," *Trans. Moscow Math. Soc.* **19**, 197 (1968).
23. J. P. Crutchfield, J. D. Farmer, and B. A. Huberman, "Fluctuations and Simple Chaotic Dynamics," *Phys. Rep.* **92** (1982).
24. C. de Boor, *A Practical Guide to Splines* (Springer, New York, 1978).
25. International Mathematical and Statistical Library, Houston, TX.
26. B. P. Belousov, *Ref. Radiates. Med.* **1958**, 145 (1959).
27. A. M. Zhabotinskii, *Dokl. Akad. Nauk. SSSR* **157**, 392 (1964).
28. J.-M. Wersinger, J. M. Finn, and E. Ott, "Bifurcation and 'Strange' Behavior in Instability Saturation by Nonlinear Three-Wave Mode Coupling," *Phys. Fluids* **23**, 1142 (1980); and J.-M. Wersinger (private communication).
29. J.-C. Roux, R. H. Simoyi, and H. L. Swinney, "Observation of a Strange Attractor," *Physica* **8D**, 257 (1983).
30. N. H. Packard, J. P. Crutchfield, J. D. Farmer, and R. S. Shaw, "Geometry From a Time Series," *Phys. Rev. Lett.* **45**, 712 (1980).
31. J. Guckenheimer (private communication).
32. K. Matsumoto and I. Tsuda, "Noise-Induced Periodicity," *J. Stat. Phys.* **31**, 87 (1983).
33. J. Guckenheimer, "Noise in Chaotic Systems," *Nature* **298**, 358 (1982).

FRACTAL BASIN BOUNDARIES IN NONLINEAR DYNAMICAL SYSTEMS

Celso Grebogi, Steven W. McDonald, Edward Ott, and James A. Yorke
University of Maryland, College Park, Maryland

Abstract. *The structure of the boundaries separating basins of attraction for different attractors is investigated.*

A common situation in nonlinear dynamical problems is the existence of multiple attractors. By an *attractor* we mean a dynamical state to which the system orbit asymptotes with time. If more than one attractor exists, then the one to which an orbit asymptotes is determined by its initial conditions. The set of initial conditions that yield a given time asymptotic attractor is called the *basin of attraction* of that attractor.

As an example of the above definitions, consider the simple case of a point particle moving under the influence of friction in a potential $V(x)$ as shown in Fig. 1a. For almost any initial condition, the orbit will eventually come to rest in either of the two stable fixed points $x = x_0$ or $x = -x_0$. Figure 1b shows schematically the basins of attraction of these two fixed point attractors. Thus, any initial condition originating in the crosshatched region eventually comes to rest at $x = x_0$, while any initial condition in the black region eventually comes to rest at $x = -x_0$. An initial condition exactly on the boundary of these two regions approaches the unstable fixed point $x = 0$ as $t \to \infty$ ($t \equiv$ time). The boundary separating these two basins of attraction is the stable manifold of the unstable fixed point, $x = 0$. This boundary is a smooth curve. The main point of this chapter is that basin boundaries can have a different character; namely, they can be nondifferentiable (fractal) curves or surfaces. (Note that, although the two

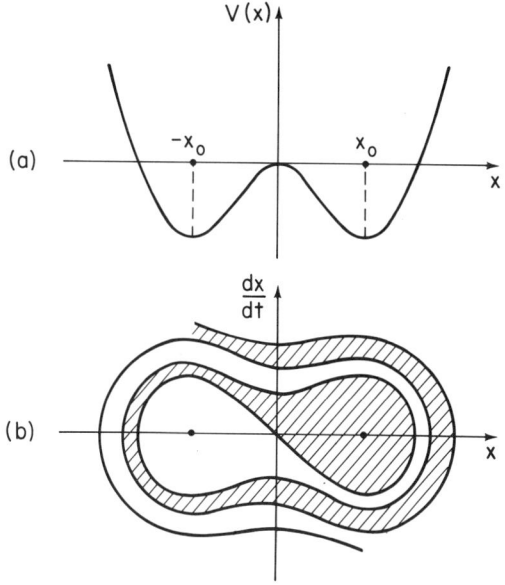

Figure 1. (a) The potential $V(x)$. (b) Basins of attraction for the two attractors $x = +x_0$ and $x = -x_0$.

attractors of the example in Fig. 1 are fixed points, attractors can also be periodic orbits, chaotic orbits, etc.)

To illustrate the possibility of fractal basin boundaries in the simplest possible context, and also to facilitate an analytical calculation, we shall consider a particular two-dimensional noninvertible map. While this example may appear to be somewhat contrived, we emphasize that the phenomena found have also been demonstrated in more-general examples. The fractal basin boundaries which we describe here are nondifferentiable curves. Still, worse boundaries may occur, one which are not curves (i.e., cannot be parametrically represented as $x = f_1(t)$, $y = f_2(t)$ with f_1 and f_2 continuous) and result due to horseshoes in the dynamics. For such an example see Ref. 2.

The map which we consider is[1]

$$\theta_{n+1} = 2\theta_n \bmod 2\pi \qquad (1a)$$

$$z_{n+1} = \lambda z_n + \cos \theta_n \qquad (1b)$$

where we take $2 > \lambda > 1$ and $0 \le \theta \le 2\pi$. For almost every initial condition the orbit generated by Eqs. (1) will asymptote to either $z = +\infty$ or $z = -\infty$ as $n \to +\infty$. We regard $z = +\infty$ and $z = -\infty$ as two different attractors and ask what is their basin boundary. As we shall show, there is a curve $z = f(\theta)$ such that orbits initialized in $z > f(\theta)$ go to $z = +\infty$, while those initialized in $z < f(\theta)$ go to $z = -\infty$. Thus, $z = f(\theta)$ is the basin boundary. To find $f(\theta)$, we note that, since forward iterates of Eqs. (1) diverge form $z = f(\theta)$, backward iterates approach it. Thus, to find $f(\theta)$, we adopt the following procedure:

1. Pick a point $\theta = \theta_o$ for which we wish to find the corresponding value of z on the boundary.
2. Using Eq. (1a), iterate θ_0 forward N steps generating $(\theta_0, \theta_1, \theta_2, \ldots, \theta_N)$.
3. Using Eq. (1b) and the θ orbit generated in step 2, pick a value z_N and interate back to z_0.
4. Now let $N \to \infty$. z_0 will then approach $f(\theta_0)$, thus determining the basin boundary.

In this way we obtain

$$z = f(\theta) = -\sum_{l=1}^{\infty} \lambda^{-(l+1)} \cos(2^l \theta) \qquad (2)$$

Since $\lambda > 1$, the sum in Eq. (2) converges absolutely and uniformly. On the other hand, consider $df/d\theta$. Formally taking the derivative of Eq. (2), we obtain

$$\frac{df}{d\theta} = \frac{1}{2} \sum_{l=1}^{\infty} \left(\frac{2}{\lambda}\right)^{(l+1)} \sin(2^l \theta). \qquad (3)$$

Since $\lambda < 2$, this sum diverges, and thus $f(\theta)$ is nondifferentiable. In fact, the curve [Eq. (2)] has an infinite length and a fractal dimension between one and two.

Figure 2 shows a picture of the basin boundary obtained from Eqs. (1).

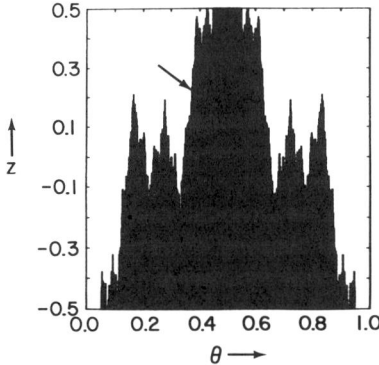
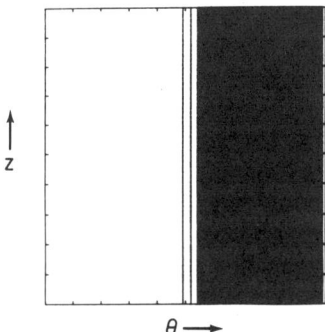

Figure 2. The black region is the basin of attraction for $z = +\infty$, and the blank region is the basin of attraction for $z = -\infty$. The arrow indicates a region to be magnified and shown in Fig. 3.

Figure 3. A blow up of a small section of the boundary shown in Fig. 2.

Note the spiky structure, suggestive of the nondifferentiable nature of the curve [demonstrated in Eq. (3)]. Figure 3 shows a magnification of a small region of the boundary. Note the linearlike structure. Further successive blowups of points on the boundary also demonstrate similar one-dimensional linear structure. In addition, our other examples of fractal basin boundaries also show this type of structure upon magnification. Thus we believe that this structure is a typical property of fractal basin boundaries for dynamical systems.

To conclude we believe that typical dynamical systems will often exhibit fractal basin boundaries and that their structure will often be of the form shown in Fig. 3.

This work was supported by the U.S. Department of Energy (Office of Basic Energy Sciences).

REFERENCE

1. C. Grebogi, E. Ott, and J. A. Yorke, *Phys. Rev. Lett.* **50**, 935 (1983).
2. C. Grebogi, S. W. McDonald, E. Ott, and J. A. Yorke, *Phys. Lett.* *A***99**, 415 (1983).

EVOLUTION FROM COHERENCE TO TURBULENCE IN PLASMAS

A. Y. Wong, P. Y. Cheung, and T. Tanikawa
Department of Physics, University of California, Los Angeles, California

Abstract. Highly monochromatic plasma waves are observed to evolve into turbulent spectra with spreads in frequencies and wavenumbers. In the weakly turbulent regime parametric decay and mode-coupling processes account for the spreading in frequency and wavenumber spectra. Certain parametric processes satisfying linear dispersion relations exhibit period doubling as a route to turbulence. In the strong turbulent regime, where $E^2/4\pi nT > (k\lambda_D)^2$, the route to turbulence proceeds via plasma-wave collapse. The turbulent spectra are a result of sharp spatial inhomogeneities (cavitons) and the interactions between waves trapped inside cavitons. The relationship between caviton collapse and the formation of phase-space granulations is described.

As working definitions, we shall define a coherent plasma wave as having narrow widths in frequency and wavenumber spaces $\Delta\omega/\omega$, $\Delta k/k \ll 1$; a turbulent state is one in which these bandwidths are large. The purpose of this chapter is to review past plasma experiments in which an initial coherent wave evolves into a turbulent state.

We shall consider the evolution to turbulence in two regimes: weak and strong turbulence. The dividing line is whether the dispersion relation is significantly modified or not; in the case of Langmuir turbulence this division is represented by the ratio of the self-consistent field energy density normalized to the thermal energy density, $E^2/4\pi nT$, is equal to the ratio $(k/k_D)^2$, where k_D is the Debye wavenumber.

WEAKLY TURBULENT REGIME

In this regime the dispersion relation derived from linear theories is still applicable[1] to the frequency and wavenumber matching condition, for example, in the weakly nonlinear parametric decay process. In this regime the turbulence is homogeneous in space and time and wave–particle interactions are usually treated by resonant quasilinear theory. A characteristic of this regime is that the nonlinear energy transfer between waves is toward long wavelengths ($k \to 0$). We shall illustrate by an experiment that the cascading behavior in a parametric decay process of an initially coherent wave generates successively broader sidebands, as illustrated in Fig. 1. A simple laboratory experiment[2] is used to illustrate this behavior using the amplitude of the externally excited coherent wave as a parameter.

The experiment is performed in a double-plasma (DP) source[3] of 40 cm diameter and 50 cm length operating with argon at 0.5 mTorr. Typical plasma parameters as measured with Langmuir probes are electron temperature

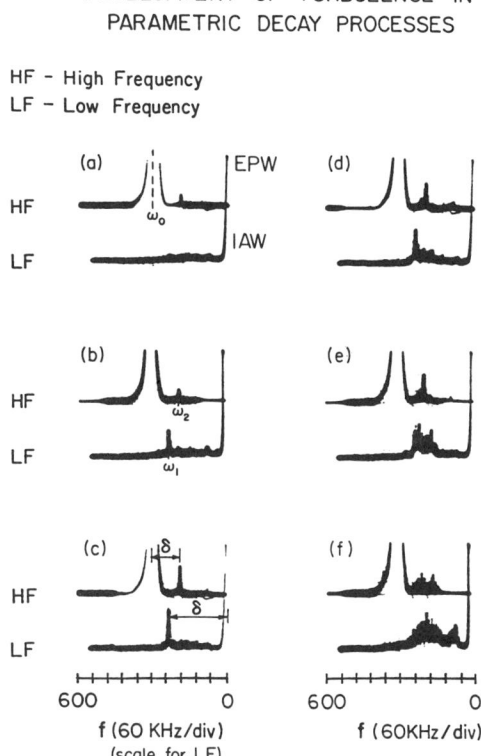

Figure 1. Evolution of the high-frequency and low-frequency spectra in a parametric decay process as the pump power is increased from threshold $E_0^2/16\pi n T_e = 2 \times 10^{-4}$ to 4.5×10^{-4}.

$T_e \simeq 2$ eV, electron density $n_e \simeq 10^9$ cm^{-3}, ion temperature $T_i \simeq 0.2$ eV, and ion fluctuation level $\delta n_i/n_i \lesssim 0.1\%$. No electron or ion beam is present in the plasma. The pump rf field with frequency $f_0 = 400$ MHz $\gtrsim f_{pe}$ is provided by a parallel-plate capacitor (5-cm-diameter grids, 3-cm spacing) immersed into the center of the plasma. The electrodes are coupled capacitively to the rf power amplifier in order to avoid drawing dc currents from the plasma. Ion waves are detected with movable Langmuir probes, while the electric fields of electron plasma waves are measured with a high-impedance rf probe. Spatial correlation between probe signals yields the **k** spectrum. The rf pump can be pulsed on with a rise time of $t_r = 2\,\mu$s in order to measure growth rates.

Both the high-frequency (near ω_{pe}) and low-frequency ($0 < \omega < \omega_{pi}$) spectra of the parametric decay instability have been observed. For low pump field strengths the high-frequency spectrum shows a single line at $\omega = \omega_0$, while the low-frequency ion wave spectrum consists of a low level of background noise extending up to ω_{pi}. As the pump field strength is increased (Fig. 1a), we observe at a threshold $E_0^2/16\pi n T_e \simeq 2 \times 10^{-4}$ a second high-frequency line $\omega_2 < \omega_0$ and simultaneously a low-frequency oscillation ω_1; the sum of the excited oscillation frequencies is equal to the pump frequency, that is, $\omega_1 + \omega_2 = \omega_0$. As the pump power is increased beyond threshold (Figs. 1b–1f), electron- and ion-wave spectra broaden; the frequency and wavenumber matching conditions continue to hold. The width of the ion-wave spectrum increases from $\Delta\omega/\omega \simeq 10^{-2}$ to 0.5 when the pump power is increased from the threshold value to twice the threshold.

PARAMETRIC DECAY AND PERIOD-DOUBLING ROUTE TO TURBULENCE

We shall illustrate using drift waves[4] the connection between parametric decay and the route to turbulence through period doubling. Drift waves have an approximately linear dispersion relation, which allows high-frequency modes to decay into successively lower-frequency modes. In particular, the decay into modes at exactly half its initial frequency has the lowest threshold and the highest growth rates. This is the same as period doubling, and explains how successive parametric decay leads to turbulence via the period doubling scheme.[5] We wish to point out that only in systems with linear dispersion relation is this scheme possible.

The drift wave is an electrostatic mode occurring in a plasma with a density gradient across a zeroth-order magnetic field. This instability is observed as a low-frequency density and potential perturbation with its maximum amplitude in the region of maximum density gradient of the plasma. In the cylindrical plasma of a Q device, drift waves are essentially ion acoustic waves propagating azimuthally, almost perpendicular to the axial magnetic field ($k_\perp \gg k_z$). The periodicity of the wave in the azimuthal

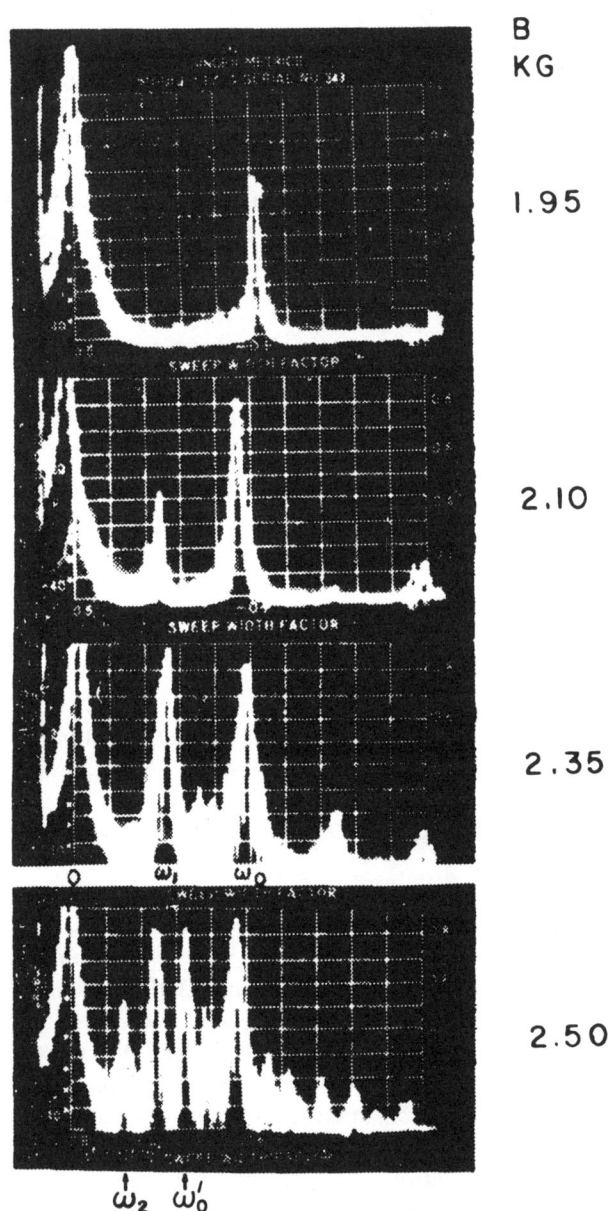

Figure 2. Spectra showing the occurrence of degenerate three-mode parametric coupling in an unstable plasma as the magnetic field is increased. The density gradient drift mode at ω_0 is the self-excited pump and that at ω_1 is the parametrically excited mode. The peak at the extreme left is the zero-frequency marker. Local density $\simeq 2 \times 10^{10}$ cm^{-3}, $(1/n)(dn/dr) \simeq 1$ cm^{-1}; relative fluctuation amplitude $n_1/n_0 \simeq 0.06$; $(B_0 = 1.95$ kG) $k_\perp = 1.18$ cm^{-1}, $k_z = 2.5 \times 10^{-2}$ cm^{-1}.

direction gives rise to resonant modes with the integral relationship $k_\perp = m/r_0$, where m is the integral azimuthal mode number and r_0 is the radius at which the mode is observed. Drift waves satisfy approximately the linear dispersion relation $\omega = k v_D$ and, therefore, meet the matching conditions for propagating waves in parametric coupling: $\omega_0 = \omega_1 + \omega_2$, $k_0 = k_1 + k_2$ (for the degenerate case $\omega_1 = \omega_2$, $k_1 = k_2$). In addition, because the pump mode and the excited modes propagate with the same phase velocity and travel continuously in the azimuthal direction, the coupling among these modes is particularly strong.

We now present data in a Q device where the spectra of drift waves are monitored by Langmuir probes. In Fig. 2 as the magnetic field is increased from 1.95 to 2.35 kG (the top three traces), the $m = 4$ drift mode first goes unstable, and as its amplitude increases it excites two $m = 2$ modes at frequency $\omega_1 = \omega_0/2$, exactly half the frequency of the $m = 4$ mode. As the magnetic field is increased to $B = 2.5$ kG, an additional set of period-doubling schemes appear: ω_0' decays into $\omega_2 = \omega_0'/2$.

In the stable regime (Fig. 3) a pump mode at $m = 4$ decays into $m = 1$ and $m = 3$ modes ($\omega_0 = \omega_1 + \omega_3$), and simultaneously decays into two $m = 2$ modes ($\omega_0 = 2\omega_2$). The $m = 2$ mode further decays into two $m = 1$ modes,

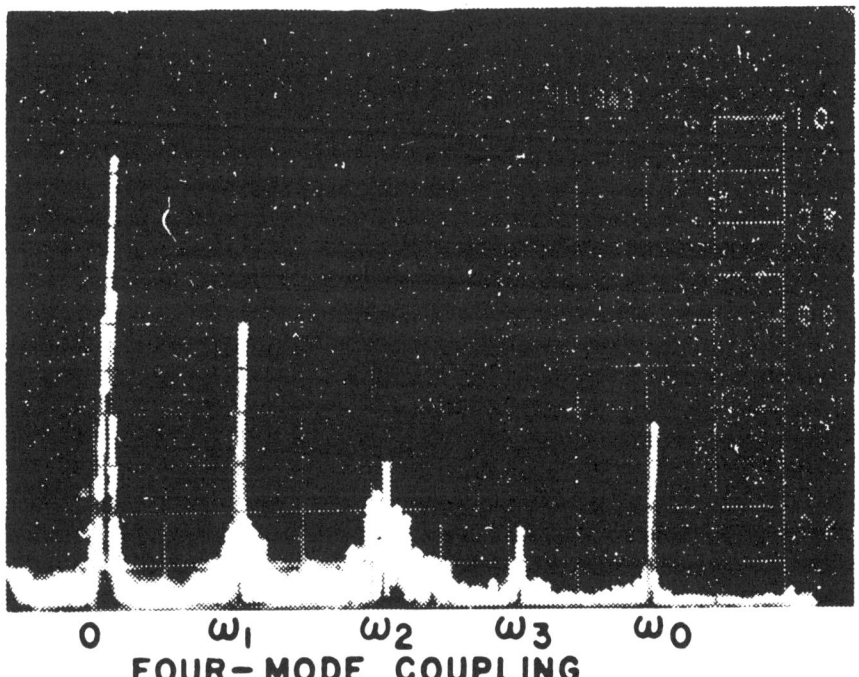

Figure 3. Spectra showing nondegenerate parametric coupling with drift waves near the radial edge of the plasma. The externally excited mode at ω_0 is the pump and those at ω_1, ω_2, and ω_3 are the decayed modes from the pump.

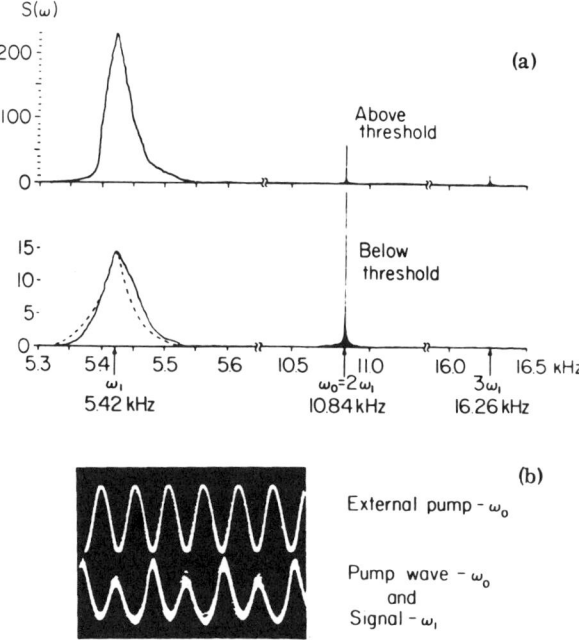

Figure 4. (a) Representative power spectra illustrating the enhancement from fluctuation about 5.42 kHz as the pump power at 10.84 kHz is increased. The theoretical curve (dashed) is obtained from Eq. (2) in Wong et al.[6] The ordinate has been expanded in the lower trace. (b) Oscillogram showing the signal on the excitation grid (upper trace) and the wave detected by a Langmuir probe near threshold. The lower trace representing the simultaneous occurrence of the self-consistent pump and the excited coherent signal can be Fourier decomposed as $\phi_0 \cos(2\omega_1 t) + \phi_1 \cos(\omega_1 t + \alpha)$, where $\alpha \approx 0$. The analysis for three coherent modes has shown that optimum coupling should take place for this condition of nearly zero phase shift.

which are the lowest-frequency modes permitted by the periodicity condition. A separate experiment[6] has shown that the decay of an excited $m = 2$ mode into two $m = 1$ modes (Fig. 4) is the preferred process with the lowest threshold.

STRONG TURBULENCE

We are interested in strong turbulence because it occurs frequently in plasmas under excitation by rather modest driving fields. Even weakly nonlinear phenomena can lead to strong turbulence through a series of cascade decay processes. The "Langmuir condensate," large-amplitude fields at long wavelengths, is the end product of a series of decays starting from high-frequency and short-wavelength modes. The accumulation of energy in this "condensate" acts as the source of strong plasma turbulence.[7]

In this strongly turbulent regime the self-consistent field E is large, that is, $E^2/4\pi nT \gg (k\lambda_D)^2$, and the linear dispersion relation is significantly altered. This strong turbulent regime is quite easily accessible because of two reasons: first, the self-consistent field is much larger than the externally imposed field by virtue of the plasma collective effects; second, the wavenumber k is small, for example, in situations where very energetic beams or long-wavelength rf fields are used. As will be shown in this section, both the background density and velocity distributions are modified such that the dispersion derived from linear theories is no longer valid and a one-to-one relationship between ω and k does not exist. We shall illustrate the evolution from a coherent Langmuir wave to a strongly turbulent regime by the process of spatial collapse or broadening in k space. There is also frequency broadening: a result of wave trapping inside density cavities and the process of parametric decay among these modes. The spatial collapse is the process in which plasma waves become increasingly localized because the ponderomotive forces of the wave fields create density cavities that confine these waves and enhance their amplitudes. This process continues until the caviton shrinks to the size of several Debye lengths at which stage particle–caviton interactions prevent further collapse. The collapse phenomenon, which preserves the number of plasmons, proceeds at a rate proportional to the instantaneous amplitude of the electric field. The final amplitude of the electric field is much higher than the initial amplitude, the amplitude being inversely proportional to the caviton size. We shall present for the first time how a collapsed caviton leads to the generation of clumps in phase space[8-10] as a consequence of particle–caviton interactions.[11,12]

We shall describe two beam-plasma experiments, one for a warm beam and another for a cold beam. The difference between these two experiments is that the growth rate and saturated amplitude of the plasma waves due to the cold beam are much higher and correspondingly the turbulence is stronger.

Caviton Generation and Collapse in the Presence of Two Counterstreaming Warm Beams

A symmetric experimental arrangement is used in which two counterstreaming warm beams produce oppositely propagating electrostatic plasma waves.[13] Standing waves in electric fields are produced in the central interaction region, where cavitons are generated and their collapse is studied. This experimental arrangement is particularly relevant to recent theoretical studies of cavitons[14] since warm beams are expanded before they reach the central region.

The experiment is performed in a triple-plasma device (Fig. 5). The chambers are electrically insulated and separated from each other by a pair of transparent grids (90% transparency). Plasmas in the source chambers

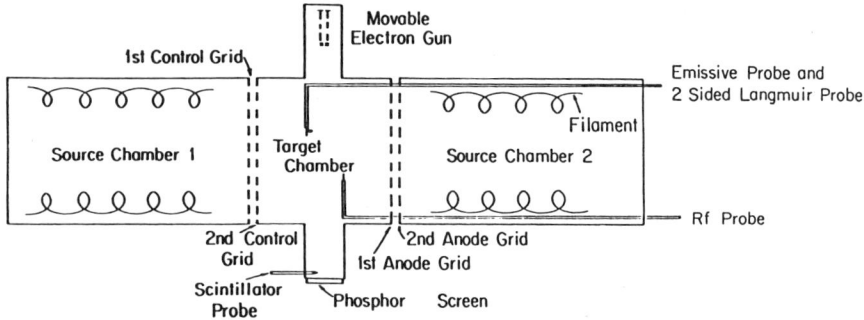

Figure 5. Schematic of the triple-plasma device, which is used for the beam–plasma interaction experiment with two counterstreaming warm electron beams.

are created by ionization of the argon gas. Potentials applied to the grids extract electron beams from the two opposite source chambers. These counterstreaming high-energy beams excite unstable plasma waves in the target chamber. They also ionize the neutral gas, thus creating the plasma in the target chamber. The operating conditions for the experiment are $T_e = 2$ eV, $T_e/T_i = 10$, $n_0 = 5 \times 10^8$ cm^{-3}. The electron neutral collision frequency ν_{en} is equal to $10^{-3} \omega_{pe}$. The beam characteristics are $n_b/n_0 = 1-5\%$, $v_b/v_{eT} = 5-10$, and $\Delta v_b/v_b = 0.05-0.10$, where n_b is the beam density, v_b is the beam velocity, Δv_b is the beam velocity spread, and v_{eT} is the background electron thermal velocity. The scaled thermal spread $S = (\Delta v_b/v_b)(6n_0 v_{eT}^2/n_b v_b^2)^{1/3} \lesssim 0.3$ (Ref. 15) characterizes a warm beam. It increases to a value much greater than unity in the interaction region as the beams are dispersed and merge with the background distribution.

Langmuir probes were used to measure electron distribution functions and density fluctuations. The rf electric-field amplitudes were measured with an rf probe and a diagnostic electron beam. The rf probe[16] allows relative measurements of the electric-field amplitude. Its absolute value can be measured from the lateral spread of the diagnostic electron beam ($E_D = 5-9$ keV, $I_D = 100$ nA, $n_D = 10^4$ cm^{-3}), which propagates perpendicular to the electric field. The spatial and the temporal evolution of the Langmuir wave electric field can be observed by moving the diagnostic electron beam along an axis parallel to the axis of the machine (Fig. 5).

Figure 6c represents the electron distributions near the grids (positions A and B) and at the center of the target chamber (position C). Near the two grids, the electron beams are clearly visible. However, owing to various dissipative processes, for example, ionization and wave–particle interaction, the two beams quickly thermalize: at the center of the chamber, where the collapse process takes place, they have disappeared.

As the two counterstreaming beams are dense and warm, the unstable

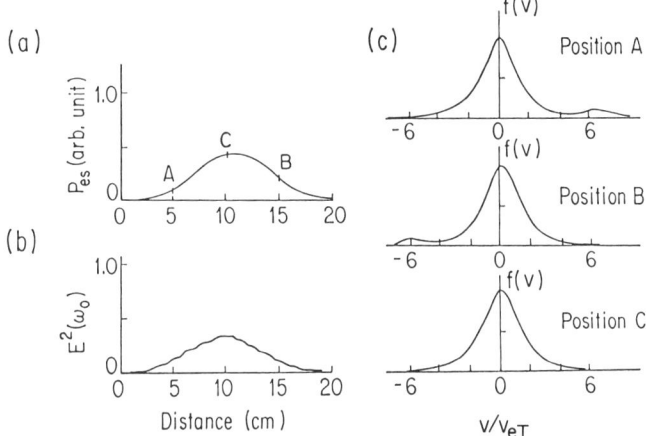

Figure 6. (a) Spatial profile of the total electrostatic power $P_{es}(x)$ generated by the two counterstreaming beams. The total spatial scan covers the whole target chamber except for a 2-cm region near the grids. (b) Spatial profile of $E^2(\omega_0)$ generated by the two counterstreaming beams. The bandwidth of the detector for this measurement is 30 kHz. (c) Electron distribution functions measured near the two extracting grids (positions A and B) and in the middle of the target chamber (position C).

wave spectrum is broadband: the bandwidth of the excited wave is $\Delta\omega/\omega_{pe} = 0.15$. As the two beams are generated and propagated in the target chamber in a symmetric way from the two grids, it is not surprising that the total electrostatic wave-energy density $P_{es}(x)$ [$P_{es}(x) = \int E^2(\omega, x)d\omega$, where $E^2(\omega, x)$ is the spatial intensity measured along the axis] is also symmetric (Fig. 6a).

As shown in Fig. 6a, $P_{es}(x)$ has a very smooth profile. However, by using a very-narrow-bandwidth detector to investigate the individual component of the unstable waves, that is, $E^2(\omega_0, \Delta\omega, x)$ where $\Delta\omega/\omega_0 \ll 1$, interesting results were revealed. Ripples were found to be developed (Fig. 6b) in the E^2 profile, indicating that a standing-wave pattern was already developed by the interaction of antiparallel propagating waves. The $P_{es}(x)$ profile in Fig. 6a is the result of standing-wave patterns of all the unstable waves; consequently, a smooth profile was obtained.

In order to follow the collapse phenomena, an rf pulse (frequency = 200 MHz, rise time <80 ns) was applied to the extracting grids to both source chambers. The frequency of this pulse was chosen to be that of the highest noise level. The amplitude was chosen such that only a monochromatic wave would grow at the expense of the beam-excited wideband noise. The phase coherence of a single wave is essential to the collapse process in this warm-beam situation. Measurements of the density profile and the electric-field amplitude at 100, 300, and 500 ns after the pump turn-on are presented in Fig. 7. The smooth electric field, which is observed

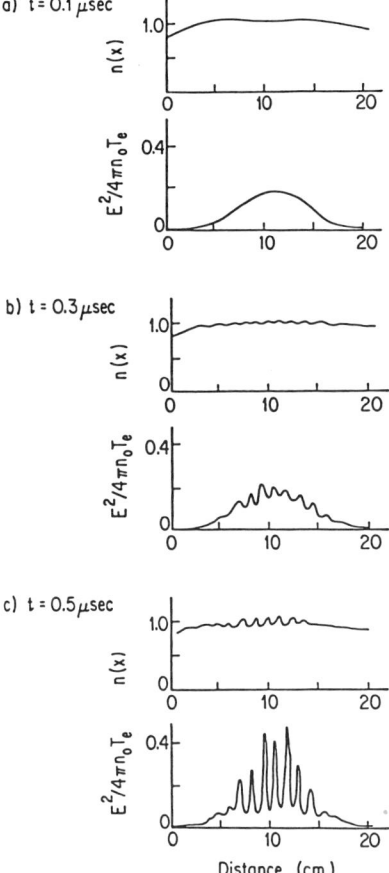

Figure 7. Time development of the spiky electric-field structures and the associated density cavities.

at time $t = 100$ ns, starts to develop oscillations at time $t = 300$ ns. At time $t = 500$ ns, the spiky electric-field structure is already formed. The measurement presented in Fig. 7 has been obtained with the rf probe. The electric-field profile obtained with the diagnostic beam is in agreement with the probe measurement. As the electric-field amplitude increases, the density profile shows a ripple ($t = 300$ ns), which becomes deeper at time $t = 500$ ns. It is also to be noted that the lowest density does not correspond exactly to the maximum of the electric field. This effect has also been noted by Morales and Lee in their simulation.[17] The reason is that the cavity-digging process involves not only the ponderomotive force but also ion motion which has a certain inertia. Since the electric field in the target chamber arises from the superposition of two Langmuir waves of wave vector $\mathbf{k}_0$ and $-\mathbf{k}_0$, the separation of the peaks of the localized field is equal to $\lambda_0/2 = \pi/k_0$. Measurement of the phase of the adjacent peaks shows that they are out of phase by 180°. The phase measurement has been inferred

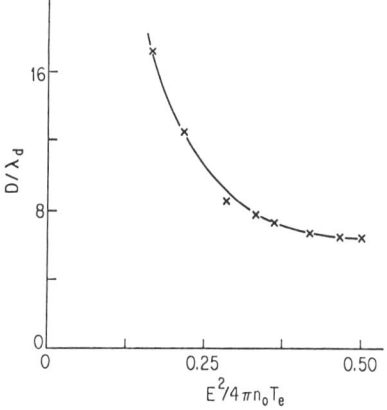

Figure 8. Dependence of the full width at half maximum D of the electron plasma waves versus the electric-field strength. The decrease of D with increasing the electric-field strength clearly shows the collapse process. The value of D is normalized to the Debye length, which has been estimated by using the final electron temperature $T_e = 4.5\,\text{eV}$.

from the Lissajous figure generated by the diagnostic beam, when a vertical deflection modulated at the same frequency as the Langmuir wave frequency is applied to it.[16]

At the beginning, the separation of the peaks of the localized electric field is equal to $\lambda_0/2$. The full width at half maximum (FWHM) of the localized structures as a function of the square of the peak electric field is plotted in Fig. 8. The measured values of both the FWHM and the electric field were taken during the time evolution of the localized structure. One notes that the FWHM decreases as E^2 increases. In the steady state, which is reached in about 500 ns after the pump is turned on, the FWHM is 7 Debye lengths.

Electron acceleration by the intense localized electric-field structure has also been observed.[12] The electron distribution, measured after the formation of the localized electric fields, shows that an increase in the population of energetic electrons with velocity up to four or five times the thermal velocity (Fig. 9). This acceleration of electrons is related to the formation of clumps as discussed in a later section.

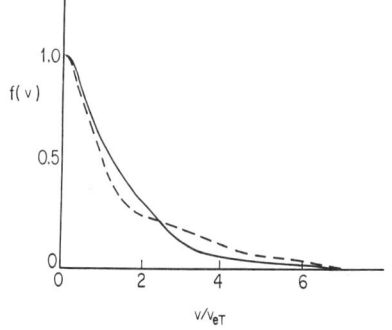

Figure 9. The electron distribution functions measured in the absence of the counterstreaming beams (the solid line) and measured after the formation of the spiky fields in the presence of the counterstreaming beams (the broken line). It is clearly seen that there is a formation of a high-energy tail in the latter case. The curves are symmetric with respect to $v = 0$.

Caviton Generation and Strong Plasma Turbulence in the Presence of a Cold Electron Beam

We will now turn our attention to an experimental situation where the strong growth rate and saturation amplitude of electron plasma waves (EPW) excited by a cold beam combine to create a much more significant perturbation in the plasma.[11] The excitation by a cold beam is unique in that it first creates from the natural plasma fluctuations a very coherent electron plasma wave which then evolves into strong turbulence by modifying the zeroth-order background plasma. Under our experimental conditions the spatial growth rate due to a cold beam is seven times that due to a warm beam when the beam particles are the same.

The experiments are performed in a large (1.8 m long and 1.8 m wide) vacuum chamber (Fig. 10). The background plasma is uniform and unmagnetized, and is produced independently of the electron beam by pulsed filament discharge in argon gas. Each experiment is performed in an afterglow plasma with typical plasma parameters of $T_e = 1.5$ eV and $n = 2.3 \times 10^9$ cm^{-3}. The electron beam, which is also pulsed, has a typical pulse duration of $\tau_b = 13 \mu s$, and is injected into the background plasma approximately 700 μs after the background plasma is shut off. The beam has a diameter of approximately 4 cm and is produced by a hot-cathode source. It has a beam voltage of $V_b = 800$ V, a beam spread of $\Delta v_b / v_b \simeq 2\%$, where Δv_b is the beam velocity spread, and beam densities of $n_b / n_0 = 0.2$–4%. Since the scaled thermal spread S is approximately 10^{-2}, the experimental situation is describable in terms of a cold-beam plasma interaction in which the spatial growth rate is very high ($k_i / k_0 \simeq 1$) and the initial unstable

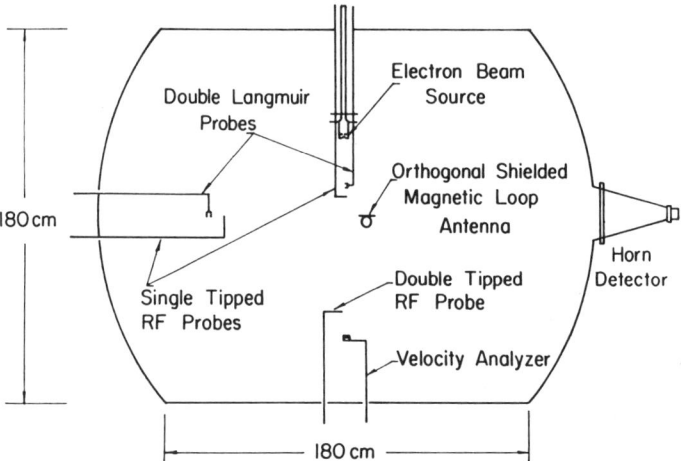

Figure 10. Schematic diagram of the experimental setup for the beam-plasma interaction experiment with a cold electron beam.

wave has a narrow frequency bandwidth ($\Delta\omega/\omega \leq 3\%$). Electrostatic potential and electric-field fluctuations are measured by high-frequency probes that have either a single wire tip (3 mm long, 0.5 mm diameter) or two wire tips closely spaced together. Beam characteristics are measured by double-sided Langmuir probes and a multigrid electrostatic velocity analyzer. A further check on the beam parameters can be made by the test-wave or free-streaming technique; the beam is modulated externally with a small signal at a frequency ω_t much higher than the plasma frequency. The measured wavenumber $k_t = \omega_t/v_b$ and the spatial decay of the test wave due to phase mixing are directly related to the beam spread Δv_b. Electromagnetic radiation is measured by shielded magnetic loops ($k_{em}d \ll 1$, where d is the diameter of the loop and k_{em} is the free-space wavenumber) located approximately 30 cm away from the interaction region, and a microwave horn with a cutoff frequency of 900 MHz mounted on one end of the vacuum chamber. All probes are removed from the interaction region when the emission data are being taken. While fundamental emissions are picked up only by the magnetic loops, emissions at higher harmonics are received by both the loops and the horn and the signals are carefully examined using filters and cross checked to eliminate the possibility of spurious signals generated within the detection system.

At the beam densities ($n_b/n_0 < 0.5\%$), the electron beam excites propagating Langmuir waves with resonant wavenumber $k_0 = 2\pi/\lambda_0 = 1.6 \text{ cm}^{-1}$ and a frequency spread of $\Delta\omega/\omega_{pe} \leq 3\%$ ($\omega_{pe}/2\pi \simeq 430 \text{ MHz}$). As the beam density increases, the wave intensity, E_{es}^2, becomes strong enough to create a density cavity in the region where it saturates. Experimental results for the case $n_b/n_0 = 2\%$ are systematically displayed in Figs. 11a–11c and show the temporal evolution of (a) the wave intensity of the beam-driven electrostatic waves, E_{es}^2, and the corresponding density cavities; (b) the spectral density of the wave intensity $E_{es}^2(\omega)$ as measured by the probe located in the middle of the density cavity at $x/\lambda_0 = 6$ (x is the distance from the beam injection point); and (c) the electromagnetic power radiated per unit frequency interval, $P_{em}(\omega)$.

The bandwidth of the electromagnetic radiation received by the microwave horn and the magnetic loops (which do not perturb the main interaction region) gives an indication of the bandwidth of the electrostatic instabilities excited over the spatial extent of the electron beam. The electromagnetic radiation comes from the conversion of electrostatic waves to electromagnetic waves through local density gradients and from the beating of two electrostatic waves to form electromagnetic waves. High-frequency probes sample the bandwidth of electrostatic waves only at the location where the probes are situated.

At $\omega_{pi}\tau = 10$ (ω_{pi} is the ion plasma frequency and τ is the time after the beam is turned on), a density cavity of the size $|\delta n/n_0| \simeq 20\%$ has already formed (Fig. 11a). Density measurements are obtained from complete swept probe characteristics. The time resolution is achieved with a PAR boxcar integrator.

At $\omega_{pi}\tau = 20$, the density cavity has become deeper with $|\delta n/n_0| \simeq 40\%$ and broader with $\Delta x/\lambda_0 \simeq 3$ (Δx is the spatial width of the cavity). It has a steep slope on the side nearer the beam injection point and a much gentler slope on the other side, matching the spatial growth and decay pattern of the unstable wave. The predominant electrostatic fluctuations are centered around the unperturbed plasma frequency of $\omega_{pe}/2\pi = 430$ MHz (Fig. 11b). The electromagnetic emission spectra are discrete and narrow with the fundamental emission at $\omega \simeq \omega_{pe}$ ($\omega/2\pi \simeq 450$ MHz) and a frequency spread of $\Delta\omega/\omega \simeq 8\%$ (Fig. 11c). Harmonic emissions are first observed at this time at $\omega \simeq 2\omega_{pe}$ and $\omega \simeq 3\omega_{pe}$ and $P_{em}(\omega_{pe})/P_{em}(2\omega_{pe}) \simeq 20$ dB.

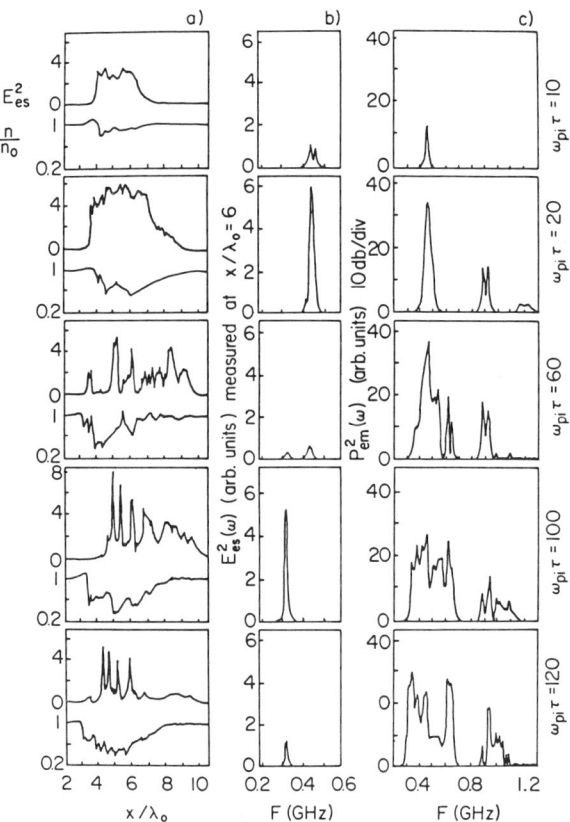

Figure 11. (a–e) Temporal evolution of (a) the wave intensity, E_{es}^2, and the corresponding density cavities; (b) the spectral density of the wave intensity, $E_{es}^2(\omega)$, as measured by the probe located in the middle of the density cavity at $x/\lambda_0 = 6$; and (c) the electromagnetic power radiated per unit frequency interval, $P_{em}(\omega)$. (d) Space-time evolution of wave intensity, E_{es}^2. At later time of the beam pulse, strong turbulence effects begin to modulate the field intensity spatially, forming spiky, localized fields. (e) Two-dimensional density profile created by a cold electron beam interacting with homogeneous plasma at $\omega_{pi}\tau = 30$. All the above data are for the case $n_b/n_0 = 2\%$ and $V_b = 800$ V. (Continued on next page.)

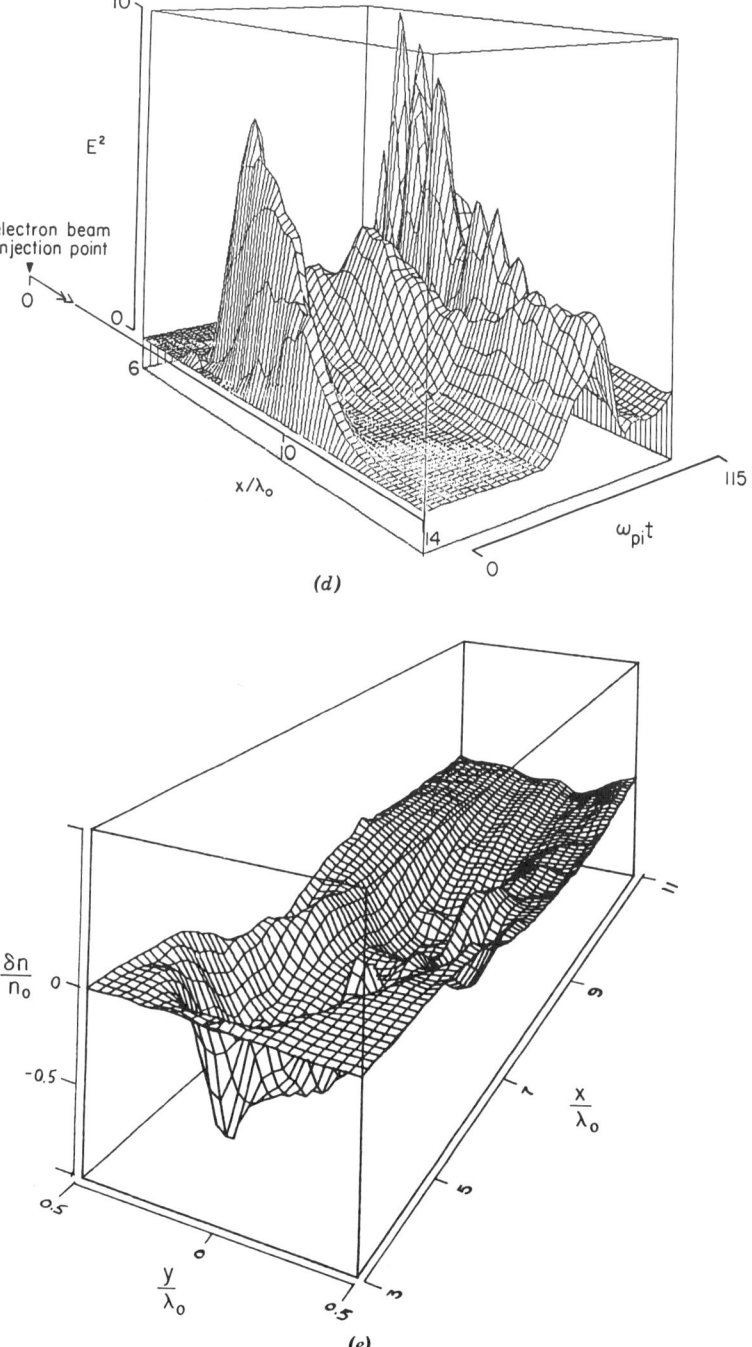

Figure 11. (d-e).

At $\omega_{pi}\tau = 60$, E_{es}^2 has now developed into spatially localized peaks and small density modulation begins to appear inside the large cavity (Fig. 11a). The original electrostatic waves at 430 MHz have decreased in intensity and new waves begin to emerge at a much lower frequency of 320 MHz (Fig. 11b). This shifting down in frequency is also evident in the electromagnetic wave signals as emission spectrum near ω_{pe} has now considerably broadened with emission occurring below ω_{pe} and a discrete peak is also observed near 640 MHz (Fig. 11c).

At $\omega_{pi}\tau = 100$, E_{es}^2 has developed into very localized peaks with a spatial extent (full width at half maximum) of $12\lambda_D$ ($k_0\lambda_D \simeq 0.03$, where λ_D is the Debye length) (Fig. 11a). The most unstable electrostatic waves within the cavity are now at 320 MHz (Fig. 11b). Use of $\omega_{pe} - \bar{\omega}_{pe} = \omega_{pe}[1 - (\bar{n}/n_0)^{1/2}]$ with $\bar{\omega}_{pe}/2\pi = 320$ MHz and $\bar{n}$ the perturbed density gives $\bar{n}/n_0 \simeq 50\%$, which agrees reasonably well with the actual depth of the density cavity.

At $\omega_{pi}\tau = 120$, with the beam still on, E_{es}^2 has actually decreased in intensity while the spiky structures still remain (Fig. 11a). The original electromagnetic emissions near ω_{pe} and their harmonics have decreased in amplitude and appreciable emissions are fundamental ($\bar{\omega}_{pe}$) and harmonics ($2\bar{\omega}_{pe}$, $3\bar{\omega}_{pe}$) radiation of the new waves excited within the cavity. The space-time evolution of E_{es}^2 is summarized in the three-dimensional picture in Fig. 11d, demonstrating the development of a spiky field structure from a broad envelope. This type of data is meant to capture the qualitative feature of the caviton. Ideally, the absolute amplitude of the field and its spatial distribution have to be measured by an array of diagnostic electron beams. This new diagnostic technique is now under development in our laboratory. In Fig. 11e, the two-dimensional density profile at $\omega_{pi}\tau = 30$ is shown. The deep density cavity at $x/\lambda_0 \simeq 4$ is clearly seen. The width of the density cavity in the y direction is essentially determined by the beam diameter.

We can also derive from the data above a rate of collapse to be compared with theory. Figure 12 shows the initial increase in the electrostatic field in a caviton with time. When the caviton reaches a certain dimension of $\sim 10\lambda_D$, the field starts to decrease as a result of particle–caviton interaction. This behavior is to be compared with the computer simulation of collapse including kinetic effects by Anisimov et al.[18,19] (Fig. 13). The reduction in the E field after collapse is called "burnt-out" and is a consequence of allowing particles to interact with the caviton field. The functional form of the collapse $|E| \propto (t_0 - t)^{-1}$ observed in the experiment is the same predicted by the Zakharov equations derived from fluid equations[20] up to $t = 0.8 t_0$ when kinetic effects become important.

We would like to comment on the experimental technique used in our studies of cavitons. In the case of beam–plasma interactions, the beam starts at a particular spatial point and the plasma instability saturates at a fixed distance from the starting point. The location of the initial cavitation is, therefore, well defined. Our experimental technique of repetitive sampling at various distances will, therefore, pick out peaks that are strong and

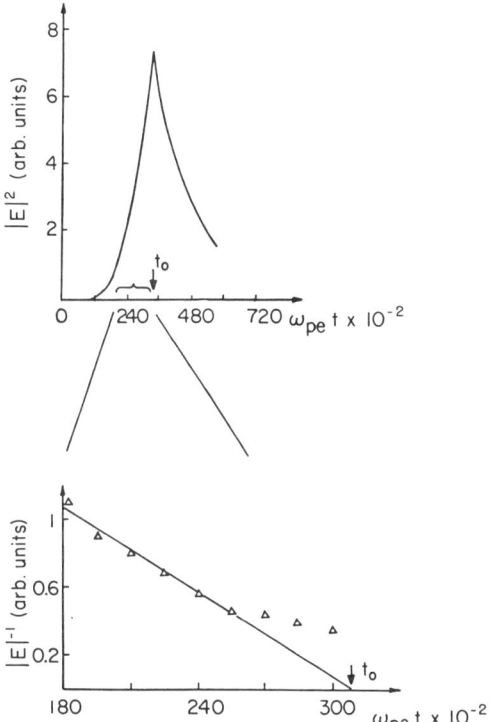

Figure 12. Time dependence of wave intensity $|E_{es}|^2$ (top figure) and $|E_{es}|^{-1}$ (bottom figure) at a spatial location inside the density cavity. The solid line in the bottom figure corresponds to $|E| \propto (t_0 - t)^{-1}$.

recurrent spatially and temporally as the experiment is repeated in an identical fashion. This point might not be fully understood by theoreticians who deal with homogeneous plasmas with no particular location for the excitation source.

It is interesting to note that in spite of plasma fluctuations from shot to shot the final state remains the same. This implies a form of attractor to which all nonlinear development gravitate.[5] In spite of this reproducibility in the caviton experiment, it is worthwhile to examine each single event in order to uncover those nonlinear states that are not reproducible. Such experiments are currently in progress.

In summary, this experiment has shown how the wavenumber spectrum broadens due to spatial collapse while the frequency spectrum expands as a result of the formation of deep density cavities.

We shall present a separate experiment to demonstrate how trapped modes arise inside a density cavity.

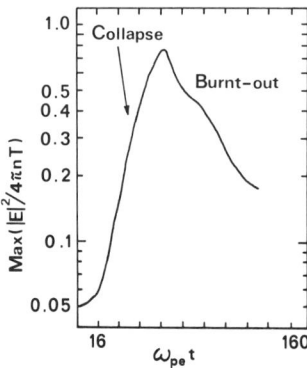

Figure 13. Normalized field energy in logarithmic scale at the center of the density cavity versus time. This is the result of two-dimensional particle-in-cell simulation depicted in Anisimov et al.[18] The ion to electron mass ratio $M/m = 100$ is used for this simulation. The processes of the Langmuir wave collapse and its burnt-out are clearly seen.

PLASMA WAVE TRAPPING IN CAVITONS

The previous section has revealed the spatial development of cavitons. The frequency broadening of the initial coherent electron-plasma wave is inferred from the observed electromagnetic spectra. We will now present experimental results that support the concept that the broadening of the electron-plasma wave spectrum is a result of the increase in the number of eigenstates inside a density cavity. In order to delineate the physics, we shall confine our attention to a single caviton.

The temporal and spatial evolution of a caviton is describable by the set of Zakharov equations,

$$\frac{2i}{\omega}\frac{\partial E}{\partial t} + \frac{3v_{eT}^2}{\omega^2}\frac{\partial^2 E}{\partial x^2} + \left(1 - \frac{\omega_{pe}^2(x,|E|^2)}{\omega^2} + i\frac{\nu_e}{\omega}\right)E = E_0 \quad (1)$$

$$\frac{\partial^2 n_1}{\partial t^2} - C_s^2\frac{\partial^2 n_1}{\partial x^2} + \nu_i\frac{\partial n_1}{\partial t} = n_0 C_s^2 \frac{\partial^2}{\partial x^2}\left(\frac{|E^2|}{16\pi n_0 T_e}\right) \quad (2)$$

where E is the envelope of the high-frequency field, E_0 is the amplitude of the pump field, n_1 is the ion-density perturbation, and ν_e, ν_i are the respective damping rates of electron and ion waves.

This set of equations has been solved for the density profile observed in our experiments and verified by analytic results using a parabolic density cavity.[21,22]

The initial single caviton is created by a capacitor rf-driven field of frequency f_0 ($k_0 = 0$) at the resonant location x_0, where $\omega_{pe}(x_0) = 2\pi f_0$.[23] After the single caviton is created, the plasma response to different excitation frequencies is monitored by a diagnostic electron beam. It was found that the enhanced responses occur at frequencies below f_0, the initial excitation frequency. According to Fig. 14, at least two resonances have

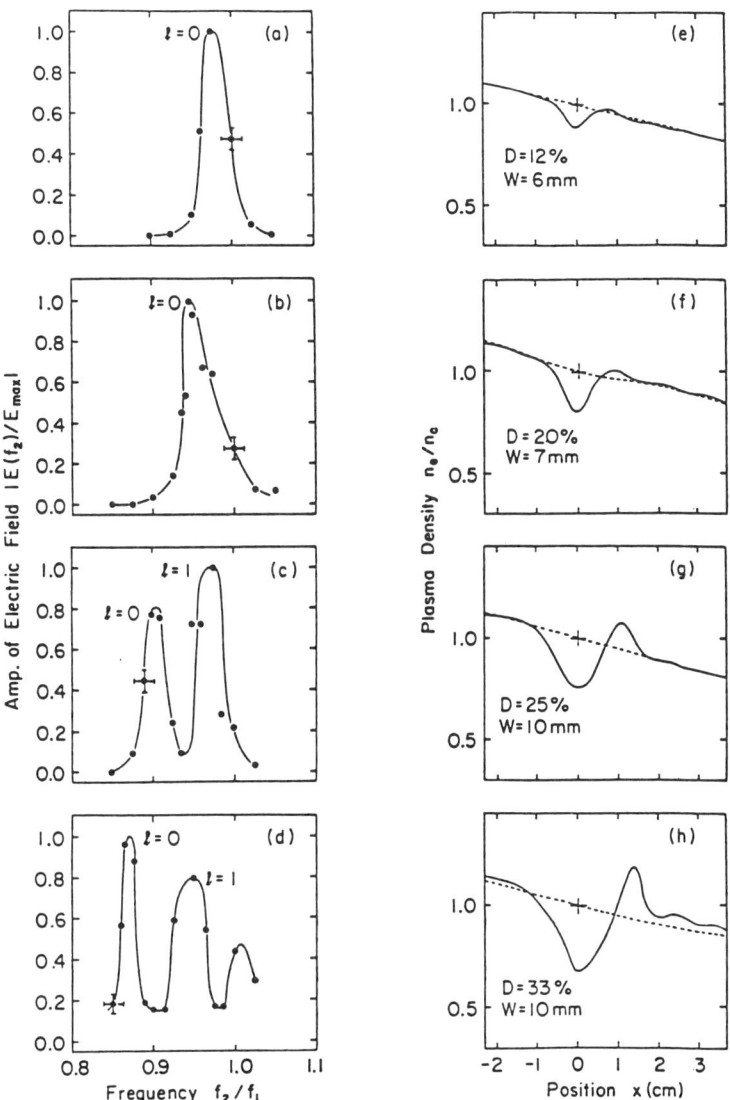

Figure 14. Response functions, $|E(f_2)|$, to different excitation frequencies for preexisting density cavities of different depth and width (a)–(d). Corresponding density profiles are shown in (e)–(h). The depth ($D = |\delta n/n_c|$) and the full width at half maximum depth W of the density cavities are indicated in the figures. In these experiments each density cavity is created by a strong excitation pump at frequency f_1 ($= 200$ MHz); then, the second excitation pump at frequency f_2 (variable), which is too weak to create a density cavity by itself, is applied to measure the plasma response. The plasma density is normalized to the critical density for the cavity creating pump, that is, $n_c = \pi m f_1^2 / e^2$. It is clearly seen that the peak position corresponding to $l = 0$ state shifts toward the lower frequency side as the density cavity deepens. When the density cavity is sufficiently deep [(g) and (h)], the peak corresponding to $l = 1$ state is observed in the response function [(c) and (d)].

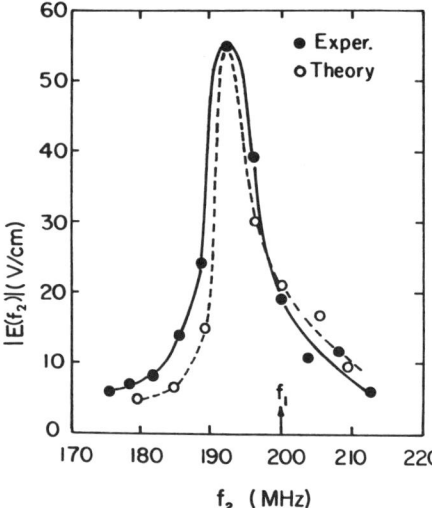

Figure 15. Comparison between theory and experiment for the response function, $|E(f_2)|$. The experimental data (solid circles) are taken for the preexisting density cavity of depth $D \simeq 10\%$ and width $W \simeq 10$ mm. Theoretical points (open circles) are obtained by numerically solving the Zakharov equations (1) and (2) with the preexisting density cavity similar to that in the experiment.

been observed inside a caviton of $|\delta n/n_0| \simeq 30\%$, which we have identified as $l = 0$ and 1 modes corresponding to the ground state and the first excited state in an equivalent potential well determined by the caviton density profile. A comparison between theory and experiment for $l = 0$ mode is presented in Fig. 15; the theoretical points are computed from the Zakharov equations using the exact density profile. Experimentally, after the termination of the external pump field, E_0, the ground state has been found to have the longest lifetime because it has the slowest leakage rate out of the confining cavity. The single resonance at the original location is replaced by two or more resonances each with a finite width as a consequence of the caviton formation. Therefore, while the caviton undergoes the spatial-collapsing process, it acquires a range of frequencies in addition to the initial monochromatic mode. It is worth pointing out that these trapped states can interact among themselves producing difference and sum frequencies. If the difference frequency lies in the ion-acoustic-frequency range, it generates ion acoustic waves, which propagate out of the cavity. The sum frequencies can convect out as electromagnetic radiation as in the cold-beam experiment described in the previous section.

CAVITONS AND CLUMPS

Once localized, large-amplitude fields are formed in cavitons, they can generate clumps in phase space, which are analogous to the clumps discussed by Dupree and co-workers.[8-10] The main difference is that we are

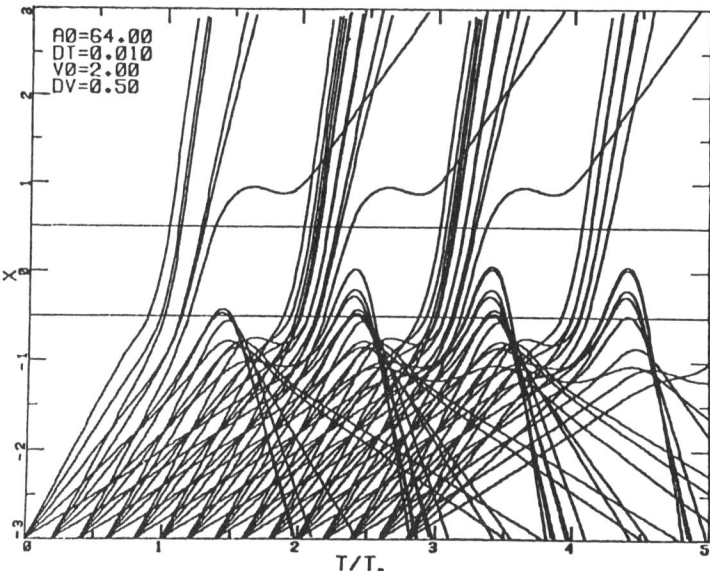

Figure 16. An $x - t$ diagram describing the interaction between charged particles and single caviton. The caviton region, where strong oscillating E fields exist, is between two straight lines at $x = -0.5$ and 0.5. The particles are accelerated or decelerated depending upon their phase as they approach the caviton. Particle bunching, creating a clump in phase space, is obvious after the particles pass through the caviton region.

considering cavitons and clumps under rather strong excitations. Consider one such caviton interacting with background thermal particles (Fig. 16).[12] For those particles that approach the caviton at the proper phase so that they are accelerated during their passage, the velocity vectors are bunched closer together at the same spatial location, hence forming a clump in phase space. The particles in this clump in phase space experience the same self-consistent field as they move through the plasma. The lifetime of these clumps can be extended through repeated bindings by such cavitons. The formation of cavitons and clumps might be a common occurrence inside a plasma and might have escaped detection by standard experimental methods, which look at ensemble averages. Plasma might be a strongly turbulent medium and much more "granular" in appearance than is generally perceived.

Future work should consider the self-consistent interaction between clumps and cavitons. Can the clumps produced by cavitons regenerate new cavitons at other locations by virtue of their retaining phase information in a collisionless medium? Do cavitons communicate with one another through particles and waves?

The development from linear to strongly nonlinear regimes in a plasma is summarized in Fig. 17.

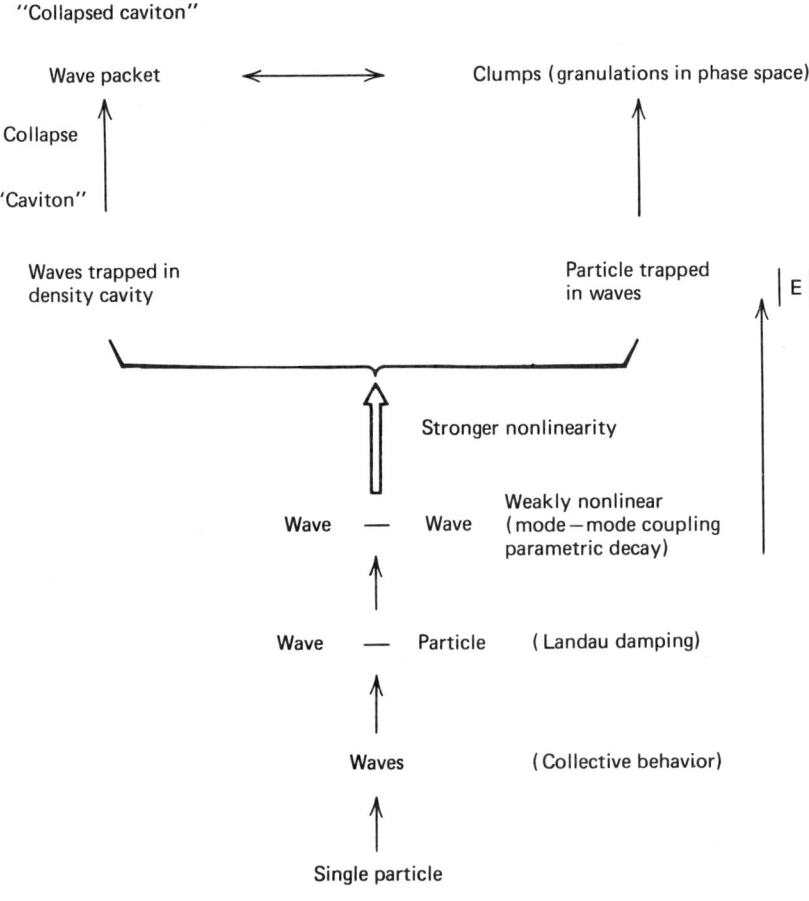

Figure 17. Development of strong nonlinearity in a plasma. The amplitude of the pump field, $|E|$, increases in the direction of the vertical arrows. When $|E|$ is sufficiently small, the plasma stays in the linear regime. When $|E|$ becomes very large, the plasma shows a strong nonlinearity characterized by "cavitons" and "clumps".

ACKNOWLEDGEMENT

This work was supported by the National Science Foundation.

REFERENCES

1. K. Papadopoulos, in *Diagnostics for Fusion Experiments*, E. Sindoni and C. Wharton, eds. (Pergamon, Oxford, 1979), pp. 355–365.

2. R. Stenzel and A. Y. Wong, *Phys. Rev. Lett.* **28**, 274 (1972).
3. H. Ikezi and R. J. Taylor, *Phys. Rev. Lett.* **22**, 923 (1969).
4. F. Hai and A. Y. Wong, *Phys. Fluids* **13**, 672 (1970).
5. E. Ott, *Rev. Mod. Phys.* **53**, 655 (1981).
6. A. Y. Wong, M. V. Goldman, F. Hai, and R. Rowberg, *Phys. Rev. Lett.* **21**, 518 (1968).
7. A. A. Galeev, R. Z. Sagdeev, V. D. Shapiro, and V. I. Shevchenko, Space Institute Report II, The Academy of Sciences of the USSR, 1977, p. 353.
8. R. H. Berman, D. J. Tetreault, T. H. Dupree, and T. Boutros-Ghali, *Phys. Rev. Lett.* **48**, 1249 (1982).
9. T. H. Dupree, C. E. Wagner, and W. M. Manheimer, *Phys. Fluids* **18**, 1167 (1975).
10. T. H. Dupree (private communication, 1983).
11. P. Y. Cheung, A. Y. Wong, C. B. Darrow, and S. J. Qian, *Phys. Rev. Lett.* **48**, 1348 (1982).
12. A. Y. Wong, P. Leung, and D. Eggleston, *Phys. Rev. Lett.* **39**, 1407 (1977).
13. P. Leung, M. Q. Tran, and A. Y. Wong, *Plasma Phys.* **24**, 567 (1982).
14. G. D. Doolen, D. F. DuBois, H. A. Rose, and B. Hafizi, LANL Report LA–UR–83–605, 1983; *Phys. Rev. Lett.* **51**, 335 (1983).
15. T. M. O'Neal and J. H. Malmberg, *Phys. Fluids* **11**, 1754 (1968).
16. H. C. Kim, R. L. Stenzel, and A. Y. Wong, *Phys. Rev. Lett.* **33**, 886 (1974).
17. G. J. Morales and Y. C. Lee, *Phys. Fluids* **19**, 690 (1976).
18. S. I. Anisimov, M. A. Berezovskii, M. F. Ivanov, I. V. Petrov, A. M. Rubenchik, and V. E. Zakharov, Reprint N167, Institute of Automation and Electrometry, Siberian Branch, USSR Acad. Sci., Novosibirsk, 1981.
19. S. I. Anisimov, M. A. Berezovskii, V. E. Zakharov, I. V. Petrov, and A. M. Rubenchik, Reprint N198, Institute of Automation and Electrometry, Siberian Branch, USSR Acad. Sci., Novosibirsk, 1982.
20. L. N. Shchur and V. E. Zakharov, Preprint 1981–16, The Academy of Sciences of the USSR, L. D. Landau Institute for Theoretical Physics, Chernogolovka, 1982.
21. T. Tanikawa, A. Y. Wong, and D. L. Eggleston, UCLA Plasma Physics Group Report 702, 1983.
22. A. Y. Wong, *Physica Scripta* **T2/1**, 262 (1982).
23 D. L. Eggleston, A. Y. Wong, and C. B. Darrow, *Phys. Fluids* **25**, 257 (1982).

COHERENCE IN CHAOS AND THE ZAKHAROV MODEL

G. D. Doolen, D. F. DuBois, and H. A. Rose
Los Alamos National Laboratory, Los Alamos, New Mexico

B. Hafizi
Science Applications Inc., Plasma Research Institute, Boulder, Colorado

Abstract. The chaotic nature of "caviton turbulence" is studied in one-dimensional many-Fourier mode numerical simulations of the dissipative Zakharov equations in which the system is driven to modulational instability by energy injection into long-wavelength modes. Just above this instability threshold the system evolves, with hysteresis, into a variety of stable space-time patterns of cavitons. For stronger driving the system becomes chaotic along various routes; the cavitons remain quite coherent with lifetimes long compared to the shortest Lyapunov time.

In this chapter we report on numerical studies in one spatial dimension of a many-mode Fourier representation of the nonlinear partial differential equations known as the Zakharov equations.[1] These studies demonstrate the coexistence of coherent spatial structures—cavitons—and temporal chaos. The level of chaos, as measured by the largest positive Lyapunov exponent, is found to be essentially independent of the method of energy injection into a fixed number of long-wavelength modes of the system for driving levels well above the chaotic threshold. On the other hand, the lifetime of the cavitons—density cavities with trapped electrostatic fields—is generally long compared to the Lyapunov time. The behavior near the chaotic threshold depends sensitively on the method of driving and on parameters such as the length of the system; at least in the threshold regime the cavitons do not evolve independently.

There have been a number of many-mode numerical simulations of the Zakharov equations,[2] but to our knowledge this is the first quantitative study of the chaotic nature of "caviton turbulence." All of the interesting nonlinear phenomena are observed to occur above the linear threshold for modulational instability of the system driven by the long-wavelength modes. These phenomena include hysteresis effect with stationary solutions (fixed points), limit cycles, two-frequency behavior, and transitions to chaos.

The Zakharov equations in dimensionless form in Fourier representations are[1]

$$\{i[\partial_t + \nu_e(k)] - k^2\}E_k(t) = \sum_{k'} n_{k'} E_{k-k'} + S_k(t) \tag{1}$$

$$[\partial_t^2 + 2\nu_i(k)\partial_t + k^2]n_k(t) = -k^2 \sum_{k'} E^*_{-k'}(t) E_{k-k'}(t) \tag{2}$$

supplemented by the complex conjugated Eq. (1) for $E^*_k(t)$. These equations are a useful model of Langmuir turbulence and are discussed elsewhere.[1,5] Here $E_k(t)$ is the slowly varying envelope of the electrostatic-field oscillation at the background plasma frequency and $n_k(t)$ is the Fourier component of the density fluctuation induced by the ponderomotive force, which is the right-hand side of Eq. (2). (We have $n_{k=0} = 0$.) The dissipation rates $\nu_e(k)$ and $\nu_i(k)$ represent the coupling to the suppressed particle degrees of freedom. If $\nu_e = \nu_i = S_k = 0$, the system is Hamiltonian and possesses analytic single-caviton solutions.[1] With dissipation (and driving) the following modified Liouville relation in solution space holds:

$$\sum_k \left(\frac{\partial \dot{E}_k}{\partial E_k} + \frac{\partial \dot{E}^*_k}{\partial E^*_k} + \frac{\partial \dot{n}_k}{\partial n_k} + \frac{\partial \dot{u}_k}{\partial \dot{u}_k} \right) = -2 \sum_k [\nu_e(k) + \nu_i(k)] \tag{3}$$

Here $u_k = -k^{-2}\dot{n}_k$ is the conjugate Hamiltonian variable to n_k. We consider only cases in which the right-hand side of Eq. (3) is negative, which ensures the contraction of volumes in solution space.

Equations (1) and (2) were numerically solved using a split-time-step spectral method in which linear terms are advanced in Fourier space and nonlinear terms in real space. Periodic boundary conditions in a box of length L are imposed and from 64 to 1024 Fourier modes were used to ensure adequate resolution of spatial structures. Tests for spatial and temporal resolution were employed,[4] and the dependence on box size was studied.

Numerical experiments were carried out for three methods of driving or energy injection:

1. Coherent source drive: $S_k(t) = \nu_e W_0^{1/2}$ for $|k| \leq k_{DR}$ and $S_k = 0$ for $|k| > k_{DR}$. If $k_{DR} = 0$, this represents a capacitor model drive where $|E_0| = W_0^{1/2}$ is the amplitude of the spatially uniform, vacuum capacitor field driven at the plasma frequency ω_p.

2. Coherent beam drive: $\nu_e(k) = -\nu_{DR}$ for $|k| \leq k_{DR}$ and $\nu_e(k) = \nu_e =$ const for $|k| > k_{DR}$. This might represent a system excited by counterstreaming electron beams where the Langmuir waves with $|k| \leq k_{DR}$ are unstable with growth rates ν_{DR}.

3. Noise source drive: $S_k(t) = (2\nu_e W_0)^{1/2} \xi_k(t)$ for $|k| \leq k_{DR}$ and $S_k = 0$ for $|k| > k_{DR}$ and $\xi_k(t)$ are delta-correlated complex white-noise sources of unit rms amplitude. This model is useful for comparison with statistical theories and can be thought of as a directly driven Langmuir condensate.[5]

The measure of chaos in our work will be the Lyapunov exponent, which measures the exponential rate of separation of two solutions with nearby initial conditions. Formally, the largest Lyapunov exponent is given by[6]

$$L_1 = \lim_{t \to \infty} \lim_{D(0) \to 0} \left\{ t^{-1} \ln\left[\frac{D(t)}{D(0)}\right] \right\} \quad (4)$$

where $D(t)$ is a measure of the distance between solutions "one" and "two":

$$D(t) = \left[\sum_k D_k^2(t) \right]^{1/2}; \quad D_k^2(t) = |\Delta E_k|^2 + (\Delta n_k)^2 + (\Delta u_k)^2 \quad (5)$$

where $\Delta E_k(t) = E_1(t) - E_2(t)$, etc., represents the difference at time t of two solutions differing slightly in their initial conditions.

When the variables $E(x, t)$ and $n(x, t)$ evolve from almost flat initial conditions, the most modulationally unstable mode ($k \simeq W^{1/2}$) evolves into a set of steepened cavitons. Here W is the total electrostatic energy per unit length, $W = \sum_k |E_k(t)|^2 = L^{-1} \int dx |E(xt)|^2$. In Fig. 1 we show this stage of development for case (ii) with five driven modes. An x, t plot of the contours of equal $|E(x, t)|^2$ is given in Fig. 2, which shows the persistence of the caviton "trajectories" over times long compared to the Lyapunov time. The amplitudes of these cavitons oscillate up and down in time in a random manner; the spacing between peaks is roughly the Lyapunov time, L_1^{-1}. It appears that these motions are not independent and neighboring cavitons exchange energy. A large number of such simulations have been carried out and yield interesting data on spectral width in k space and the number of cavitons versus driving or turbulence level. These will be discussed elsewhere; here we comment that both of these quantities scale as $\bar{W}^{1/2}$ in the near-threshold regime.

In Fig. 3 we summarize results for the Lyapunov exponents L_1 vs $\bar{W}$, where the bar denotes a time average, for various methods of driving and with fixed background dissipation. We note that for $W \gg \nu_e$ the level of chaos is independent of the *method* of driving but depends on the *number* of driven modes. We have verified this for several values of the background dissipation. Note, however, in all cases energy is injected into the low

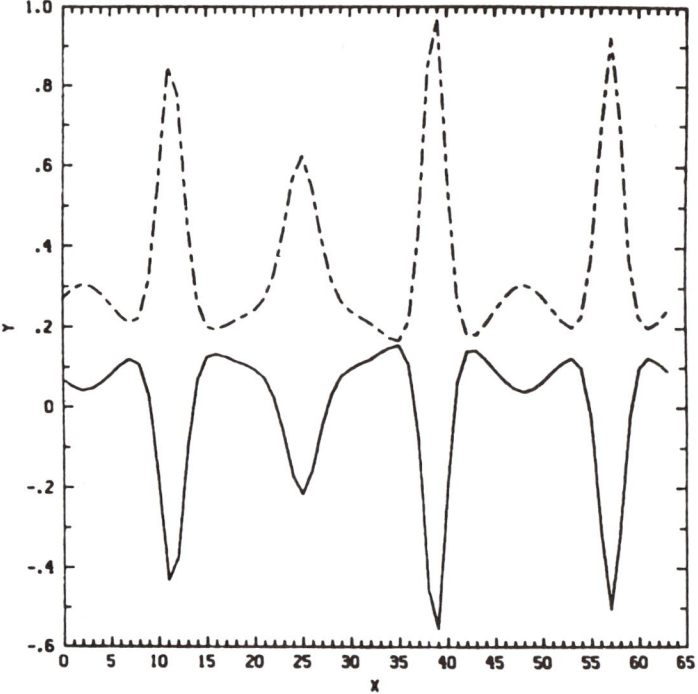

Figure 1. Instantaneous spatial patterns of $|E(x, t)|^2$ and $n(x, t)$ for the case of a five-mode ($k_{DR} = 0.2$), coherent-beam drive [case (ii)], $\nu_{DR} = -0.025$, $\nu_e = 0.05$, $\nu_i = 0.5k$ and $L = 64$, $\bar{W} = 0.13$.

wavenumbers, $|k| < k_{DR}$ and that in the range of driving independence $W^{1/2} \gg k_{DR}$. Here we present data only for the weak driving where $W < 1$.

The caviton lifetime is obtained from contour plots such as Fig. 2 and measuring the length (in time) of identifiable "caviton mountain chains." A major result of this work is that the cavitons can persist much longer than the Lyapunov time—$\tau_{cav} L_1 \gg 1$. This, then, is a relatively simple system in which spatial coherence and temporal chaos can coexist.[7] The largest Lyapunov exponent itself is not a good measure of the coherence time of the cavitons.

A finite threshold for chaos has been observed for both the coherent-source- and noise-source-driven cases (see Fig. 3). In the latter case, it is interesting to note that the intrinsic chaotic behavior characterized by exponential separation of solutions is clearly observable in the presence of strong extrinsic noise which drives the turbulence. Note that when we compare nearby orbits, we do so for a *specific* realization of the noise-drive-time series.

We have only an incomplete picture of the threshold regime at present; we will here present some results for the case of a single-mode coherent

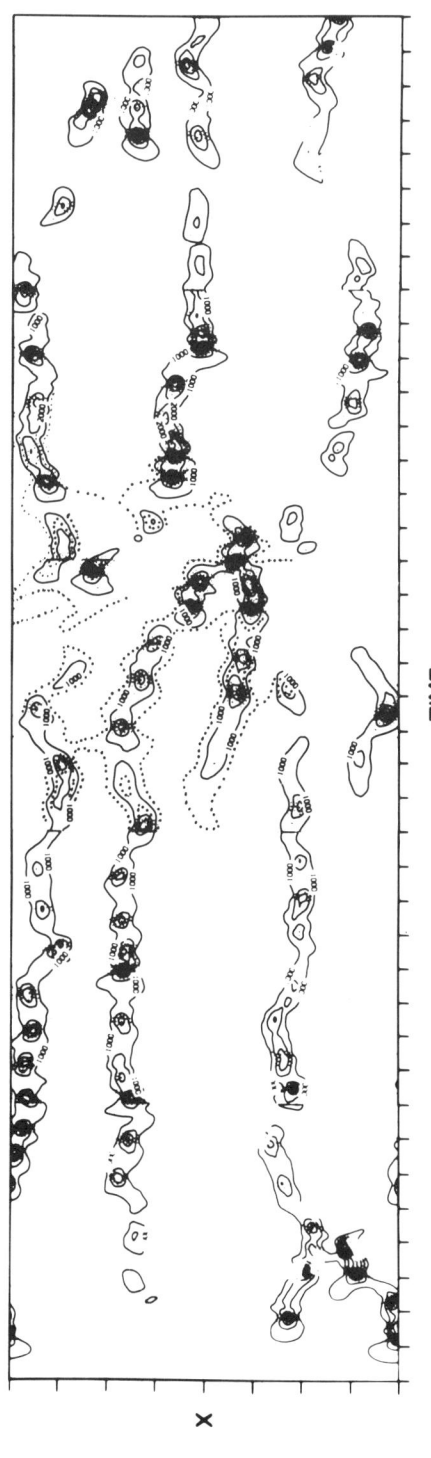

Figure 2. Contours of equal $|E(xt)|^2$ versus x and t for the parameters of Fig. 1. The length of the system in x is 64 and the spacing between divisions on the time scale are $\Delta t = 20$, which is approximately the Lyapunov time L_1^{-1} for this case.

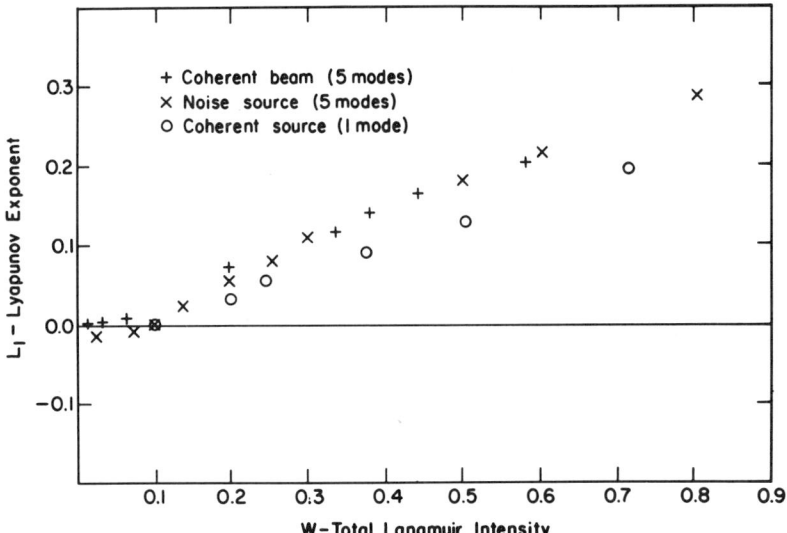

Figure 3. Lyapunov exponent L_1 versus total energy $\bar{W}$ for several methods of energy injection: × noise driven, + beam driven, both with five driven modes ($k_{DR} = 0.2$); and ○ coherent-source drive with one driven mode. In all cases $\nu_e = 0.1$, $\nu_i = 0.1k$.

drive [case (i) with $k_{DR} = 0$]. For drive strengths W_0 below the linear modulational instability threshold, $W_0(MI) = \nu_e$, a trivial fixed point (FP) or stationary solution is observed with $|E_k| = W_0^{1/2} \cdot \delta_{k0}$ and $n_k = 0$. Above this threshold we observe complicated hysteresis effects. Among the possible states for $W_0 > W_0(MI)$ is bifurcation to another nontrivial FP with stationary spatial patterns shown in Fig. 4a for $W_0 = 0.14$ ($\nu_e = 0.1$) and characterized by zero momentum. In this stationary case n is adiabatically related to $|E|^2$; $n = -|E|^2$. For larger $W_0 = 0.145$ we observe a stable limit cycle in which $W(t)$, after initial transients, oscillates sinusoidally about a mean value of 0.104 with a frequency of 0.50 in our dimensionless units [multiply by $\frac{2}{3}(m_e/m_i)\omega_{pe}$ to obtain physical units]. The contour plots of $|E(xt)|^2$ and $n(xt)$ are shown in Figs. 4b and 4c. These show a periodic structure moving at an average speed of $v \simeq 0.94$ (times the ion sound speed). The spacing in time between repeated structures in $|E|^2$ agrees with the limit cycle period of $W(t)$. We remark that $n(x, t)$ is *not* adiabatically related to $|E(xt)|^2$ in this case. This stable limit cycle is characterized by a finite momentum and apparently arises from a symmetry-breaking momentum instability.

The stable limit cycle evolves into a two-frequency aperiodic state, which we show for $W_0 = 0.150$. In Fig. 4d we see the two frequencies in the time behavior of $W(t)$, the high frequency ω_H is the continuation of the stable limit cycle frequency and a new low-frequency modulation ω_L ap-

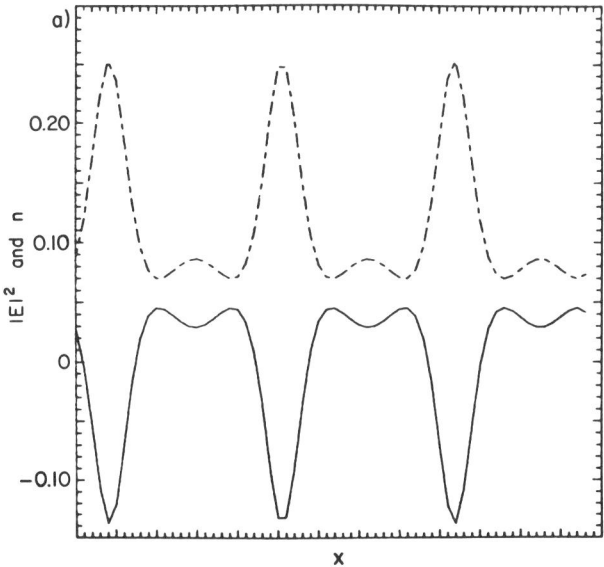

Figure 4. (a) Behavior near the modulational threshold $W_0(MI) = \nu_e = 0.1$ and chaotic threshold $W(\text{chaos}) \sim 0.15$ for a single-mode coherent-source drive with $\nu_e = 0.1$, $\nu_i = 0.1k$, and $L = 64$: instantaneous spatial patterns for $|E(x, t)|^2$ and $n(xt)$ for a coherent stationary (FP) solution $W_0 = 0.140$.

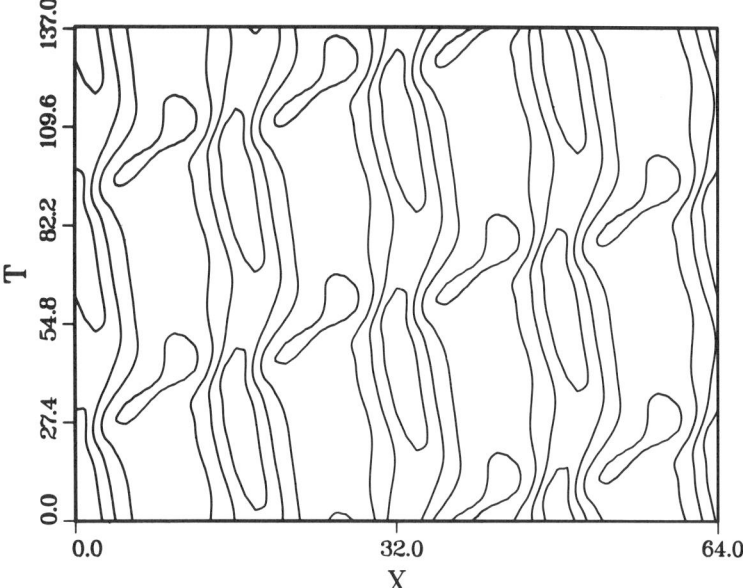

Figure 4. (b) Contours of $|E(x, t)|^2$.

(*Continued on next page*)

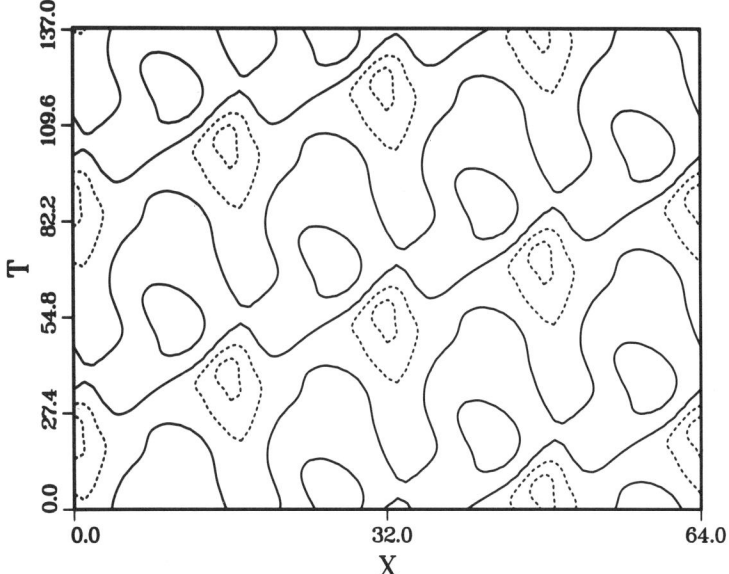

Figure 4. (c) Contours of $n(x, t)$.

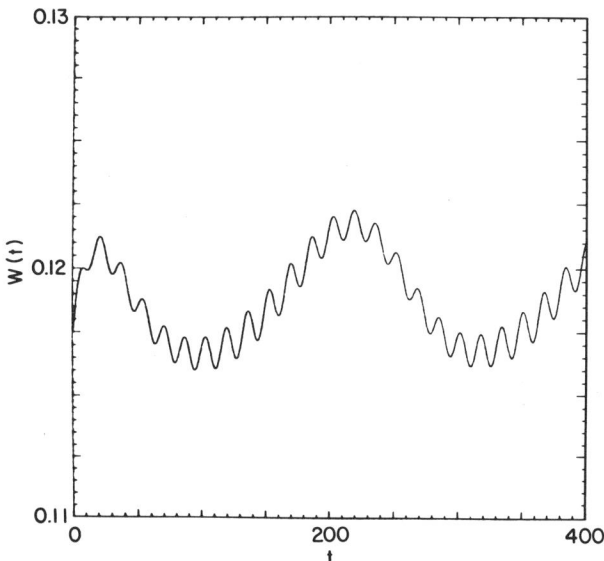

Figure 4. (d) $W(t)$ versus time for a two-frequency, aperiodic state at $W_0 = 0.145$ which is slightly chaotic.
(*Continued on next page*)

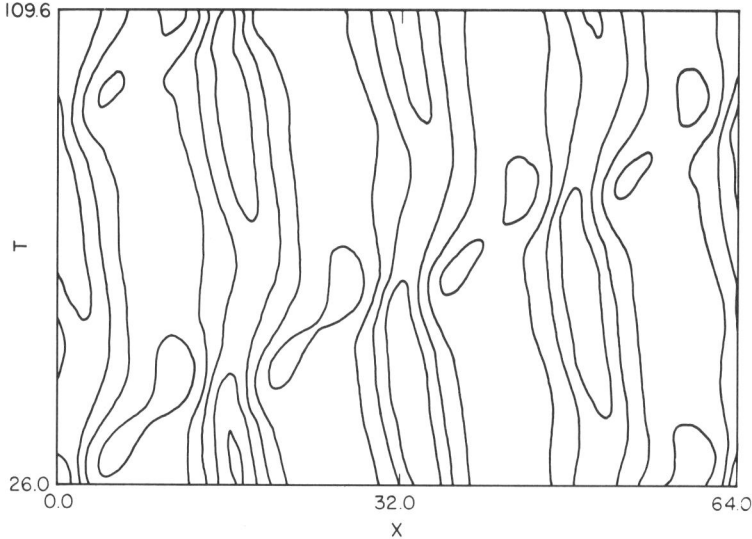

Figure 4. (*e*) Contours of $|E|^2$ for this aperiodic state.

pears with $\omega_H/\omega_L \simeq 13$. This case appears to be slightly chaotic. In the corresponding contour plot shown in Figs. 4*e* we observe that the periodicity of the structures observed in Figs. 4*b* is broken. A long-time continuation of this plot reveals that the pattern *appears* to repeat after about 13 limit cycle periods. We have not yet determined whether these frequencies are incommensurate.

In the regime near the chaotic threshold with a coherent-source drive we have observed a sensitive dependence of the various transitions on the system length L. The comparison between solutions in two different length boxes is complicated by hysteresis effects. However, we have been able to establish a dependence on box size for a particular class of solutions. As W_0 increases above $W_0(MI)$ which is independent of L, there is the sequence of transitions discussed above: from a fixed point to a momentum-carrying limit cycle at $W_0 = W_0(LC)$; from the limit cycle to a more-complicated, two-frequency, but nonchaotic, state at $W_0 = W_0(TF)$; and then a transition to chaos at $W_0 = W_0(C)$. The transition values—$W_0(LC)$, $W_0(TF)$, $W_0(C)$—depend on the box size with the corresponding transition values being smaller for the larger box. For example, $W_0(LC) = 0.136$, $W_0(C) = 0.18 \pm 0.005$ for $L = 128$, number of modes = 128; whereas $W_0(LC) = 0.155$, $W_0(C) = 0.20 \pm 0.005$ for $L = 55$, number of modes = 64. The lengths of the boxes were chosen so that in the fixed-point regime both boxes had the same repeated-unit caviton spatial pattern with seven unit cavitons in $L = 128$ and three unit cavitons in $L = 55$. There are some classes of solutions (involving large structures) that are observed in the large box and not in the small one, so we cannot make any general statements at this point relating the behavior of solutions. However, for the special class of solutions mentioned above we can say that chaos occurs for weaker driving in the larger box. We also observe that for driving levels well into the chaotic region the results for spatial-averaged quantities (spectra, energy levels, Lyapunov exponents, etc.) appear to be independent of box size presumably because the box size is large compared to the basic correlation length of the turbulent fields. We are currently studying the nature of the transition to chaos in the regime of large dynamic aspect ratio, $L\sqrt{W_0} \gg 1$, to see if this behavior implies that for very large values of this ratio the system will go from the trivial FP for $W_0 < W_0(MI)$ directly to chaos at $W_0 = W_0(MI)$.

Russell and Ott[11] have made a detailed study of the rich structure of the chaotic threshold regime in a three-mode truncation of the nonlinear Schrödinger equation, which is the adiabatic limit of the Zakharov equations with $n = |E|^2$. We have observed that near the chaotic threshold only a few modes (three or *more*) are in fact excited; however, the many-mode system self-consistently determines which modes these are—for example, the specific modes depend on the level of driving. As noted above, the adiabatic approximation can break down even in the threshold regime.

Preliminary results indicate that the chaotic threshold region for the noise-source-driven case (iii), does not have the bifurcation structure observed in our studies for the coherent source or in other studies where the noise is a small perturbation to a finite coherent drive.[9]

Although we have not studied the chaotic threshold for the coherent beam drive [case (ii)], we have observed a regime where arbitrarily weak beam-unstable modes grow past the modulational instability threshold creating a "burst" of cavitons, which depletes the driven modes. The high-k caviton modes then damp away and the driven modes again grow. In Fig. 5a $\ln[D(t)]$ and in Fig. 5b $W(t)$ and $W_{k=0} \equiv |E_0(t)|^2$ are displayed for a case where $k_{DR} = 0$. This example illustrates a connection between the traditional concept of the genesis of turbulence by instabilities (in this case the modulational instability) and the more-recent concept of stochastic instability. It appears that the local in time leading positive Lyapunov exponent, $L_1(t)$, can be identified with either the growth rate of the beam-unstable mode or a modulationally unstable mode k_m, for the local value of $W_{k=0}$. The modulational growth rate does not appear in $L(t)$ until $W_{k_m} \simeq W_{k=0}$. A related analysis of L_1 can be obtained from an average of $D_k(t)$, the difference spectrum,[10] which shows peaks at $k = 0$ and k_m. For example, in the interval $0 \le t \le 320$, the slope of the curve in Fig. 5a is about 0.004, corresponding to the beam-unstable mode at $k = 0$, while for $320 < t < 350$, $W_{k=0} \simeq 0.5$ and the modulational instability growth rate at $k = 0.5$ is given by $k(2W_{k=0} - k^2)^{1/2} - \nu_e(k) = 0.33$, which is close to the measured slope, 0.30 [an inspection of data not presented here indicates $E(k = 0.5)$ was rapidly growing in this time interval]. It is the nonlinearity

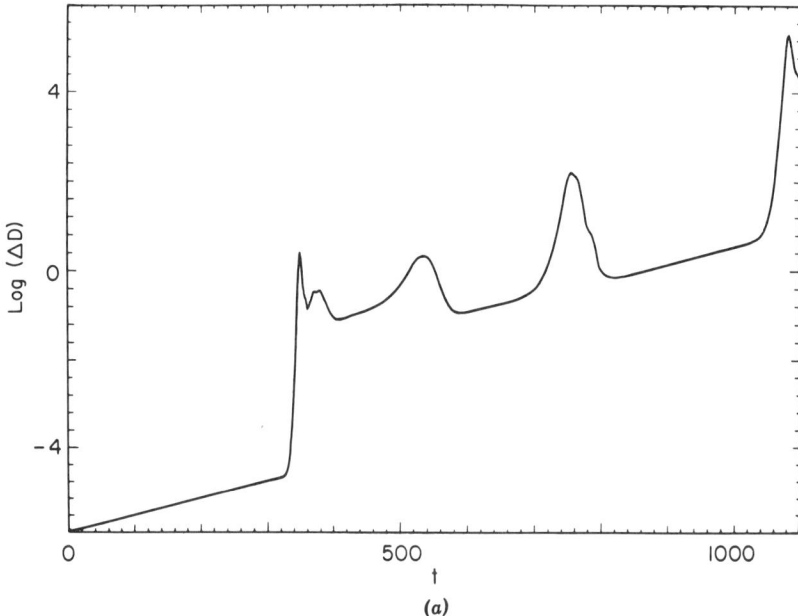

Figure 5. Illustration of caviton "burst" phenomena: (a) The logarithm of $D(t)$ vs time t for a single-mode coherent-beam drive with $\nu_e = 0.1$, $\nu_i = 0.1k$, and $\nu_{DR} = -0.004$, $k_{DR} = 0$, and $L = 64$. (*Continued on next page*)

of the dynamics that allows for the "turning on and off" of the two types of instabilities, and determines the time-average stochastic instability of the leading Lyapunov exponent. This process also appears to occur for stronger driven cases in which these instabilities act to produce local bursts of cavitons, which occur throughout the system and produce a less intermittent time signature for box-averaged quantities.

The Zakharov system is a particularly simple multidimensional system in which to study the evolution of a well-defined linear instability—the modulational instability—to a nonlinear state involving well-characterized coherent cavitons; this nonlinear state may be chaotic or not. Similar phenomena of coherence in chaos appear in other settings; for example, spatial rolls and chaos transitions in Bénard convection. In the Zakharov system transitions to chaos along with hysteresis have been observed. In the chaotic regime caviton structures are observed to live for times long compared to the time scale of the largest Lyapunov exponent. We often find it necessary to integrate for several hundred time units [which are $\frac{3}{2}(mi/me)\omega_{pe}^{-1}$] for a system to reach its asymptotic (attracting) state; for a physical mass ratio this is long compared to the time of most particle-in-cell simulations. For shorter times misleading coherent transients may occur,

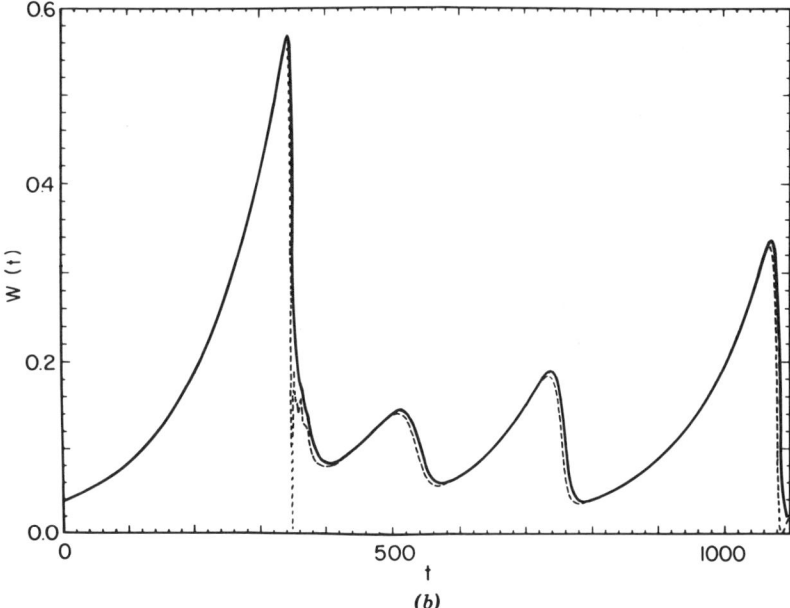

Figure 5. (b) For the same case $W(t)$, solid curve and $W_{k=0}(t) = |E_{k=0}(t)|^2$, dotted curve, versus time t.

which ultimately lead to a chaotic attractor. The reverse scenario may also occur. The study of coherent caviton structures in the Zakharov model may provide some insight into analogous structures such as "clumps" and "holes" observed in simulations of Vlasov plasmas; the cavitons are analogous to self-consistent BGK modes in the Vlasov model, which in turn may be related to "holes" in phase space.[12] Our results indicate that the caviton lifetime is not directly related to the shortest e-folding time for exponential separation of solutions.

Further studies of the important problem of the dependence of these threshold scenarios on box size are feasible, and extension of these studies to the case of collapsing cavitons is of considerable interest but may present formidable computational problems.

We are grateful to Professor M. V. Goldman for making available to us an early version of the computer code. This research was supported by the U.S. Department of Energy, in particular the work of one of us (B.H.) was supported in part under contract DE-AC03-76 ETS3057 with Science Applications, Inc.

REFERENCES

1. V. E. Zakharov, Zh. Eksp. Teor. Fiz. **62**, 1745 (1972) [*Sov. Phys.–JETP* **35**, 908 (1972)]. Two review articles: S. G. Thornhill and D. ter Haar, *Phys. Rep.* **43C**, 43 (1979) and L. I. Rudakov and V. N. Tsytovich, *Phys. Rep.* **40**, 1 (1979).

2. J. J. Thomson, F. J. Faehl, and W. L. Kruer, *Phys. Rev. Lett.* **31**, 918 (1973); R. N. Sudan, N. R. Pereira, and J. Denavit, Phys. Fluids **20**, 271 (1977); L. M. Degtyarev, R. Z. Sagdeev, G. I. Solovev, V. D. Shapiro, and V. I. Shevchenko, *Fiz. Plazmy* **6**, 485 (1980) [*Sov. J. Plasma Phys.* **6**, 283 (1981)]; B. Hafizi, J. C. Weatherall, M. V. Goldman, and D. R. Nicholson, *Phys. Fluids* **25**, 392 (1982).

3. Recently, studies of the relation between spatial coherence and temporal chaos have been carried out for the sine-Gordon system by K. Nozaki, *Phys. Rev. Lett.* **49**, 1883 (1982) and by A. Bishop, K. Fesser, P. Lomdahl, W. Kerr, M. Williams, and S. Trullinger, *Phys. Rev. Lett.* **50**, 1095 (1982).

4. The limiting factor in determining the spatial resolution is the width in k space of the density spectrum $|n_k|^2$, which is generally much wider than the $|\varepsilon_k|^2$ spectrum.

5. D. F. DuBois and H. A. Rose, *Phys. Rev.* **A24**, 1476 (1981).

6. G. Bennetin, L. Galgani, and J. Strelcyn, *Phys. Rev. A* **14**, 2338 (1976).

7. The Langmuir cavitons do not possess the topological stability of sine-Gordon kinks, for example, so that their long lifetime seems to be a nontrivial effect.

8. The nontrivial fixed point generally involves only a small number of modes—3, 5, 7, etc.—and the conditions for this stationary state can be found semianalytically.

9. J. D. Farmer, *Phys. Rev. Lett.* **47**, 179 (1981) and references therein.

10. R. H. Kraichnan, *Phys. Fluids* **13**, 569 (1970).

11. D. A. Russell and E. Ott, *Phys. Fluids* **24**, 1976 (1981).

12. T. H. Dupree, *Phys. Fluids* **25**, 277 (1982), and M.I.T. P.F.C./JA-83-1 preprint, 1983 (to be published), and earlier references therein.

Part 3

KINETIC THEORY

KINETIC THEORY OF FLUCTUATIONS IN FLUIDS FAR FROM EQUILIBRIUM

E. G. D. Cohen
Rockefeller University, New York, New York

T. R. Kirkpatrick
Institute for Physical Science and Technology, University of Maryland, College Park, Maryland

Abstract. An outline is given of the kinetic theory of a convective instability in a fluid, and the density and velocity fluctuations near it. The first convective instability in the Rayleigh–Bénard cell is considered. Results are presented for fluids of arbitrary density. The singular behavior of the fluctuations below and above the instability point and their experimental detection are discussed.

1. INTRODUCTION

In this chapter we shall sketch how a kinetic theory of a convective instability can be developed and how the fluctuations below and just above the instability can be computed. The anomalous behavior of the fluctuations near the instability can in principle be observed by microwave scattering as far as density fluctuations are concerned or laser-Doppler-velocimetry techniques as far as velocity fluctuations are concerned. We have studied in particular the first instability in the Rayleigh–Bénard cell.

Since kinetic theory is restricted to dilute gases, we have also developed a hydrodynamic theory of the fluctuations near the first instability point that is applicable to fluids of arbitrary density. We prefer, however, to present a kinetic treatment, not only because it is the first time, we believe, that such a treatment has been given but also because the structure of the theory is clearer and the derivations are more controlled than the more-formal hydrodynamic ones, since correction terms can be systematically explored and be shown to be of higher order in the density than the terms kept. We shall give our final formulas for general densities, as they follow from a hydrodynamic treatment. The kinetic theory results can then simply be obtained by taking the low-density values for the thermodynamic and transport coefficients in these expressions. It is not the purpose of this chapter to give an account of all the derivations, which are quite involved, especially above the instability point. Rather, we would like to outline the main steps in the calculations, so that the logical structure of the theory becomes clear. For the details we refer to a recent publication.[1]

Before a discussion of the theory, a very brief sketch of the basic properties of the Rayleigh–Bénard cell relevant for this paper will be given.[2]

In a Rayleigh–Bénard cell a fluid in a container is heated from below in the presence of a gravitational field (cf. Fig. 1). For sufficiently small temperature gradients the fluid will be at rest, so that the local velocity $\mathbf{u}(\mathbf{R})$ vanishes everywhere: $\mathbf{u}(\mathbf{R}) = 0$ for all $\mathbf{R}$. This is so because the buoyancy force, caused by the expansion of the hotter fluid below, cannot overcome the dissipative forces due to viscous friction and thermal conduction. The ratio of these two forces is the dimensionless Rayleigh number R:

$$R = \frac{\text{buoyancy force}}{\text{dissipative forces}} = \frac{\alpha_T g(dT/dz)}{\nu D_T} d^4$$

Here $\alpha_T = -(1/\rho)(\partial \rho/\partial T)_p$ is the thermal expansion coefficient, with ρ being the mass density, T is the absolute temperature and p is the pressure

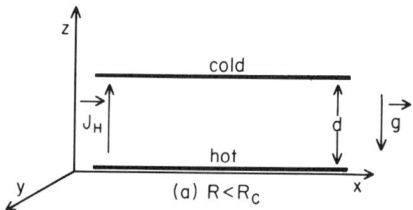

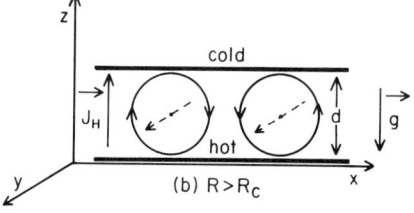

Figure 1. A fluid in a Rayleigh–Bénard cell in the presence of a heat flux ($\mathbf{J}_H$) and gravity ($\mathbf{g}$): (a) at rest (for $R < R_c$); (b) moving in cylindrical rolls with axes parallel to the y axis (for $R > R_c$). d is the distance between the plates.

of the fluid, g is the magnitude of the acceleration of gravity, dT/dz is the temperature, gradient, $\nu = \eta/\rho$ is the kinematic viscosity and $D_T = \lambda/\rho c_p$ is the thermal diffusivity, where η is the shear viscosity, λ is the thermal conductivity and c_p is the specific heat per unit mass of the fluid at constant pressure. d is the distance between the two plates at different temperatures. As long as $R < R_c$, where R_c is a critical Rayleigh number, the fluid will remain at rest. However, for sufficiently large temperature gradients, when R exceeds the critical value R_c, that is, for $R > R_c$, the buoyancy force will dominate and convection will occur. The value of R_c as well as the pattern of convection depend on the boundary conditions. We will restrict ourselves to the case of a fluid that is of finite extent in the z direction, that is, confined between two plates at $z = 0$ and $z = d$ and of infinite extent in the x and y directions. Instead of the more-realistic stick boundary conditions, which prescribe that the fluid sticks to the boundaries, we shall use the more-convenient free–free boundary conditions, which apply when the fluid can move freely at the boundaries, and allow a completely analytic treatment of the problem. The critical Rayleigh number for stick boundary conditions is about 1708, while for free–free boundary conditions it is $27\pi^4/4 = 657.4$. A more mathematical description of the occurrence of this first instability in the Rayleigh–Bénard cell at $R = R_c$ will be given in Section 4.

For the free–free boundary conditions and the geometry of Fig. 1 the convection for $R > R_c$ takes place in cylindrical rolls, whose axes are all parallel, say to the y axis (cf. Fig. 1b).

The plan of this chapter is as follows. In Section 2, we define the fluctuations and their correlations in μ-space and give a kinetic equation for the unequal time correlation function $C(1t|2)$. In Section 3, we discuss the average state of the fluid around which the fluctuations are studied. Kinetic equations are given for both the single-particle distribution function $f_1(1)$ and the nonequilibrium pair-correlation function $G_2(12)$. In Section 4, we determine explicitly $f_1(1)$ for both $R < R_c$ and $R \gtrsim R_c$ and discuss the hydrodynamic stability of this solution. In Section 5, the formal solutions to the kinetic equations for $C(1t|2)$ and $G_2(12)$ are given. In Section 6, the eigenvalue problem is treated that enables us to compute $C(1t|2)$ and $G_2(12)$ explicitly, while in Section 7 expressions for the density–density and momentum–momentum correlation functions are presented that follow from these explicit solutions for $C(1t|2)$ and $G_2(12)$. Finally, in Section 8, some concluding remarks are made.

2. FLUCTUATIONS IN μ-SPACE

2.1 Definitions

All the fluctuations we want to discuss can be defined as velocity moments of density fluctuations in μ-space, the six-dimensional phase space of one molecule:

$$\delta f(1t) = \tilde{f}(1t) - \langle \tilde{f}(1t) \rangle_{ss} = \tilde{f}(1t) - f_1(1) \tag{1}$$

Here the fluctuation $\delta f(1t)$ of the number of particles at the position $1 \equiv \mathbf{R}_1$, $\mathbf{V}_1$ in μ-space is the difference between the *actual* number of particles at 1 at time t:

$$\tilde{f}(1t) = \sum_{i=1}^{N} \delta^{(6)}[1 - x_i(t)] \tag{2}$$

where $x_i(t) = \mathbf{r}_i(t), \mathbf{v}_i(t)$ ($i = 1, \ldots, N$) is the phase of particle i, and the *average* number of particles $f_1(1)$ at 1 at time t:

$$f_1(1) = \langle \tilde{f}(1t) \rangle_{ss} \tag{3}$$

where the average is taken over the stationary state of the fluid. $\delta^{(6)}$ represents a six-dimensional Dirac delta function and $f_1(1)$ is the first distribution function of the fluid.

Obviously, the average fluctuation vanishes:

$$\langle \delta f(1t) \rangle_{ss} = 0 \tag{4}$$

However, the average correlation between two fluctuations at positions 1 and 2 in μ-space, $C(1t|2)$, defined by

$$C(1t|2) \equiv \langle \delta f(1t) \delta f(20) \rangle_{ss} \tag{5}$$

does not. Here $C(1t|2)$ is the unequal-time-correlation function, which reduces for $t = 0$ to

$$C(1|2) \equiv \langle \delta f(10) \delta f(20) \rangle_{ss} \tag{6}$$

where $C(1|2)$ is the equal-time-correlation function.

2.2. Equation for $C(1t|2)$

An equation for $C(1t|2)$ can be derived from the BBGKY-hierarchy for time-correlation functions. For low densities it reads[5,6,12]

$$\left[\frac{\partial}{\partial t} + \mathscr{L}_{ss}(1) \right] C(1t|2) = 0 \tag{7}$$

where the kinetic operator $\mathscr{L}_{ss}(1)$ is defined by

$$\mathscr{L}_{ss}(1) = \mathbf{V}_1 \cdot \frac{\partial}{\partial \mathbf{R}_1} - \int d3 \, \hat{T}(13)(1 + P_{13}) f_1(3) - \bar{T}_w(1) \tag{8}$$

Here the rate of change of $C(1t|2)$, $(\partial C/\partial t)$, has been expressed as a sum of four terms ($\mathscr{L}_{ss}C$), the first of which expresses the change of $C(1t|2)$ by streaming in μ-space, while the last two give the rate of change of $C(1t|2)$ due to binary collisions. In fact, the last two terms give the contributions to $\partial C/\partial t$ due to interparticle collisions and collisions with the walls of the fluid container, respectively.

In Eq. (8) the binary-collision operators characterizing these two types of collisions are

$$\hat{T}(13) = \delta^{(3)}(\mathbf{R}_1 - \mathbf{R}_3) \int_0^{2\pi} d\varepsilon \int_0^a db\, b |\mathbf{V}_1 - \mathbf{V}_3|(b_\sigma - 1) \tag{9}$$

and[3]

$$\bar{T}_w(1)h(1) = \int dw\, \delta^{(3)}(\mathbf{R}_1 - \mathbf{R}_w)\{\theta(\mathbf{V}_1 \cdot \hat{\mathbf{n}}) \int d\mathbf{V}_1' \theta(-\mathbf{V}_1' \cdot \mathbf{n})$$
$$\times |(\mathbf{V}_1' \cdot \mathbf{n}| K(\mathbf{V}_1, \mathbf{V}_1') h(\mathbf{R}_1 \mathbf{V}_1') - \theta(-\mathbf{V}_1 \cdot \mathbf{n})|\mathbf{V}_1 \cdot \mathbf{n}| h(1)\} \tag{10}$$

In Eq. (9) $\delta^{(3)}$ denotes a three-dimensional delta function, while b and ε are the impact parameter and the azimuthal angle, respectively, that characterize the binary collision between two particles with velocities $\mathbf{V}_1$ and $\mathbf{V}_3$, a is the range of the interparticle forces and b_σ is an operator that replaces the velocities $\mathbf{V}_1$ and $\mathbf{V}_3$ by $\mathbf{V}_1'$ and $\mathbf{V}_3'$, the velocities of the restituting collision. Furthermore, P_{13} is a permutation operator that permutes the indices 1 and 3. In Eq. (10), $\mathbf{R}_w$ denotes the position of a point on the walls, dw is an integration over the walls, $\theta(x)$ is the Heaviside step function, $\hat{\mathbf{n}}$ is a unit vector normal to the wall pointing into the fluid, and $K(\mathbf{V}_1, \mathbf{V}_1')$ is a scattering kernel that specifies the interaction between the walls and the fluid particles. For diffusive reflection of the particle at the walls one has, for instance, that $K(\mathbf{V}_1, \mathbf{V}_1') = K_D(\mathbf{V}_1, \mathbf{V}_1')$ with[3]

$$K_D(\mathbf{V}_1, \mathbf{V}_1') = |\mathbf{V}_1 \cdot \mathbf{n}| 2\pi \left(\frac{m}{2\pi k_B T_w}\right)^2 \exp\left(-\frac{mV_1^2}{2k_B T_w}\right) \tag{11}$$

Here m is the mass of a particle, $T_w \equiv T_w(\mathbf{R}_w)$ is the temperature of the wall at $\mathbf{R}_w$ and k_B is Boltzmann's constant.

The unequal-time-correlation function $C(1t|2)$, as given by Eq. (7), can be formally expressed in terms of the equal-time-correlation function (6) by

$$C(1t|2) = \exp[-\mathscr{L}_{ss}(1)t] C(1|2) \tag{12}$$

where, with Eqs. (1)–(3),

$$C(1|2) = \delta^{(6)}(1-2)f_1(1) + G_2(12) \tag{13}$$

Here the pair-correlation function $G_2(12)$ is given by

$$G_2(12) = f_2(12) - f_1(1)f_1(2) \tag{14}$$

where $f_2(12)$ is the pair-distribution function

$$f_2(12) = \left\langle \sum_{i \neq j}^N \delta^{(6)}(1 - x_i) \delta^{(6)}(2 - x_j) \right\rangle_{ss}. \tag{15}$$

Equation (12) shows that the computation of $C(1t|2)$ involves two problems: (1) a dynamical problem involving the application of the time-evolution operator $e^{-\mathscr{L}_{ss}(1)t}$ and (2) an equal-time problem involving the computation of the correlation function $C(1|2)$. As we shall see below, because of the presence of a heat flow in the stationary state, this equal-time problem is also of a *dynamical* nature. This is contrary to the case of a fluid in thermal equilibrium, where only static correlations that can be computed from equilibrium statistical mechanics occur.

3. THE AVERAGE STATE

Before the correlation functions of the fluctuations can be determined, one must first find the average state, with respect to which the fluctuations are defined.

3.1. Equation for $f_1(1)$

To lowest order in the density, one can derive from the BBGKY-hierarchy for distribution functions the following extension of the nonlinear Boltzmann equation for $f_1(1)$ in a stationary state in the presence of gravity and walls[3,5]

$$\left[\mathbf{V}_1 \cdot \frac{\partial}{\partial \mathbf{R}_1} + \mathbf{g} \cdot \frac{\partial}{\partial \mathbf{V}_1}\right] f_1(1) = \int d2\, \hat{T}(12) f_1(1) f_1(2) + \bar{T}_w(1) f_1(1) \quad (16)$$

Equation (16) for $f_1(1)$ has the same structure and interpretation as Eq. (7) for $C(1t|2)$, except that $\partial f_1/\partial t = 0$ in the stationary state.

3.2. Equation for $G_2(12)$

Similarly, one has for $G_2(12)$ the equation[4,5,13]

$$[\mathscr{L}_{ss}(1) + \mathscr{L}_{ss}(2)] G_2(12) = \hat{T}(12) f_1(1) f_1(2) \quad (17)$$

Since the right-hand side (rhs) of this equation is proportional to the temperature gradient ∇T, or the heat flux $\mathbf{J}_H$ in the fluid,* this equation expresses the fact that dynamical processes in the fluid, as occur on the left-hand side (lhs) of Eq. (17) in the kinetic operator $\mathscr{L}_{ss}(1) + \mathscr{L}_{ss}(2)$, must balance the effect of the heat flux that appears on the rhs of Eq. (17), in order to maintain $G_2(12)$ stationary.

3.3. Solution Procedure

From the foregoing it follows that in order to find $C(1t|2)$ one has to proceed as follows: first determine $f_1(1)$ from Eq. (16), then $G_2(12)$ from

---

*This follows from Eq. (19) for $f_1(1)$ and that $\hat{T}(12) f_l(1) f_l(2) = 0$.

Eq. (17), since the rhs and the lhs [cf. Eq. (8)] are determined by $f_1(1)$. Equation (13) then gives $C(1|2)$ and finally Eq. (12), $C(1t|2)$:

$$f_1(1) \to \mathcal{L}_{ss}(1) \to G_2(12) \to C(1|2) \to C(1t|2) \qquad (18)$$

4. DETERMINATION OF $f_1(1)$

$f_1(1)$ can be found from the extended Boltzmann equation (16) by using the Chapman–Enskog solution method, that is, by an expansion in the uniformity parameter $\mu = l/L_\nabla$ where l is the mean free path of the particles in the fluid and $L_\nabla = T/\nabla T$, a gradient length, is a characteristic length over which the temperature in the fluid varies. Using the fact that $f_1(1)$ vanishes outside the fluid container, one has to $\mathcal{O}(\mu)$ (Ref. 3):

$$f_1(1) \equiv f_1^{\text{Ch-E}}(1) = W(\mathbf{R}_1)[f_l(1) + f_\nabla(1)] \qquad (19)$$

In Eq. (19), $W(\mathbf{R}_1)$ is the characteristic function that vanishes outside the fluid volume Ω:

$$W(\mathbf{R}_1) = 1 \text{ if } \mathbf{R}_1 \in \Omega$$
$$= 0 \text{ otherwise} \qquad (20)$$

$f_l(1)$ is the local Maxwellian distribution function:

$$f_l(1) = n(\mathbf{R}_1)\left(\frac{m}{2\pi k_B T(\mathbf{R}_1)}\right)^{3/2} \exp\left(-\frac{mC_1^2(\mathbf{R}_1)}{2k_B T(\mathbf{R}_1)}\right) \qquad (21)$$

where $\mathbf{C}_1(\mathbf{R}_1) = \mathbf{V}_1 - \mathbf{u}(\mathbf{R}_1)$ and $n(\mathbf{R}_1)$, $\mathbf{u}(\mathbf{R}_1)$, and $T(\mathbf{R}_1)$ are the local density, velocity, and temperature in the fluid at $\mathbf{R}_1$, respectively.

$f_\nabla(1)$, the correction to $f_l(1)$ of $\mathcal{O}(\mu)$, that is, $\sim \nabla T$, is given by

$$f_\nabla(1) = \frac{1}{\bar{\Lambda}_l(1)} f_l(1) \left\{ \left(\frac{m}{k_B T}\right)(C_{1\alpha}C_{1\beta} - \tfrac{1}{3}\delta_{\alpha\beta}C_1^2)\frac{\partial C_{1\alpha}}{\partial R_{1\beta}} \right.$$
$$\left. + \left(\frac{mC_1^2}{2k_B T} - \frac{5}{2}\right)C_{1\alpha}\frac{\partial \log T}{\partial R_{1\alpha}} \right\} \qquad (22)$$

with

$$\bar{\Lambda}_l(1) = \int d3\, \hat{T}(13)(1 + P_{13})f_l(3) \qquad (23)$$

While the $\mathbf{V}_1$ dependence of $f_1(1)$ follows from Eqs. (21) and (22), the $\mathbf{R}_1$ dependence has to be determined from the nonlinear Navier–Stokes equations for $n(\mathbf{R}_1)$, $\mathbf{u}(\mathbf{R}_1)$, and $T(\mathbf{R}_1)$ that follow from the Chapman–Enskog solution (19) of the Boltzmann equation (16), *appended by boundary conditions*, that follow from the particle–wall collision term $\bar{T}_w(1)$ in Eq. (16). For the case of diffusive reflection, these conditions are the stick boundary conditions[3]

$$\mathbf{u}(\mathbf{R}_w) = 0$$
$$T(\mathbf{R}_w) = T_w(\mathbf{R}_w) \tag{24}$$

Usually, $f_1(1)$ is used to obtain expressions for the transport coefficients η and λ of the fluid, in terms of the interparticle forces. Since for a dilute gas η and λ are independent of the density, the dependence of η and λ on position is only though $T(\mathbf{R}_1)$, which has to be determined from a solution of the nonlinear Navier–Stokes equations. This hydrodynamic part needed for a complete determination of the transport coefficients in a fluid is usually not considered in the context of the kinetic theory of nonuniform gases. In our case, however, this part *is* important since we not only need to determine $f_1(1)$ formally, but we need an $f_1(1)$ that is stable. That is, we have to use the *stable* solutions of the Navier–Stokes equations for $n(\mathbf{R}_1)$, $\mathbf{u}(\mathbf{R}_1)$, and $T(\mathbf{R}_1)$ in $f_1(1)$ in order to compute the unequal- and equal-time correlation functions of interest.

Therefore, in order to determine the stable stationary-state solution $f_1(1)$, one has to consider the hydrodynamic stability problem associated with the Rayleigh-Bénard cell. That is, since we expect the long-wavelength, or slowly varying, modes to be the important modes near a hydrodynamic instability, we can discuss the stability of $n(\mathbf{R}_1)$, $\mathbf{u}(\mathbf{R}_1)$, and $T(\mathbf{R}_1)$ [and thus of $f_1(1)$] by using hydrodynamic stability theory.[2] Only a very general outline of the solution to this problem will be given here. However, some of the results are needed in the sequel for the computation of the fluctuations near the first instability in the Rayleigh–Bénard cell.

The condition for stability of a stationary solution of the nonlinear Navier–Stokes equations is that an arbitrary infinitesimal perturbation $A(\mathbf{R}_1 t)$

$$A(\mathbf{R}_1 t) = \sum_j A_j(\mathbf{R}_1) e^{-\omega_j t} \tag{25}$$

decays in time, that is, that $Re\,\omega_j > 0$ for all j. Here j is a general eigenfunction index that can be discrete or continuous. In our case one considers five infinitesimal disturbances of n, $\mathbf{u}$, and T simultaneously, so that $A(\mathbf{R}_1 t)$ is really a five-component vector. The time evolution of the $A(\mathbf{R}_1 t)$ can be obtained from the nonlinear Navier–Stokes equations linearized around the stationary solution of these equations, whose stability one investigates. As long as the temperature gradients are small, one expects that the stable stationary solution for n, $\mathbf{u}$, and T follows from the Navier–Stokes equations:

$$\mathbf{u} = 0$$
$$\frac{d}{dz_1}(nk_B T) = \rho g \tag{26}$$
$$\frac{d}{dz_1}\left(\lambda_B \frac{dT}{dz_1}\right) = 0$$

for a fluid at rest. Here $\rho = nm$ and λ_B is the low-density thermal conductivity as given by the Chapman–Enskog solution of the Boltzmann equation.

In order to discuss the stability of this solution, that is, whether the fluid at rest as given by the Eqs. (26) is indeed stable, one considers an eigenvalue problem of the form

$$\mathbf{H}(\mathbf{R}_1)\mathbf{\Psi}^R_{i,\mathbf{k}}(\mathbf{R}_1) = -\omega_i(\mathbf{k})\mathbf{\Psi}^R_{i,\mathbf{k}}(\mathbf{R}_1) \qquad (27)$$

Here **H** is a five-by-five matrix that can be read off from the linearized Navier–Stokes equations around the steady state, given by the Eq. (26). Since **H** is a nonsymmetric matrix, one has to consider the right and left eigenvalue problem with eigenvectors $\mathbf{\Psi}^R_{i,\mathbf{k}}(\mathbf{R}_1)$ and $\mathbf{\Psi}^L_{i,\mathbf{k}}(\mathbf{R}_1)$, respectively. Here we have labeled the eigenvalues and eigenvectors by a wavenumber index $\mathbf{k}$, and by an index i that refers to one of the five eigenmodes of the matrix **H** for a fixed $\mathbf{k}$. The eigenvalue problem given by Eq. (27) is completely specified by giving the boundary conditions on the $\mathbf{\Psi}_{i,\mathbf{k}}(\mathbf{R}_1)$. It can be solved by using the fact that there are two small parameters in the problem: l/d and d/L_∇. The first parameter, l/d, is the ratio of the mean free path, l, and a distance that represents the length scale over which the infinitesimal perturbations vary. We characterize this length by d for the Bénard problem. The condition $l/d < 1$ is then needed in order that a hydrodynamic theory of stability can be applied. The second parameter, d/L_∇, represents the relatively slow variation of the hydrodynamic fields compared to the spatial variation of the infinitesimal perturbations. Using these small parameters the eigenvalue problem can then be solved if we give the boundary conditions on the temperature and velocity fluctuations, $\delta T(\mathbf{R}_1 t)$ and $\delta \mathbf{u}(\mathbf{R}_1 t)$, respectively. One often considers, as we did, instead of the stick boundary conditions [Eq. (24)], the free–free boundary conditions[1,2,8,9,16]:

$$\delta T(\mathbf{R}_w t) = 0$$

$$\delta u_z(\mathbf{R}_w t) = \frac{\partial}{\partial z_1}\delta u_x(\mathbf{R}_w t) = \frac{\partial}{\partial z_1}\delta u_y(\mathbf{R}_w t) = 0 \qquad (28)$$

Nontrivial solutions are then obtained only for certain $\omega_i(\mathbf{k})$ ($i = 1,\ldots,5$): the five hydrodynamic modes *in the presence of* ∇T. These hydrodynamical modes reduce to the usual ones, familiar from the linearized Navier–Stokes equations around (total) equilibrium, when one sets $\nabla T = 0 = g$. As in the latter case, there are here two sound modes (which we neglect), one viscous mode, and two visco-heat modes.*

We shall only give the eigenvalues of the two visco-heat modes here and refer for further details to the literature[1,17]:

---

*The visco-heat modes correspond to combinations of one viscous mode and the heat mode in an equilibrium fluid, due to the presence of gravity and a temperature gradient.

$$\omega^<_\pm(\mathbf{k}) = \frac{(\nu + D_T)k^2}{2} \left\{ 1 \pm \left[1 - \frac{4\nu D_T}{(\nu + D_T)^2}\left(1 - \frac{R(\mathbf{k})}{R_c}\right) \right]^{1/2} \right\} \quad (29)$$

In Eq. (29), $R(\mathbf{k})$ is a $\mathbf{k}$-dependent Rayleigh number, defined by

$$\frac{R(\mathbf{k})}{R_c} = \frac{\alpha_T g dT/dz_1}{\nu D_T} \frac{k_\parallel^2}{k^6} \quad (30)$$

where $k_\parallel^2 = k_x^2 + k_y^2$ and the superscript $<$ refers to the fact that $R < R_c$. While k_x and k_y are continuous variables, k_z is a discrete variable, in that $k_z = n\pi/d$ ($n = 1, 2, \ldots$) due to the boundary conditions at $z = 0$ and $z = d$. For sufficiently large temperature gradients $R(\mathbf{k})/R_c > 1$, so that $\text{Re}\,\omega_-(\mathbf{k})$ is no longer positive and not all infinitesimal perturbations of all $\mathbf{k}$ decay anymore in time. The critical value of dT/dz_1, or of R, above which this is going to occur for the first time is determined by[2]

$$R = R_c = \frac{27\pi^4}{4} = 657.4 \quad (31)$$

for which $\omega_-(\mathbf{k}) \to 0$ for $\mathbf{k} = \mathbf{k}_c$, with $\mathbf{k}_c \equiv (k_{z_c} = \pi/d, k_{\parallel c} = \pi/d\sqrt{2})$. Thus for $R < R_c$, the stable solutions of the nonlinear Navier–Stokes equations are given by those of the Eqs. (26): $\mathbf{u}_{nc}(\mathbf{R}_1) = 0$, $n_{nc}(\mathbf{R}_1)$, $T_{nc}(\mathbf{R}_1)$, where the subscript nc refers to the nonconvective state of the fluid [i.e., $\mathbf{u}_{nc}(\mathbf{R}_1) = 0$].

However, for $R > R_c$ this nonconvective solution is no longer stable and a new stable solution, that of parallel cylindrical rolls, establishes itself. Very near the instability point, that is, for small $R/R_c - 1 > 0$, the deviations from the previous, nonconvecting hydrodynamic fields, are given by[14]

$$\Delta\rho(x_1, z_1) = -\frac{\rho}{T}\Delta T(x_1, z_1)$$

$$\Delta T(x_1, z_1) = -9\sqrt{3}\frac{d\pi^3}{R}\frac{dT_{nc}}{dz_1}\sqrt{\frac{R}{R_c} - 1}\sin\left(\frac{\pi z_1}{d}\right)\cos\left(\frac{\pi x_1}{d\sqrt{2}}\right)$$

$$+ \frac{d}{\pi}\frac{dT_{nc}}{dz_1}\left(\frac{R}{R_c} - 1\right)\sin\left(\frac{2\pi z_1}{d}\right)$$

$$u_z(x_1, z_1) = -\frac{D_T}{d}2\pi\sqrt{3}\sqrt{\frac{R}{R_c} - 1}\sin\left(\frac{\pi z_1}{d}\right)\cos\left(\frac{\pi x_1}{d\sqrt{2}}\right)$$

$$u_x(x_1, z_1) = -\frac{D_T}{d}4\sqrt{3/2}\sqrt{\frac{R}{R_c} - 1}\cos\left(\frac{\pi z_1}{d}\right)\sin\left(\frac{\pi x_1}{d\sqrt{2}}\right)$$

$$u_y(x_1, z_1) = 0 \quad (32)$$

The stability of this solution can be investigated by linearizing the nonlinear Navier–Stokes equations around this stationary solution: $n = n_{nc} + \Delta\rho/m$, $T = T_{nc} + \Delta T$, and $\mathbf{u}$, as given by the Eqs. (26) and (32), and ordering the equations so obtained according to three small parameters: l/d, d/L_∇, and

$\varepsilon = (R/R_c - 1)^{1/2}$. With free–free boundary conditions on the hydrodynamical fields, a similar eigenvalue problem occurs as before for $R < R_c$.

Thus, again, five hydrodynamic modes are obtained that are the continuation of those for $R < R_c$ to $R \geqslant R_c$ with eigenvalues $\omega_i^>(\mathbf{k})$ and eigenvectors $\Psi_{i,\mathbf{k}}^{>R,L}(\mathbf{R}_1)$. Here i characterizes either two sound modes (that are again neglected), a viscous mode, or two visco-heat modes, one of which, $\omega_i^>(\mathbf{k}) \to 0$ for $\mathbf{k} = \mathbf{k}_c$ and $R \to R_c$. Since the $\mathrm{Re}\,\omega_i^>(\mathbf{k})$ are positive for all $\mathbf{k}$ when $R > R_c$, the stable solution for $f_1(1)$ follows from the Eqs. (19)–(22), (26), and (32).

The solutions below and above R_c could be denoted by $f_1^<(1)$ for $R < R_c$ and $f_1^>(1)$ for $R \geqslant R_c$ and, consequently, there are also two corresponding operators $\mathscr{L}_{ss}^<$ and $\mathscr{L}_{ss}^>(1)$, given by Eq. (8). In the following we shall omit the superscripts $<$ and $>$ as well as R, L in order not to overburden the notation.

5. FORMAL DETERMINATION OF $G_2(12)$ AND $C(1t|2)$

Using the stable solutions for $f_1(1)$ described in the previous section and Eq. (17), yields to $\mathcal{O}(\mu)$ for $G_2(12)$ the expression:

$$G_{2_{n.eq.}}(12) = \frac{1}{\mathscr{L}_{ss}(1) + \mathscr{L}_{ss}(2)} [\hat{T}(12)(1 + P_{12}) W(\mathbf{R}_1) W(\mathbf{R}_2) f_1(1) f_\nabla(2)] \quad (33)$$

As pointed out above, one should distinguish between $G_{2_{n.eq.}}^<(12)$ and $G_{2_{n.eq.}}^>(12)$ according to whether $f_1^>(1)$ or $f_1^<(1)$ are used in (33).* Equations (19)–(22) for $f_1(1)$, Eq. (33) for $G_2(12)$ together with Eqs. (12) and (13) give for $C(1t|2)$:*

$$C(1t|2) = e^{-\mathscr{L}_{ss}(1)t} [\delta^{(6)}(1-2) f_1(1) + G_{2_{n.eq.}}(12)] \quad (34)$$

$G_{2_{n.eq.}}(12)$ as well as $C(1t|2)$ can be obtained from the Eqs. (19)–(22), (33), and (34) by a spectral decomposition of the kinetic operator $\mathscr{L}_{ss}(1)$. The relevant eigenmodes of this operator will be discussed in the next section.

6. EIGENMODES OF $\mathscr{L}_{ss}(1)$

The right eigenvalue problem defined by the operator $\mathscr{L}_{ss}(1)$ reads:

$$\mathscr{L}_{ss}(1) \theta_{i,\mathbf{k}}^R(1) = -\omega_i(\mathbf{k}) \theta_{i,\mathbf{k}}^R(1) \quad (35)$$

---

*To the particular solution of Eq. (17) for $G_2(12)$ given by Eq. (33) one should, in principle, add the solution of the homogenous equation $[\mathscr{L}_{ss}(1) + \mathscr{L}_{ss}(2)] G_2^h(12) = 0$. However, in the low-density approximation considered here, where the difference in position of two colliding particles is neglected, this solution is zero. In general, although $G_2^h(12)$ will not vanish, it will be of very short range, like in thermal equilibrium. It can, therefore, always be neglected when one is interested in the long-range behavior of $G_2(12)$.

We are only interested here in those eigenmodes of $\mathscr{L}_{ss}(1)$ whose eigenvalues vanish when $\mathbf{k} \to 0$, that is, the long-wavelength eigenmodes with small $\omega_i(\mathbf{k})$. This follows from the fact that these modes are the analogs in kinetic theory of the hydrodynamic eigenmodes discussed in Section 4 and thus determine the long-time, long-wavelength behavior of the fluid. In particular, it is one of these long-wavelength (or hydrodynamic) modes that has an eigenvalue which vanishes at the convective instability. Now for these modes the relevant $\mathbf{V}_1$ dependence of the eigenfunctions follows from the collisional invariants in a binary collision and is therefore given by a linear combination of 1, $\mathbf{V}_1$, and V_1^2, with $\mathbf{R}_1$-dependent coefficients. The $\mathbf{R}_1$ dependence of these coefficients is given by the components of the eigenvector $\Psi_{i,\mathbf{k}}^R(1)$ of the *same* eigenvalue problem that was discussed in Section 4. Thus the eigenvalues $\omega_i(\mathbf{k})$ are identical to those found in Section 4 and the eigenfunctions $\theta_{i,\mathbf{k}}^{R,L}(1)$ can be expressed in terms of the components of the $\Psi_{i,\mathbf{k}}^{R,L}(\mathbf{R}_1)$ discussed in Section 4, both for $R < R_c$ and $R \gtrsim R_c$. In fact, the eigenfunctions are bilinear combinations of the five collisional invariants 1, $\mathbf{V}_1$, V_1^2 and the five components of the $\Psi(\mathbf{R}_1)$. Also $\theta_{i,\mathbf{k}=\mathbf{k}_c}^{>R,L}(1) = \theta_{i,\mathbf{k}=\mathbf{k}_c}^{<R,L}(1) + \mathcal{O}[(R/R_c - 1)^{1/2}]$ or: the eigenfunctions just above and below the instability point are identical to lowest order in ε, that is, to $\mathcal{O}[(R/R_c - 1)^0]$. Using the (approximate) completeness of these "kinetic" hydrodynamic modes, Eq. (34) for $C(1t|2)$ reveals immediately that the decay of the correlation function $C(1t|2)$ with time will be given by the *single* modes $\omega_i(\mathbf{k})$, in that terms proportional to $e^{-\omega_\pm(\mathbf{k})t}$, and so on, , will occur. Similarly, $G_{2_{n.eq.}}(12)$ will depend on the contribution of *pairs of modes* $\omega_i(\mathbf{k}) + \omega_j(\mathbf{k})$, in that terms proportional to $1/[\omega_\pm(\mathbf{k}) + \omega_\pm(\mathbf{k})]$, etc., will appear. These contributions to $G_{2_{n.eq.}}(12)$ are called mode-coupling contributions, since two hydrodynamic modes are involved. We note that near $R = R_c$, $\omega_-(\mathbf{k} = \mathbf{k}_c) \approx 0$, which implies that $G_2(12)$ will become singular at $R = R_c$ since $G_2(12)$ will have contributions proportional to $[2\omega_-(\mathbf{k}_c)]^{-1}$. Furthermore, the structure of the theory presented here indicates that this is so for any instability since there always exists an $\omega_i(\mathbf{k})$ that goes to zero near the instability point.*

7. RESULTS

Although explicit expressions for $C(1t|2)$ and $G_2(12)$ have been obtained in terms of the "kinetic" hydrodynamic modes discussed above,[1] we will only give results here for the density–density and velocity–velocity correlation functions just below and just above the instability. Then only the con-

---

*This singularity of $G_2(12)$ does not necessarily imply that also all velocity moments of $G_2(12)$ are singular. For example, although in the Rayleigh–Bénard cell both the density–density and the velocity–velocity correlation functions become singular for $R = R_c$, in the circular Couette flow near the Taylor instability, only the velocity–velocity correlation functions become singular, while the density–density correlation function remains regular.[15]

tributions of $\omega_-(\mathbf{k})$ and $\theta_{-,\mathbf{k}}(1)$ will play a role, since they are much larger than those of the other hydrodynamic modes.

Unequal-Time-Correlation Functions

(a) The density–density correlation function $M_{\rho\rho}(\mathbf{R}_1\mathbf{R}_2 t)$ is defined by the relation

$$M_{\rho\rho}(\mathbf{R}_1\mathbf{R}_2 t) \equiv \langle \delta\rho(\mathbf{R}_1 t)\delta\rho(\mathbf{R}_2 t)\rangle_{ss} = m^2 \int d\mathbf{V}_1 \int d\mathbf{V}_2 C(1t|2) \qquad (36)$$

where $\delta\rho(\mathbf{R}t)$ is the fluctuation in the mass density at $\mathbf{R}, t$. Since the spectral decomposition of the operator $\mathscr{L}_{ss}$ is performed in $\mathbf{k}$ space, it is convenient first to calculate the Fourier transform of $M_{\rho\rho}(\mathbf{R}_1\mathbf{R}_2 t)$, which is defined by

$$M_{\rho\rho}(k_z, k'_z, \mathbf{k}_\|, t) = \int d\mathbf{R}_{12} \int_0^d dz_1 \int_0^d dz_2 e^{-i\mathbf{k}_\| \cdot \mathbf{R}_{12\|}}$$
$$\times \sin(k_z z_1)\sin(k'_z z_2) M_{\rho\rho}(\mathbf{R}_1\mathbf{R}_2 t) \qquad (37)$$

where $\mathbf{k}_\| = (k_x, k_y)$ and $\mathbf{R}_{12} = (R_{12x}, R_{12y})$ with $\mathbf{R}_{12} = \mathbf{R}_1 - \mathbf{R}_2$. One then finds that the $M_{\rho\rho}(k_z, k'_z, \mathbf{k}_\|, t)$ are only different from zero if $k_z = k'_z$ and have contributions proportional to $e^{-\omega_-(\mathbf{k})t}$, which will decay slower and slower for $\mathbf{k} \cong \mathbf{k}_c$ when $R \to R_c$ either from below or from above, while the amplitudes will diverge. This slow decay near the instability point is analogous to the *critical slowing down* near a (gas–liquid) critical point. For $R < R_c$, $k_z = k'_z = \pi/d$, $\mathbf{k}_\| \cong \mathbf{k}_{\|c}$ and R near R_c, we obtain[1]

$$M_{\rho\rho}(k_z = k'_z = \pi/d, \mathbf{k}_\| \cong \mathbf{k}_{\|c}, t) \cong \frac{\rho k_B T \mathbf{k}_{\|c}^2}{D_T(\nu + D_T)k_c^6}\left(\alpha_T \frac{dT}{dz_1}\right)^2$$
$$\times \frac{1}{[(1-R/R_c) + \frac{4}{3}(\mathbf{k}_\| - \mathbf{k}_{\|c})^2/\mathbf{k}_{\|c}^2]} \exp\left[\frac{-\nu D_T k_c^2}{(\nu + D_T)}\left(\left[1 - \frac{R}{R_c}\right]\right.\right.$$
$$\left.\left.+ \frac{4}{3}\frac{(\mathbf{k}_\| - \mathbf{k}_{\|c})^2}{\mathbf{k}_{\|c}^2}\right)t\right] \qquad (38)$$

Here we have used the fact that for R near R_c one can write:

$$1 - \frac{R(k_z = \pi/d, \mathbf{k}_\| \cong \mathbf{k}_{\|c})}{R_c} \cong \left(1 - \frac{R}{R_c}\right) + \frac{4}{3}\frac{(\mathbf{k}_\| - \mathbf{k}_{\|c})^2}{\mathbf{k}_{\|c}^2} \qquad (39)$$

The slow decay of $M_{\rho\rho}$ near the instability point, as given by Eq. (38) has been experimentally verified by Allain et al.[10] For $R \gtrsim R_c$, we obtain a similar result except that $(1 - R/R_c)$ is replaced by $(R/R_c - 1)f(\cos\theta)$ where[1]

$$f(\cos\theta) = \frac{1}{P} \sum_{\sigma=\pm 1} \frac{(1 + \sigma\cos\theta)^2}{[(5 - \sigma\cos\theta)^3 - \frac{27}{4}(1 - \sigma\cos\theta)]}$$
$$\times \left(\frac{27}{4}(1 - \sigma\cos\theta) + P(5 - \sigma\cos\theta)^2 + \frac{3}{P}(1 - \sigma\cos\theta)(5 - \sigma\cos\theta)\right) \qquad (40)$$

Here $P = \nu/D_T$ is the Prandtl number and θ is the angle between $\mathbf{k}_\parallel$ and the x axis. The singularity in the amplitude of $M_{\rho\rho}$ for $\mathbf{k} = \mathbf{k}_c$ when $R \to R_c$ is due to the contribution of $G_{2_{n.eq.}}$ (12) to the equal-time-correlation function [cf. Eqs. 13, 33 and 43].

(b) The momentum–momentum correlation function $M_{p_ip_j}(\mathbf{R}_1\mathbf{R}_2 t)$ is defined by the relation

$$M_{p_ip_j}(\mathbf{R}_1\mathbf{R}_2 t) \equiv \langle \delta p_i(\mathbf{R}_1 t) \delta p_j(\mathbf{R}_2 t) \rangle_{ss}$$

$$= m^2 \int d\mathbf{V}_1 \int d\mathbf{V}_2 V_{1i} V_{2j} C(1t|2) \qquad (41)$$

where $\delta p_i(\mathbf{R}t)$ is the ith component ($i = x, y, z$) of the momentum fluctuation at $\mathbf{R}, t$. One can define its Fourier transform $M_{p_ip_j}(k_z, k'_z, \mathbf{k}_\parallel, t)$ by Eq. (37) with the subscripts $\rho\rho$ replaced by p_ip_j and $\sin(k_z z_1)$ or $\sin(k'_z z_2)$ replaced by $\cos(k_z z_1)$ and $\cos(k'_z z_2)$, respectively, where $i, j = x$ or y. These correlation functions also exhibit the analog of critical slowing down for $\mathbf{k} \simeq \mathbf{k}_c$ and $R \to R_c$ and their amplitudes also diverge in this limit.[1]

7.2 Equal-Time-Correlation Function

The equal-time-correlation function $M_{\rho\rho}(\mathbf{R}_1\mathbf{R}_2) \equiv M_{\rho\rho}(\mathbf{R}_1\mathbf{R}_2 t = 0)$ exhibits singular behavior with longer and longer spatial correlations when $R \to R_c$.

We shall quote the results for the density–density correlation function $M_{\rho\rho}(\mathbf{R}_1\mathbf{R}_2)$; the velocity–velocity correlation function will display an analogous behavior.

Using the results for $M_{\rho\rho}(k_z, k'_z, \mathbf{k}_\parallel, t)$, $M_{\rho\rho}(\mathbf{R}_1\mathbf{R}_2)$ can be found as the inverse Fourier transform of $M_{\rho\rho}(k_z, k'_z, \mathbf{k}_\parallel, t = 0)$ from the formula

$$M_{\rho\rho}(\mathbf{R}_1\mathbf{R}_2) = \frac{2}{d} \sum_{n=1}^{\infty} \sum_{m=1}^{\infty} \sin\left(\frac{n\pi z_1}{d}\right) \sin\left(\frac{m\pi z_2}{d}\right)$$

$$\times \int \frac{d\mathbf{k}_\parallel}{(2\pi)^2} e^{i\mathbf{k}\cdot\mathbf{R}_{12\parallel}} M_{\rho\rho}(k_z, k'_z, \mathbf{k}_\parallel, t = 0) \qquad (42)$$

1. $R < R_c$. For $M_{\rho\rho}(k_z, k'_z, \mathbf{k}_\parallel, t = 0)$ our explicit result for $R < R_c$ is

$$M_{\rho\rho}(k_z, k'_z, \mathbf{k}_\parallel, t = 0)$$

$$= \delta_{k_z, k'_z} \left\{ \rho^2 k_B T \chi_T \frac{(\gamma - 1)}{\gamma} + \frac{\rho k_B T k_\parallel^2}{D_T(\nu + D_T k^6)} \frac{(\alpha_T dT/dz_1)^2}{[1 - R(k_z, k_\parallel)/R_c]} \right\} \qquad (43)$$

here $\gamma = c_p/c_v$ is the ratio of specific heats, χ_T is the isothermal compressibility, and $R(k_z, k_\parallel)/R_c$ is given by Eq. (30). We next examine the spatial decay of $M_{\rho\rho}(\mathbf{R}_1\mathbf{R}_2)$ near $R = R_c$.

Using the approximation (39) in Eq. (43) and inserting the result into Eq. (42) yields $M_{\rho\rho}(\mathbf{R}_1\mathbf{R}_2)$ for $R \lesssim R_c$ and $R_{12\parallel} \gg d$:

$$M_{\rho\rho}(\mathbf{R}_1\mathbf{R}_2) \cong \rho k_B T \left(\alpha_T \frac{dT}{dz_1}\right)^2 \frac{\sqrt{6}}{27} \frac{d}{\pi^2} \sin\left(\frac{\pi z_1}{d}\right) \sin\left(\frac{\pi z_2}{d}\right)$$

$$\times \frac{1}{D_T(\nu + D_T)} \frac{1}{(1 - R/R_c)^{1/2}} \frac{e^{-R_{12\|}/\xi_\|}}{\sqrt{\pi k_\| R_{12\|}}} \cos\left(\frac{\pi R_{12\|}}{d\sqrt{2}} - \frac{\pi}{4}\right) \quad (44)$$

The correlation length $\xi_\|$ is given by

$$\xi_\| = \frac{2d}{\pi} \sqrt{\frac{2}{3}} \frac{1}{(1 - R/R_c)^{1/2}} \quad (45)$$

The long-range contribution to the equal-time density–density correlation function given by Eq. (44) is due to the mode-coupling contribution $1/[\omega_-(k) + \omega_-(k)]$ to $G_{2n.eq.}$. The contribution $\exp[-R_{12\|}/\xi_\|]/\sqrt{R_{12\|}}$ represents an Ornstein–Zernike-like behavior in *two* dimensions[7] and the cosine term reflects the existence of the rolls. Although $M_{\rho\rho}$ is considered here for $R \lesssim R_c$ and there are no rolls present yet in the *average* state, the fluctuations already "feel" their imminent appearance for $R \gtrsim R_c$. $M_{\rho\rho}$ becomes singular and the correlation length tends to diverge with $R \to R_c$.[1,17]

2. $R \gtrsim R_c$. Here we only give how Eq. (44) for $M_{\rho\rho}(\mathbf{R}_1\mathbf{R}_2)$ for $R \lesssim R_c$ is modified, when $R \gtrsim R_c$, in two special cases.

  (i) *Across the rolls:* $R_{12_x} \gg d$ and $R_{12_y} = 0$. $M_{\rho\rho}(\mathbf{R}_1\mathbf{R}_2)$ is given by Eqs. (44) and (45) with $(1 - R/R_c)$ replaced by $R/R_c - 1$, R_{12} by R_{12_x} and $\xi_\|$ by ξ_x, respectively.
  (ii) *Along the rolls:* $R_{12_x} = 0$ and $R_{12_y} \gg d$. $M_{\rho\rho}(\mathbf{R}_1\mathbf{R}_2)$ is given by Eqs. (44) and (45), with $(1 - R/R_c)$ replaced by $(R/R_c - 1)f(P)$, $R_{12\|}$ by R_{12_y} and $\xi_\|$ by ξ_y. Here $f(P)$ is a function of the Prandtl number $P = \nu/D_T$ only, and is given by Eq. (40) with $\theta = 0$. For gases $(P \approx 1)$, $f(P) \approx 0.80$, while for liquids $(P \approx 5)$, $f(P) \approx 0.34$.

8. DISCUSSION

1. Near $R = R_c$ the correlation functions exhibit very-long-range correlations owing to mode-coupling contributions of two ω_- modes to $G_{2n.eq.}$.

2. The singular behavior of the correlation functions when the instability point is approached either from below or above is similar to that near a critical point as given by mean-field theory. For instance, in the latter case the correlation length $\xi \sim 1/\sqrt{|1 - T/T_c|}$ (T_c is the critical temperature), which is similar to the behavior of $\xi_\|$ or ξ_x, ξ_y near $R = R_c$ [cf. Eq. (45)] as a function of $|1 - R/R_c|$. However, the *length* scale is quite different: while near the critical point this scale is of the order of many σ (σ is the range of the interparticle forces $\sim 10^{-8}$ cm) near the instability point the scale is of the order of many times the distance between the plates d ($d \sim 0.1$ cm).

3. As a direct consequence of this, there will be "instability opalescence" near $R = R_c$, just as there is critical opalescence near $T = T_c$. However, the electromagnetic radiation to detect this directly is in the cm-wave range rather than in the visible light range.

Alternatively, one could try to detect the behavior of the velocity-velocity correlation function by using laser-Doppler-velocimetry.

4. Because the results presented here for the correlation functions near the convective instability are analogous to the results found near a critical point in the mean-field approximation, one might think that there are important corrections to our results when one is close enough to the instability point, since this is so for the critical point. We have examined this question[1] using both kinetic theory and a hydrodynamic theory and found that important corrections to our results only occur in an experimentally inaccessible region close to R_c, that is, when $|R/R_c - 1| \lesssim 10^{-6}$. Similar conclusions had been reached before by others[8,9,16] using fluctuating hydrodynamics.

5. The unequal- and equal-time density–density correlation functions have been computed before for $R \ll R_c$, when the effects of gravity and the walls can be ignored. In that case, the expression (43) for $M_{\rho\rho}(k_z, k'_z, \mathbf{k}_\parallel, t = 0)$ reduces to the corresponding formula given earlier.[11]

6. In this chapter we have only considered the treatment of a convective instability (in **R** space) within the framework of kinetic theory. We have left open the question whether there are kinetic instabilities in addition to those considered here, that is, whether an instability in **V** space can also occur in addition to the instability in **R** space.

ACKNOWLEDGMENT

Work was performed under NSF Grants CHE77-16308 and MCS80-17781.

REFERENCES

1. T. R. Kirkpatrick and E. G. D. Cohen, "Pair Correlation Function Near a Convective Instability," *Phys. Lett.* **88A**, 44–47 (1982), and "Kinetic Theory of Fluctuations Near a Convective Instability," *J. Stat. Phys.* **33**, 639–664 (1983).
2. S. Chandrasekhar, *Hydrodynamic and Hydromagnetic Stability*. W. Marschall and D. H. Wilkinson, eds. (Clarendon Press, Oxford, England, 1961), Chaps. I and II.
3. J. R. Dorfman and H. van Beijeren, "The Kinetic Theory of Gases," Part B in *Statistical Mechanics*, B. Berne, ed. (Plenum, New York, 1977), pp. 65–179.
4. M. H. Ernst, B. Cichocki, J. R. Dorfman, J. Sharma, and H. van Beijeren, "Kinetic Theory of Nonlinear Viscous Flow in Two and Three Dimensions," *J. Stat. Phys.* **18**, 237–270 (1978).
5. M. H. Ernst and E. G. D. Cohen, "Nonequilibrium Fluctuations in μ-Space," *J. Stat. Phys.* **25**, 153–180 (1981).

6. M. H. Ernst and J. R. Dorfman, "Nonanalytic Dispersion Relations in Classical Fluids I. The Hard Sphere Gas," *Physica* **61**, 157–181 (1972).
7. M. E. Fisher, "Correlation Functions and the Critical Region of Simple Fluids," *J. Math. Phys.* **5**, 944–962 (1964).
8. M. E. Graham, "Hydrodynamic Fluctuations Near the Convective Instability," *Phys. Rev. A* **10**, 1762–1784 (1974).
9. M. E. Graham and H. Pleiner, "Mode-Mode Coupling Theory of the Heat Convection Threshold," *Phys. Fluids.* **18**, 130–140 (1975).
10. C. Allain, H. Z. Cummins, and P. Lallemand, "Critical Slowing Down Near the Rayleigh–Bénard Convective Instability," *J. Phys. Lett.* **39**, 473–477 (1978).
11. T. R. Kirkpatrick, E. G. D. Cohen, and J. R. Dorfman, "Light Scattering by a Fluid in a Nonequilibrium Steady State: II. Large Gradients," *Phys. Rev. A* **26**, 995–1014 (1982).
12. J. A. Krommes and C. Oberman, "Anomalous Transport Due to Long Lived Fluctuations in Plasma: Part I. A General Formalism for Two-Time Fluctuations," *J. Plasma Phys.* **16**, 193–227 (1976).
13. A. Onuki, "On Fluctuations in μ-Space," *J. Stat. Phys.* **18**, 475–499 (1978).
14. A. Schlüter, P. Lortz, and F. Busse, "On the Stability of Steady Finite Amplitude Convection," *J. Fluid Mech.* **23**, 129–144 (1965).
15. R. Schmitz and E. G. D. Cohen, (unpublished).
16. J. Swift and P. C. Hohenberg, "Hydrodynamic Fluctuations at the Convective Instability," *Phys. Rev. A* **15**, 319–328 (1977).
17. V. Zaitsev and M. Shliomis, "Hydrodynamic Fluctuations Near the Convection Threshold," *Sov. Phys. JETP.* **32**, 866–870 (1971).

KINETIC DESCRIPTION OF A CHAOTIC MOTION IN A CLASSICAL CONSERVATIVE SYSTEM WITH TWO DEGREES OF FREEDOM

T. Y. Petrosky

Center for Studies in Statistical Mechanics, University of Texas, Austin, Texas

Abstract. Motion of an elastic pendulum in the vicinity of unperturbed separatrix has been investigated by a kinetic description of nonequilibrium statistical mechanics. New canonical variables of the pendulum, which are well-defined and continuous at the separatrix, have been introduced. A kinetic equation of the Fokker–Planck type for the momentum distribution function has been derived from basic principles in the weak coupling limit. This equation gives the Arnol'd–Chirikov diffusion coefficient of the energy of the unperturbed pendulum. In higher-order approximation in a coupling constant, the existence of a logarithmic singularity has been pointed out.

1. INTRODUCTION

Since the discovery of diffusionlike process in a numerical simulations of Hamiltonian systems,[8] the existence of dissipative process in conservative systems with few degrees of freedom has been the object of controversy.[4,5,7,15] Chirikov's numerical investigations of the "whisker mapping," an approximation of the Poincaré mapping of a weakly driven pendulum, suggest that in the vicinity of the unperturbed separatrix a diffusion coefficient (i.e., second moment) of the energy exists. This suggests that the motion around the separatrix of nonintegrable small systems may be similar to dissipative processes in systems with many degrees of freedom.

On the other hand, Channon and Lebowitz[4] and Karney[10] have discovered long-time correlations in numerical simulations of the area preserving Hénon mapping and the standard mapping, respectively. Since the latter mapping is locally equivalent to the whisker mapping, this suggests that the diffusionlike process of the pendulum around the separatrix may differ from ordinary diffusion present in large systems with weak coupling or low density (e.g., a gas). Because of the difficulty of the small devisor, that is, the resonance effect, in perturbation theory,[15] a theoretical understanding of the diffusionlike process in small systems is still far from satisfactory.

The purpose of this chapter is to offer a systematic theoretical treatment of this diffusionlike process, starting from the Liouville equation. The classical approach to dynamical systems is to study individual trajectories. However, when the motion starts from an unstable region of phase space, different initial conditions, which lie arbitrarily close to each other in phase space, may yield exponentially diverging trajectories or ones with qualitatively distinct types of behavior. In this situation the concept of deterministic motion along trajectories ceases to be a physically meaningful idealization.[13] This makes it necessary to go to a probabilistic description of physical states in terms of Gibbs distribution function.

The method on which we rely is the perturbation theory developed by Prigogine and his colleagues[2,18,20] for nonequilibrium statistical systems. This theory has been first introduced to treat systems with an infinite number of degrees of freedom, and is applicable when the spectrum of the Liouville operator is continuous. By using a specific example, we show the spectrum is continuous in the unstable region of a small system, so we may apply this theory even to systems with few degrees of freedom. A characteristic distinction of this perturbation theory from others (e.g., the Kolmogorov–Arnol'd–Moser theory[1,11,14]) is that it is applicable in the resonance region, as it collects the most diverging term in time of the perturbation series.

To illustrate this method, we consider an autonomous system, an elastic pendulum with two degrees of freedom, the rod of which is made from a harmonic spring that obeys Hooke's law. We focus our attention on the vicinity of the separatrix of the unperturbed pendulum. One of the difficulties encountered in this region of phase space is that the action-angle variables of the pendulum diverge at the separatrix. In Section 2, we show that there is a set of canonical variables that are well defined and continuous at the separatrix and make the unperturbed Hamiltonian of the pendulum cyclic. In Section 3 we use these variables to construct a whisker mapping similar to the one obtained by Chirikov.[5] From this we may investigate the distribution of singular points in phase space (i.e., stable and unstable periodic points) and estimate the width of the "stochastic layer" around the separatrix. In Section 4 Prigogine's perturbation theory is used to describe the motion in the stochastic layer. We show that there exists well-defined first and second moments of the observables, for example, the energy of the unperturbed pendulum, in the limit of weak coupling between the pendulum and the spring. We then construct a Markovian kinetic equation, of the Fokker–Planck type, for the momentum distribution function. From this we obtain the same diffusion coefficient of the energy of the unperturbed pendulum obtained by Chirikov's heuristic estimate.

At the same time, we point out that in higher-order approximations in the perturbation series there exists a logarithmic singularity at the separatrix. This is shown to come from the nonanalyticity of the interacting Hamiltonian in the new canonical variables. Consequently, we may expect that higher-order moments of the observables behave differently from the ones in large systems as has been suggested by Channon and Lebowitz[4] and Karney.[10]

In Section 5, we investigate the properties of our Fokker–Planck equation and show the existence of an equilibrium state defined by a microcanonical distribution for a given energy. We also remark that the H-theorem is satisfied in this system. We point out that the distribution function may not be decomposed into factors corresponding to each degree of freedom in general; thus, the equilibrium state of our system depends on its initial energy distribution.

In Section 6 we identify several characteristic time scales that are analogous to those present in large statistical systems (i.e., the microscopic interaction time, mean free time, and relaxation time).

2. THE ELASTIC PENDULUM

We will consider the motion of an elastic pendulum with two degrees of freedom, as shown in Fig. 1. The Hamiltonian of the system is given by

$$\tilde{H} = \frac{P_\theta^2}{2M} \frac{1}{(1+x/l)^2} + M\omega_0^2(1-\cos\theta) - \frac{x}{l}M\omega_0^2\cos\theta + \frac{p_x^2}{2m} + \frac{m\Omega^2 x^2}{2} \quad (1)$$

Here, θ denotes angular displacement of the pendulum from its stable equilibrium position at bottom, P_θ is the conjugate angular momentum of θ, x is the displacement of the spring from its natural length l, and p_x is the momentum of the spring. m is mass of the weight, M ($\equiv ml^2$) is moment of inertia of the pendulum when we neglect the vibration of the spring. Ω is the natural frequency of the spring and ω_0 ($\equiv \sqrt{g/l}$) is the natural frequency of the harmonic pendulum.

We are interested in the motion of the pendulum in the vicinity of the

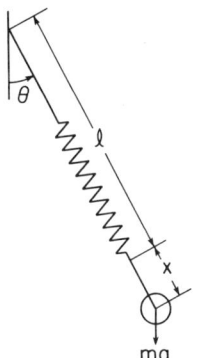

Figure 1.

separatrix of the unperturbed pendulum (i.e., the pendulum with nonvibrating rod), when the coupling between the pendulum and the spring is small. The energy of the separatrix of the unperturbed pendulum is given by

$$\tilde{H}_{sx} = 2M\omega_0^2 \qquad (2)$$

If we assume that the energy of the spring has the same order of magnitude with $\hat{H}_{sx}$ at the initial time $t = 0$, then the weak-coupling condition is given by

$$\mu \equiv \frac{\omega_0}{\Omega} \ll 1 \qquad (3)$$

In this case $|x|/l$ has an upper bound that is proportional to μ, and we may approximate the Hamiltonian, Eq. (2), to first order of μ by

$$\tilde{H} = \frac{P_\theta^2}{2M} + M\omega_0^2(1 - \cos\theta) + \frac{p_x^2}{2m} + \frac{m\Omega^2 x^2}{2} - \frac{x}{l}\left(\frac{P_\theta^2}{M} + M\omega_0^2 \cos\theta\right) \qquad (4)$$

At this stage, it is useful to introduce dimensionless quantities by

$$H \equiv \frac{\tilde{H}}{\tilde{H}_{sx}}, \qquad Y_1 \equiv \frac{P_\theta}{2M\omega_0}, \qquad \theta_1 \equiv \theta$$

$$\tau \equiv \omega_0 t, \qquad y_2 \equiv \frac{\omega_0 J}{\tilde{H}_{sx}}, \qquad \alpha_2 \equiv \alpha \qquad (5)$$

where the action-angle variables (J, α) for the spring are defined by

$$x = \sqrt{2J/m\Omega}\cos\alpha, \qquad p_x = \sqrt{2m\Omega J}\sin\alpha \qquad (6)$$

Then, the dimensionless Hamiltonian H is given from Eq. (4) by

$$H = H_0 + \varepsilon V \qquad (7)$$

where $H_0 = H_1 + H_2$ and

$$H_1 = Y_1^2 + \tfrac{1}{2}(1 - \cos\theta_1), \qquad H_2 = y_2/\mu \qquad (8)$$

and
$$\varepsilon V = -2\varepsilon(2Y_1^2 + \tfrac{1}{2}\cos\theta_1)\sqrt{\mu y_2}\cos\alpha_2 \qquad (9)$$

Here we have introduced the parameter ε to specify the order of the interaction. For $\varepsilon = 1$, we recover the Hamiltonian of the elastic pendulum, Eq. (4).

If we regard H as a new "Hamiltonian" and τ as a "time," then the variables (Y_1, θ_1) and (y_2, α_2) are canonical conjugates.

The restriction of the energies which we have imposed above is given in terms of the dimensionless quantities by

$$|H_1 - 1| \ll 1 \quad \text{and} \quad O(y_2) = \mu \qquad (10)$$

where $H_1 = 1$ is the energy of the separatrix of the unperturbed pendulum.

We now introduce new canonical variables (y_1, α_1) in which the Hamiltonian of the unperturbed pendulum H_1 becomes cyclic. Our canonical variables are defined differently for the cases of rotation, motion on the separatrix, and oscillation. However, we define them such that they and their derivatives are well defined and continuous in the limit $H_1 \to 1$. This property is essential, and is not satisfied by the ordinary action-angle variables of the pendulum that diverge at the separatrix.* Our new variables are defined through the following canonical transformations:

(i) For rotation ($0 \leq c < 1$)

$$\sin\frac{\theta_1}{2} = \pm \text{sn}\left(\frac{\alpha_1}{2}, c\right), \quad Y_1 = \pm y_1 \text{dn}\left(\frac{\alpha_1}{2}, c\right) \qquad (11)$$

(ii) For the separatrix ($c = 1$)

$$\sin\frac{\theta_1}{2} = \pm \tanh\frac{\alpha_1}{2}, \quad Y_1 = \pm y_1 \text{sech}\frac{\alpha_1}{2} \qquad (12)$$

(iii) For oscillation ($c > 1$)

$$\sin\frac{\theta_1}{2} = \frac{1}{\sqrt{c}}\text{sn}\left(\frac{\sqrt{c}}{2}\alpha_1, \frac{1}{c}\right), \quad Y_1 = y_1 \text{cn}\left(\frac{\sqrt{c}}{2}\alpha_1, \frac{1}{c}\right) \qquad (13)$$

where c is related to y_1 by

$$c = 1/y_1^2 \qquad (14)$$

and the new Hamiltonian is

$$H_1 = y_1^2 \qquad (15)$$

---

*The canonical transfomation between the dimensionless action-angle variables (j, ϕ) of the unperturbed pendulum and our variable is given, for example, for rotation, by $\phi = 2\pi\alpha_1/K(c)$ and $j = (2/\pi)[E(c)/\sqrt{c}]$, where $c = y_1^{-2}$. Note that the relation between α_1 and ϕ is essentially the same with the relation between the arguments of the Jacobi elliptic functions and the theta functions.[21]

Here sn, cn, and dn are the Jacobi elliptic functions and c in Eq. (11) and c^{-1} in Eq. (13) are square of the "modulus" (i.e., the "cross-ratio") of the elliptic function.[19,21] We assume that $y_1 \geq 0$. Each sign in Eqs. (11) and (12) corresponds to each branch of the motion outside, or on, the separatrix.

Observe that the real periods of the functions $dn(u, c)$ and $cn(u, c^{-1})$ with respect u are given by $2K(c)$ and $4K(c^{-1})$, respectively, where K is the complete elliptic integral of the first kind. The period of the rotation T_r and oscillation T_0 of the unperturbed pendulum are given by

$$T_r = 2\sqrt{c}K(c) \quad \text{and} \quad T_0 = 4K(c^{-1}) \tag{16}$$

Using the approximation $K(c) \simeq \frac{1}{2}\ln(16/c')$ for $c' \ll 1$, where

$$c' = 1 - c \tag{17}$$

is the square of the complementary modulus of the elliptic function for $0 \leq c \leq 1$ [cf. Eq. (A5a) in the Appendix], we obtain the following unified expression of the period of rotation T_r and the half-period of oscillation $T_0/2$ in the vicinity of the separatrix:

$$T(c') = \ln\left(\frac{16}{|c'|}\right) \tag{18}$$

Note that c' is approximately the distance of the energy of the pendulum from the separatrix, that is, $c' \simeq H_1 - 1$ for $|c'| \ll 1$. The rotational motion corresponds to $c' > 0$.

Let us now return to the perturbed system, Eq. (7). Using the new variables, we obtain

$$H = H_0 + \varepsilon V = y_1^2 + \frac{y_2}{\mu} - \varepsilon y_1^2 [F(\alpha_1, c) + c - 2]\sqrt{\mu y_2} \cos \alpha_2 \tag{19}$$

where

$$F(\alpha_1, c) = \begin{cases} 6\,dn^2\left(\dfrac{\alpha_1}{2}, c\right), & \text{for } 0 \leq c \leq 1 \tag{20a} \\[2mm] 6\,cn^2\left(\dfrac{\sqrt{c}}{2}\alpha_1, \dfrac{1}{c}\right), & \text{for } c > 1 \tag{20b} \end{cases}$$

The Jacobi elliptic functions are generally periodic functions of α_1, and have an infinite number of Fourier components. This implies that between the pendulum and the spring there is an infinite number of resonance interactions. For example, for the rotational case we have the following expansion:

$$dn^2\left(\frac{\alpha_1}{2}, c\right) = \frac{E}{K} + \Delta l \sum_{n=-\infty}^{+\infty}{}' \frac{2l_n}{\sinh(2l_n K')} e^{il_n \alpha_1} \tag{21}$$

Here

$$\Delta l \equiv \frac{\pi}{2K}, \quad l_n = \frac{n\pi}{2K} \tag{22}$$

where n is an integer and $K' = K(c')$. $E = E(c)$ is the complete integral of the second kind. The prime on the summation sign stands for $n \neq 0$. For $c \to 1$, Eq. (21) reduces to the Fourier integral.

Combination of the Fourier expansion, Eq. (21), and $\cos \alpha_2$ in Eq. (19) gives the resonance values of c' as solutions of the resonance equation $l_n(\partial H_1/\partial y_1) \pm (\partial H_2/\partial y_2) = 0$.

Sufficiently close to the separatrix, that is, $|c'| \ll 1$, we may solve the resonance equation explicitly with the approximation, Eq. (A5a). Then, the resonance values of c' are given by

$$|c'| \simeq 16 e^{-2\pi\mu n} \tag{23}$$

where n is any integer. Because $\text{cn}^2(u, \hat{c})$ with the square of the modulus $\hat{c}$ is related to $\text{dn}^2(u, \hat{c})$ by

$$\hat{c}\, \text{cn}^2(u, \hat{c}) = \text{dn}^2(u, \hat{c}) - \hat{c}' \tag{24}$$

we have the same result, Eq. (23), for the oscillatory case. The result, Eq. (23), shows that the resonance values of c' are densely distributed around the separatrix $c' = 0$. In the next section, we will show that these resonance interactions give rise to an infinite number of singular motions (i.e., stable and unstable periodic motions) around the unperturbed separatrix.

3. THE WHISKER MAPPING AND THE STOCHASTIC LAYER

To find the singular solutions, we first construct a discrete mapping for a Poincaré surface of section on a surface, say $\theta_1 = 0$ (i.e., $\alpha_1 = 0$). Periodic solutions are found as fixed solutions of the Poincaré mapping.

To construct the mapping we first estimate an increment $\Delta c'$ of c' in a period T of the mapping. T may be approximated by the unperturbed period of the pendulum, Eq. (18), under the weak-coupling condition, $\mu \ll 1$. $\Delta c'$ is approximately equal to the increment of the energy ΔH_1 of the unperturbed pendulum [cf. Eq. (17)]. Therefore, $\Delta c'$ is given by

$$\Delta c' \simeq \int_{-T/2}^{T/2} d\tau \frac{dH_1}{d\tau} = \int_{-T/2}^{T/2} d\tau [H_1, H] \tag{25}$$

where $[,\,]$ is the Poisson bracket.

To evaluate the right-hand side of Eq. (25), we substitute Eqs. (15) and (19), and approximate the time evolution of $y_i(\tau)$ and $\alpha_i(\tau)$, where $i = 1, 2$, by unperturbed solutions. Without any loss of generality, we may choose the origin of the time $\tau = 0$ such that $\alpha_1(0) = 0$. Then, we approximate $y_1 = 1$ (i.e., $c = 1$), since we are interested in the energy H_1 very close to the unperturbed separatrix. After these approximations, we have the same integrand of Eq. (25) for the rotational and oscillatory cases. The integrand is proportional to the factor $(d/d\tau) \text{sech}^2 \tau$. This factor varies slowly, except around the origin $\tau = 0$, and that damps very quickly for large $|\tau|$. On the other hand, the integrand of Eq. (25) includes also a rapidly oscillating factor $\cos[(\tau/\mu) + \alpha_2(0)]$. This implies that the contribution of the integration in Eq. (25) comes from only a small interval of τ around $\tau = 0$. Therefore, we may replace the limit of the integration $\pm T/2$ by infinity.

We may interpret this fact that the two degrees of freedom collide in a short interaction time at $\tau = 0$ where the centrifugal force becomes maximum. The analogy of the collision time and the mean-free time of this system to large statistical systems will be discussed in more detail in Section 6.

It is worth noting that the final expression obtained by the above procedure is nothing but the Melnikov function Δ_ε in the lowest approximation, which gives us a measure of the distance between the stable and the unstable invariant manifolds of the hypabolic point, when transverse homoclinic orbits appears.[9,12]

Substituting the Fourier integral of $(d/d\tau)\mathrm{sech}^2 \tau$ into Eq. (25), and integrating by τ, we ultimately obtain the following result,*

$$\Delta c' \simeq 12 \pi \varepsilon \sqrt{\frac{y_{20}}{\mu}} \frac{e^{-\pi/2}}{\mu} \sin \alpha_2(0) \qquad (26)$$

The increment of the phase of the spring $\Delta \alpha_2$ in a period of the mapping is approximately obtained by the unperturbed solution $\alpha_2(\tau) = (\tau/\mu) + \alpha_2(0)$, and by the unperturbed period T in Eq. (18). Then, we have $\Delta \alpha_2 \simeq (1/\mu) \ln(16/|c'|)$.

After each iteration of the mapping, the value of y_{20} in Eq. (26) changes. This value can be calculated as a function of c' and α_2 by using the energy conservation $H(y_1, y_2, \alpha_1 = 0, \alpha_2) = E$. Then, we see that each increment of y_2 in a period T is proportional to the small factor μ. Therefore, we may neglect this increment in our approximations.

Now, we can construct the Poincaré mapping by using the above results. Denoting the quantities after an iteration of the mapping by $\bar{c}'$ and $\bar{\alpha}_2$, we arrive at the following difference equation:

$$\begin{cases} \bar{c}' = c' + W \sin \alpha_2 \\ \bar{\alpha}_2 = \alpha_2 + \dfrac{1}{\mu} \ln\left(\dfrac{16}{|\bar{c}'|}\right) \end{cases} \pmod{2\pi} \qquad (27)$$

where

$$W \equiv 12 \pi \varepsilon \sqrt{\frac{y_{20}}{\mu}} \frac{e^{-\pi/2\mu}}{\mu} \qquad (28)$$

In the second equation in Eq. (27), we have substituted $\bar{c}'$ rather than c', in

---

*We can obtain the same results, Eq. (26), by evaluating the Melnikov–Arnol'd (MA) integral, which is represented by the old canonical variables (Y_1, θ_1) in Eq. (7).[5] Note that in estimating the MA integral the oscillating contributions on time cancel each other in our system because of a symmetric structure of the interaction, Eq. (9), of $\cos(\theta_1 + \alpha_1)$ and $\cos(\theta_1 - \alpha_1)$.

order to make the mapping canonical. This substitution is consistent in our approximation.

This mapping, Eq. (27), is just the "whisker mapping," which has been derived and discussed by Chirikov[5] in his study of a driven pendulum. Here we summarize his findings:

(i) There exists an infinite number of fixed solutions (i.e., solution with period one) at

$$\alpha_2 = 0 \text{ or } \pi, \text{ and } |c'_n| = 16 e^{-2\pi\mu n} \quad (29)$$

[Note that $|c'_n|$ is just the resonance value, Eq. (23).]

(ii) For $c' > 0$ (i.e., rotational case), these fixed solutions are all unstable on $\alpha_2 = \pi$, and stable on $\alpha_2 = 0$ for $c' > c_h^{(1)}$ and unstable for $c' < c_h^{(1)}$, where

$$c_h^{(1)} = \frac{W}{4\mu} \quad (30)$$

For $c' < 0$ (i.e., oscillatory case), the structure is the same for increasing $|c'|$, except that the role of the axes $\alpha_2 = 0$ and $\alpha_2 = \pi$ interchanged. Consequently, we have the region around the unperturbed separatrix, $|c'| < c_h^{(1)}$, where all fixed solutions are unstable. We may expect that the motion inside this region is very erratic.

(iii) Above the threshold value $c_h^{(1)}$, there is another threshold value c'_0 that gives the boundary which separates the erratic from the regular motion. The value of c'_0 is evaluated from the "overlapping criterion" of the resonance regions; erratic motion begins to occur if the separatrices surrounding adjacent stable fixed point touch. This value is given by

$$c'_0 = \frac{W}{\mu} \quad (31)$$

For $c_h^{(1)} \leq |c'| \leq c'_0$, the motion is still erratic because of the overlapping effect of the resonances. However, because there are stable fixed points in this region, we can observe some systematic motion around these stable fixed points. In the Poincaré mapping, we see a mixed structure of erratic points and islands.*

Because trajectories cannot cross each other, the motion in the erratic regions is confined in a thin layer, $|c'| \leq c'_0$, around the separatrix. We call this layer the "stochastic layer."

Relying on this knowledge, we may construct a kinetic equation that describes the time evolution inside the stochastic layer; this is the subject of the next section.

---

*Period-doubling sequence exists in the stochastic layer. Analytical argument shows that solutions with period two appear in $|c'| \leq c_h^{(1)}$. For $c' > 0$ ($c' < 0$), stable period-one solution bifurcate to stable period-two solution at $c' = c_h^{(1)}$ and $\alpha_2 = 0$ ($c' = -c_h^{(1)}$ and $\alpha_2 = \pi$). The stable period-two solution changes to the unstable one at $|c'| = c_h^{(2)} \simeq 0.4 c_h^{(1)}$ (T. Y. Petrosky, 1982, unpublished).

4. KINETIC EQUATION IN THE STOCHASTIC LAYER

In this section, we derive a kinetic equation that describes the long-time behavior of the momentum distribution function in the stochastic layer. To derive this equation, we apply the perturbation theory for the Fourier component of the distribution function that has been developed by Prigogine and his colleagues.[2,3,18,20] When we apply this method to our system, however, we immediately encounter the following two difficulties. First, the periodicity of the Hamiltonian, Eq. (19), on α_1 depends on the momentum y_1 through c. This is the direct consequence of the nonlinearity of the original unperturbed Hamiltonian, Eq. (8). As a consequence of this dependence, the derivative of the Hamiltonian with respect to y_1 affects the basis of the Fourier expansion, for example, $\exp(il_n\alpha_1)$ in Eq. (21).

The second difficulty is related to an analytic property of the Jacobi elliptic functions in Eq. (20). These functions are not analytic at $c = 1$, and have a branch point through the logarithmic singularity of the complete elliptic function K at this point.[21] Because of this singularity, the derivative of each Fourier component in Eq. (21) with respect to y_1 diverges in the limit $c \to 1$.

These difficulties may be alleviated by expanding the Jacobi elliptic functions in a series involving c' and a function of $\ln c'$ about c'. For sufficiently small values of c', the first two terms in this expansion coincide with the formal Taylor expansion.[16] For example, $\text{dn}^2(u, c)$ can be approximated around a point $c = c_a$, where $c'_a = 1 - c_a \ll 1$, by

$$\text{dn}^2(u, c) = \text{dn}^2(u, c_a) + \left[\frac{\partial}{\partial c} \text{dn}^2(u, c)\right]_{c=c_a} (c - c_a) \qquad (32)$$

In the Appendix we will prove that the diverging terms that appear in the Fourier expansion of the second term of Eq. (32) cancel each other, and we may approximate Eq. (32) to first order of $(c - c_a)$ by

$$\text{dn}^2(u, c) \simeq \Delta k \sum_{n=-\infty}^{+\infty}{}' \frac{2k_n}{\sinh \pi k_n \left(1 + \dfrac{c'}{4}\right)} e^{2ik_n u} - \tfrac{1}{2}(c - c_a) \qquad (33)$$

Here,

$$K_a = K(c_a), \quad \Delta k = \frac{\pi}{2K_a}, \quad k_n = \frac{n\pi}{2K_a} \qquad (34)$$

In the Appendix we will also show that correction terms of the approximation, Eq. (33), consist of contributions proportional to $c'/\ln c'$, $c'^2 \ln c'$, c'^2, and so on. Substituting Eq. (33) into Eq. (24) yields expansion for $\text{cn}^2 u$.

Therefore, we have a Hamiltonian that is well-behaved Hamiltonian at the separatrix with a fixed periodicity in α, in the weakly coupled system. We may now apply Prigogine's perturbation theory and construct a kinetic equation of the Fokker–Planck type, with well-defined first and second moments.

As a typical value of c'_a, we choose the value that gives the average period of the pendulum T_a in Eq. (18) inside the stochastic layer, that is,

$$T_a \equiv 2K(c_a) = \frac{2}{c'_0}\int_0^{c'_0} dc K(c) \simeq \ln\left(\frac{16e}{c'_0}\right) \simeq \frac{\pi}{2\mu} \qquad (35)$$

This gives the value $c'_a = c'_0/e \ll 1$, and $\Delta k \simeq 2\mu$. Therefore, the right-hand side of Eq. (33) reduces the Fourier integral for sufficiently small μ.

The solution of the Liouville equation is given in the perturbation series by

$$\rho(\tau) = e^{-iL\tau}\rho(0) = \frac{1}{2\pi i}\int_\Gamma dz\, e^{-iz\tau}\frac{1}{z-L}\rho(0)$$

$$= \frac{1}{2\pi i}\sum_{n=0}^\infty \int_\Gamma dz\, e^{-iz\tau}\frac{1}{z-L_0}\left(\varepsilon L'\frac{1}{z-L_0}\right)^n \rho(0) \qquad (36)$$

where $\rho(\tau) \equiv \rho(y_1, y_2, \alpha_1, \alpha_2, \tau)$ is the distribution function of the ensemble of the system in phase space normalized such that

$$\frac{1}{8\pi K_a}\int_{-2K_a}^{2K_a} d\alpha_1 \int_{-\pi}^{\pi} d\alpha_2 \int_0^\infty dy_1 \int_0^\infty dy_2 \rho(y_1, y_2, \alpha_1, \alpha_2) = 1 \qquad (37)$$

L is the Liouvillian (The Poisson bracket) defined by $L = i[H, \cdot]$. Corresponding to the decomposition of the Hamiltonian Eq. (7), we have the decomposition, $L = L_0 + \varepsilon L'$. The contour Γ, for $t > 0$, is above the real axis and goes $-\infty$ from $+\infty$.

Introducing vector notations, $\boldsymbol{\omega} = (2y_1, 1/\mu)$, $\boldsymbol{\alpha} = (\alpha_1, \alpha_2)$, and $\mathbf{n} = (k_n, m)$, we may write the unperturbed Liouvillian by $L_0 = -i\boldsymbol{\omega}\cdot\partial/\partial\boldsymbol{\alpha}$, and its eigenvalue and eigenfunction by

$$L_0\Phi_\mathbf{n}(\boldsymbol{\alpha}) = (\mathbf{n}\cdot\boldsymbol{\omega})\Phi_\mathbf{n}(\boldsymbol{\alpha}), \quad \Phi_\mathbf{n}(\boldsymbol{\alpha}) = (8\pi K_a)^{-1/2} e^{i(\mathbf{n}\cdot\boldsymbol{\alpha})} \qquad (38)$$

where we have imposed periodic boundary conditions on the eigenfunction $\Phi_\mathbf{n}(\boldsymbol{\alpha})$ with period of $4K_a$ and 2π for α_1, and α_2, respectively. For sufficiently small μ we have the continuous spectrum of L_0 around the separatrix.

We further introduce a certain matrix element,

$$\langle \mathbf{n}|L|\mathbf{n}'\rangle = \int_{-2K_a}^{2K_a} d\alpha_1 \int_{-\pi}^{\pi} d\alpha_2 \Phi_\mathbf{n}(\boldsymbol{\alpha}) L\Phi_{\mathbf{n}'}(\boldsymbol{\alpha}) \qquad (39)$$

Obviously this expression is still an operator of the momenta with the derivatives acting on everything to their right.

By the definition of L, we have

$$\langle \mathbf{n}|L_0|\mathbf{n}'\rangle = (\mathbf{n}\cdot\boldsymbol{\omega})\delta_{\mathbf{n},\mathbf{n}'} \qquad (40)$$

and

$$\langle \mathbf{n}|\varepsilon L'|\mathbf{n}'\rangle = -\varepsilon V_{\mathbf{n}-\mathbf{n}'}(\mathbf{n}-\mathbf{n}')\cdot\frac{\partial}{\partial \mathbf{y}} + \mathbf{n}'\cdot\frac{\partial}{\partial \mathbf{y}}(\varepsilon V_{\mathbf{n}-\mathbf{n}'}) \qquad (41)$$

where $\mathbf{y} = (y_1, y_2)$ and we have written the interaction in Eq. (19) in the Fourier expansion,

$$\varepsilon V(\mathbf{y}, \boldsymbol{\alpha}) = \varepsilon \sum_\mathbf{n} V_\mathbf{n}(\mathbf{y}) e^{i(\mathbf{n} \cdot \boldsymbol{\alpha})} \quad (42)$$

We define the projection operator P which projects out the momentum distribution function $\rho_0(\mathbf{y})$ from the full distribution function $\rho(\mathbf{y}, \boldsymbol{\alpha})$ by

$$\rho_0(\mathbf{y}) = P\rho(\mathbf{y} \cdot \boldsymbol{\alpha}) = \frac{1}{8\pi K_a} \int_{-2K_a}^{2K_a} d\alpha_1 \int_{-\pi}^{\pi} d\alpha_2 \rho(\mathbf{y}, \boldsymbol{\alpha}) \quad (43)$$

We also define $Q = 1 - P$ and we have

$$P = P^2, \quad Q = Q^2, \quad PQ = QP = 0 \quad (44)$$

Note that P projects out the $\mathbf{n} = 0$ component of the Fourier expansion of ρ, while Q projects out $\mathbf{n} \neq 0$ components.

Now we are ready to apply the perturbation theory to construct the kinetic equation. Since the procedure to derive the kinetic equation has been repeatedly discussed by Prigogine and his colleagues,[2,18,20] we quote here only their results which are relevant: For a weakly coupled system we may obtain a Markovian equation for the momentum distribution function:

$$\frac{\partial}{\partial \tau} \rho_0(\mathbf{y}, \tau) = \psi_2(+i0) \rho_0(\mathbf{y}, \tau) \quad (45)$$

where $\psi_2(z)$ is called the "collision operator," which is defined for the lowest order of the interaction by

$$\psi_2(z) = \varepsilon^2 \left\langle 0 \left| PL'Q \frac{1}{z - L_0} QL'P \right| 0 \right\rangle \quad (46)$$

The angle dependence of the equilibrium distribution function for $\tau \to \infty$ is given by

$$Q\rho^{eq}(\mathbf{y}, \boldsymbol{\alpha}) = \frac{1}{+i0 - L_0} QL'P\rho_0^{eq}(\mathbf{y}) \quad (47)$$

Applying this result for our system, we ultimately arrive at the kinetic equation.

$$\frac{\partial}{\partial \tau} \rho_0(y_1, y_2, \tau) = 2\pi\varepsilon^2 \sum_{m=-1}^{1}{}' \int_{-\infty}^{+\infty} dk \left(k \frac{\partial}{\partial y_1} + m \frac{\partial}{\partial y_2} \right)$$

$$\times \left| \frac{6 k y_1^2 \sqrt{y_2}}{\mu \sinh \dfrac{\pi k}{\mu} \left(1 + \dfrac{c'}{4}\right)} \right|^2$$

$$\times \delta(2ky_1 + m) \left(k \frac{\partial}{\partial y_1} + m \frac{\partial}{\partial y_2} \right) \rho_0(y_1, y_2, \tau) \quad (48)$$

where we have replaced the Fourier series by the Fourier integral and k by k/μ. Note that the right-hand side of Eq. (48) has a contribution only at the resonant point, $2ky_1 + m = 0$.

5. PROPERTIES OF THE KINETIC EQUATION AND EQUILIBRIUM STATE

We first note that the kinetic equation, Eq. (48), reduces to the one-dimensional equation by replacing the variables (y_1, y_2) by (y_1, H_0), and we have

$$\frac{\partial}{\partial \tau} \phi(y_1, H_0, \tau) = \mathscr{L}\phi(y_1, H_0, \tau) \tag{49}$$

where $\phi(y_1, H_0) = \rho_0(y_1, y_2)$ and

$$\mathscr{L}\phi = \frac{\pi}{2} \frac{\partial}{\partial y_1} \frac{U(y_1, H_0)}{y_1^3} \frac{\partial}{\partial y_1} \phi$$

$$= -\frac{\partial}{\partial y_1}[R_1(y_1, H_0)\phi] + \frac{1}{2}\frac{\partial^2}{\partial y_1^2}[R_2(y_1, H_0)\phi] \tag{50}$$

Here,

$$U(y_1, H_0) = \left| \frac{6\varepsilon k y_1^2 \sqrt{H_0 - y_1^2}}{\sqrt{\mu} \sinh \frac{\pi k}{\mu}\left(1 + \frac{c'}{4}\right)} \right|^2_{k=1/2 y_1} \tag{51}$$

and

$$R_2(y_1, H_0) = \frac{\pi U(y_1, H_0)}{y_1^3}, \quad R_1(y_1, H_0) = \frac{1}{2}\frac{\partial}{\partial y_1} R_2(y_1, H_0) \tag{52}$$

The last expression in Eq. (50) shows that our kinetic equation is nothing but the Fokker–Planck equation with the drift coefficient R_1 and the diffusion coefficient R_2.

The probability current j is defined by

$$j(y_1, H_0, \tau) = R_1(y_1, H_0)\phi(y_1, H_0, \tau) - \frac{1}{2}\frac{\partial}{\partial y_1}[R_2(y_1, H_0, \tau)\phi(y_1, H_0, \tau)] \tag{53}$$

To discuss the evolution governed by Eq. (49), we need to specify the boundary condition. Because the stochastic layer is restricted and isolated in the region $|c'| \leq c_0'$, we impose the boundary condition such that at $|c'| = c_0'$ the probability current, Eq. (53), vanishes. The inequality $|c'| \leq c_0$ restricts the domain D of y_1. Note that D depends on H_0. For given H_0, the measure (i.e., the length) of D is given by $\text{mes } D(H_0) = \gamma\sqrt{H_0 - 1} + O(\gamma^2)$, where $\gamma = 12\pi\varepsilon(1/\mu)^2 \exp(-\pi/2\mu)$.

We further assume that our kinetic equation, Eq. (49), is valid in whole region of the stochastic layer. Of course, this assumption is not valid near the boundary of the stochastic layer because of the islands found there induce regular motion. However, we expect that the effect of the islands is only significant near the boundary.

Note that in our boundary condition the second relation in Eq. (52) is just the condition of the self-adjointness of the Fokker–Planck equation, Eq. (49). Namely, the backward equation of Eq. (49) coincides with the forward equation, Eq. (49). As a consequence of this property, we can prove that there exists a unique equilibrium state as a continuous function of y_1. The equilibrium state is a constant on y_1 and is, for a given initial distribution $\phi(y_1, H_0, 0)$,

$$\phi^{eq}(H_0) = \frac{1}{\gamma\sqrt{H_0 - 1}} \int_{D(H_0)} dy_1 \phi(y_1, H_0, 0) \tag{54}$$

where the leading factor is a normalization constant.

Note that the equilibrium state still has an initial dependence of the distribution of H_0. This dependence occurs because the distribution function may not be decomposed into factors corresponding to each degree of freedom. This result certainly differs from that obtained in systems possessing many degrees of freedom, for which the momentum distribution function is in general factorizable. However, in this case the factorization property originates because we can neglect the effect of repeated interactions between the same degrees of freedom.[17] This effect, however, is dominant in systems with few degrees of freedom. Thus, even if we begin with a factorizable distribution, this property is soon destroyed by the dynamics.

We can further prove a monotonic approach to the equilibrium state, Eq. (54), as follows. Let $\phi_i(y_1)$ be an eigenfunction of the Fokker–Planck operator $\mathcal{L}$, Eq. (50), belonging to an eigenvalue λ_i and satisfying our boundary condition. Here we have dropped the argument H_0 for simplicity. We now assume that $\phi_i(y_1)$ form an orthonormal basis. Then we have

$$\lambda_i = \int_D dy_1 \phi_i^*(y_1) \mathcal{L} \phi_i(y_1) = -\frac{\pi}{2} \int_D dy_1 \frac{U(y_1)}{y_1^3} \left| \frac{\partial \phi_i(y_1)}{\partial y_1} \right|^2 \leq 0 \tag{55}$$

Therefore, if we expand the initial condition in the solution of Eq. (49), $\phi(\tau) = \exp[\mathcal{L}\tau]\phi(0)$, then we have monotonic approach zero for $\tau \to +\infty$, except for the state belonging to zero eigenvalue, which gives us the equilibrium state, Eq. (54).

We remark that the H-theorem is satisfied by our system if we introduce the "entropy" function

$$s(H_0, \tau) = -\int_{D(H_0)} dy_1 \phi(y_1, H_0, \tau) \ln \phi(y_1, H_0, \tau) \tag{56}$$

Let us now estimate the diffusion coefficient of the energy of the unperturbed pendulum H_1 and compare it with Chirikov's heuristic estimate.[5] We assume that the system has well-defined momenta y_{10} and y_{20} inside the stochastic layer at time $\tau = 0$, that is, $\rho_0(y_1, y_2, 0) = \delta(y_1 - y_{10})\delta(y_2 - y_{20})$. The diffusion coefficient of H_1 in the Fokker–Planck equation is given by

$$D_1 = \lim_{\Delta\tau \to 0} \langle [H_1(\Delta\tau) - H_{10}]^2 \rangle / \Delta\tau \qquad (57)$$

where $H_{10} = y_{10}^2$. The average $\langle H_1(\tau) \rangle$, for example, is calculated by

$$\langle H_1(\tau) \rangle = \int d\mathbf{y}\, y_1^2 \rho_0(\mathbf{y}, \tau) = \int dH_0 \int dy_1\, y_1^2 \phi(y_1, H_0, \tau)$$

Expanding Eq. (57) in the series of $\Delta\tau$, and integrating by parts under our boundary condition, we obtain

$$D_1 = 12^2 \pi \varepsilon^2 y_{20} \left(\frac{1}{\mu}\right)^2 e^{-\pi/\mu} \qquad (58)$$

where we have approximated $y_{10} = 1$ in the final expression.

We now compare this result with the "diffusion coefficient" D_1' defined by Chirikov[5] such that

$$D_1' = \frac{\overline{[\Delta H_1]^2}}{T_a} = 2 \times 12^2 \pi \varepsilon^2 y_{20} \left(\frac{1}{\mu}\right)^2 e^{-\pi/\mu} \overline{\sin^2 \alpha_2} \qquad (59)$$

Here $\Delta H_1 \simeq \Delta c'$ is given by Eq. (26) and T_a by Eq. (35). The quantity $\overline{\sin^2 \alpha_2}$ stands for the average of $\sin^2 \alpha_2$ with respect to the phase of the spring α_2. Following Chirikov,[5] we assume a *uniform random-phase* distribution of α_2. This gives us precisely $D_1' = D_1$.

However, the agreement is accidental. Because of the gravitational field, space is not uniform, so the random-phase assumption does not hold. To see this, let us observe the angle dependence of the distribution function at the equilibrium state by Eq. (47). Applying Eq. (47), we have

$$\rho^{eq}(\mathbf{y}, \boldsymbol{\alpha}) = P\rho^{eq} + Q\rho^{eq} = \phi^{eq}(H_0) + \varepsilon V(\mathbf{y}, \boldsymbol{\alpha}) \frac{d}{dH_0} \phi^{eq}(H_0) \qquad (60)$$

This is just the first two terms in the Taylor expansion of $\phi^{eq}(H) = \phi^{eq}(H_0 + \varepsilon V)$ by ε. This implies that the equilibrium distribution is microcanonical for a given energy $H = E$.

Integrating Eq. (60), we can obtain the angle distribution function $n_i^{eq}(\alpha_i)$, where $i = 1, 2$, at the equilibrium state. For α_1, we have the uniform distribution, $n_1^{eq}(\alpha_1) = 1$. This corresponds to the distribution of the original angle θ of the pendulum, $n^{eq}(\theta) \simeq (\pi/2K_a)[1 - c_a \sin^2(\theta/2)]^{-1/2}$. Therefore, the pendulum is found around $\theta = \pi$ more likely than $\theta = 0$.

For α_2, we have

$$n_2^{eq}(\alpha_2) \simeq 1 + \varepsilon\mu \cos\alpha_2 \int dH_0 \gamma(H_0 - 1) \frac{d}{dH_0} \phi^{eq}(H_0) \qquad (61)$$

The distribution is *not* uniform. The deviation from the uniformity has the order of $\varepsilon\mu$. [Note that the integral in Eq. (61) has the order of unity, owing to the normalization of $\phi^{eq}(H_0)$.] However, the second term in Eq. (61) does not give a contribution to the average of $\sin^2 \alpha_2$. This explains why our results are consistent with Eq. (58), even though our random-phase assumption was incorrect. That the nonuniformity of the distributions in α_2 is due to the gravitational field may be seen by observing that the interaction term in Eq. (19) vanishes if $6 \operatorname{sech}^2(\alpha_1/2) - 1 = 6\cos(\theta/2) - 1 = 0$ for $|c'| \ll 1$. Consequently, at this angle θ the distribution function Eq. (60) is independent of α_2 and, therefore, the spring "feels" space to be uniform. It is easily seen that at this angle the centrifugal force cancels the radial component of the gravitational force acting on the mass.

6. CONCLUDING REMARKS

We have derived the kinetic equation of the Fokker–Planck type, Eq. (49), in the vicinity of the unperturbed separatrix in the system with two degrees of freedom and have shown that for a given energy the system approaches the microcanonical distribution monotonically, thus satisfying the H-theorem.

Also, we have found that the resulting diffusion process evolves much slower than those occurring in large (e.g., gaseous) systems. This discrepancy may be more clearly described after examining the characteristic time scales found in the elastic pendulum and comparing them with those found in a dilute gas.

The smallest time scale present in the elastic pendulum is the period of the uncoupled spring, $\tau_s = 2\pi\mu$.

The next one is the time scale that characterizes the duration of the interaction between the spring and the pendulum, τ_i. Since this interaction is strongest at $\theta = 0$ and is exponentially damped in time, we may define τ_i to be the time for which the integrand in Eq. (25) decreases to $1/e$ of its maximum value. Neglecting the oscillatory factor in the integrand, we obtain $\tau_i \simeq 1$. This estimate is also supported by evaluating the singularity of the collision operator, Eq. (46), with respect to z.[20] Using the product expansion of $\sinh \pi x / \pi x$, we can perform the integration of the Fourier argument k in $\psi_2(z)$. Then, we see that $\psi_2(z)$ has simple and double poles in the lower-half plane of z at $\operatorname{Im}(z) = -2in\gamma_1$, where n is taken over all positive integers. From this we obtain $\tau_i \simeq \frac{1}{2}$. [Note that the kinetic equation, Eq. (49), is physically meaningful for time scales that are much larger than the interaction time τ_i.] We shall regard τ_i as being analogous to the collision time in a dilute gas. Similarly, we shall regard the mean period of the pendulum in the stochastic layer $\tau_m \simeq \pi/2\mu$ as being analogous to the mean-free time.

The largest time scale in our system is the time describing its relaxation to an equilibrium distribution τ_R. For the case of the reflecting boundary which we have assumed, τ_R may be approximated by a mean approaching time to the boundary by the diffusion process, starting from the middle of the stochastic layer. From this we obtain $\tau_R \simeq c_0'^2/D_1 \simeq 1/\mu^3$.

In the weak-coupling limit, these time scales are related by the inequality,

$$\tau_s \ll \tau_i \ll \tau_m \ll \tau_R \tag{62}$$

Note in particular that the above differs from the time-scale relationship in a dilute gas, for which $\tau_m \simeq \tau_R$.[17] In this sense, the diffusion process in the elastic pendulum evolves much slower than it does those occurring in large systems. It is clear the number of degrees of freedom plays an essential role in determining the rate of diffusion. A determination of the explicit nature of dependence remains an interesting problem.

Finally, we summarize our results. Using Prigogine's perturbation theory, we have shown that there exist well-defined first and second moments of observables in the vicinity of the unperturbed separatrix of the weakly coupled elastic pendulum. Even though these moments are very small as discussed just above, this result suggests that an "embryo" of dissipative process already exists in nonintegrable conservative systems with two degrees of freedom.

At the same time, however, we have shown that there exists a logarithmic singularity at the separatrix, which may affect higher-order moments of observables. This may in turn lead to motion qualitatively different from the simple diffusion processes found in dilute gaseous systems. We shall address this problem in our continuing investigations.

ACKNOWLEDGMENTS

The author wishes to thank Professor I. Prigogine for his constant interest, encouragements, and helpful suggestions during this work. He wants also to acknowledge numerous fruitful discussions with Professors R. Balescu, Cl. George, A. Grecos, G. Nicolis, L. Reichl, and W. Schieve, and Drs. C. Baesens, R. Snapp, and J. Tennyson. Part of this work has been supported by the Robert Welch Foundation (Texas).

APPENDIX: EXPANSION OF $dn^2(u, c)$ AT $c = c_a$

Here we prove the relation, Eq. (33). We start with Eq. (32). The Fourier expansion of the first term of Eq. (32) is obtained by just replacing Δl and l_n

by Δk and k_n, respectively, in Eq. (21). For the second term of Eq. (32), we have

$$\frac{\partial}{\partial c}\mathrm{dn}^2(u,c) = \frac{d}{dk}\left(\frac{E}{K}\right) - \frac{dK'}{dc}\Delta l \sum_{n=-\infty}^{+\infty}{}' \frac{4l_n^2 \coth(2l_n K')}{\sinh(2l_n K')} e^{2il_n u} + I(c) \quad (A1)$$

Here $I(c)$ is the term including the derivative of Δl and l_n by c. After a few manipulations we have

$$I(c) = -\frac{1}{K}\frac{dK}{dc}\left[\frac{\partial}{\partial x}\left(\mathrm{dn}^2(ux,c) - \frac{\Delta l}{x^2}\sum_{n=-\infty}^{+\infty}{}' \frac{2l_n}{\sinh(2l_n K'/x)} e^{2il_n u}\right)\right]_{x=1} \quad (A2)$$

Introducing a new modulus c_1, and K_1, K'_1 and E_1 by

$$\frac{K'_1}{K_1} = \frac{K'(c_1)}{K(c_1)} = \frac{K'(c)}{xK(c)}, \qquad E_1 = E(c_1) \quad (A3)$$

we can rewrite the second term in Eq. (A2) and obtain

$$I(c) = -\frac{1}{K}\frac{dK}{dc}\left\{\frac{\partial}{\partial x}\left[\mathrm{dn}^2(ux,c) - \left(\frac{K'_1}{K'}\right)\left(\mathrm{dn}^2\left[\frac{K'_1}{K'}ux,c_1\right] - \frac{E_1}{K_1}\right)\right]\right\}_{x=1} \quad (A4)$$

Using the expansion for $c' \ll 1$,

$$K = \frac{1}{2}\ln\frac{16}{c'} + \left(\frac{1}{2}\right)^2\left(\frac{1}{2}\ln\frac{16}{c'} - \frac{2}{1\cdot 2}\right)c' + \cdots \quad (A5a)$$

$$K' = \frac{\pi}{2}\left[1 + \left(\frac{1}{2}\right)^2 c' + \left(\frac{1\cdot 3}{2\cdot 4}\right)^2 c'^2 + \cdots\right] \quad (A5b)$$

$$E = 1 + \frac{1}{2}\left(\frac{1}{2}\ln\frac{16}{c'} - \frac{1}{1\cdot 2}\right)c' + \frac{1^2\cdot 3}{2^2\cdot 4}\left(\frac{1}{2}\ln\frac{16}{c'}\frac{2}{1\cdot 2} - \frac{1}{3\cdot 4}\right)c'^2 + \cdots \quad (A5c)$$

and the relations

$$\frac{dE}{dc} = \frac{E-K}{2c}, \qquad \frac{dK}{dc} = \frac{E-c'K}{2cc'} \quad (A6)$$

we have

$$c'_1 = 16\exp\left[x\ln\left(\frac{c'}{16}\right)\right] + O(c'^2 \ln c') \quad (A7)$$

and

$$\frac{d}{dc}\left(\frac{E}{K}\right) = -\frac{1}{2} - \frac{1}{2c'K^2} + O\left(\frac{1}{\ln c'}\right) \quad (A8)$$

$$-\frac{1}{K}\frac{dK}{dc} = \frac{1}{2c'K} + O(1) \quad (A9)$$

Substituting Eqs. (A5)–(A9) into Eq. (A4) and expanding the dn^2 function by c', we obtain

$$I(c) = \frac{1}{2c'K^2} + \frac{1}{2} - \frac{1}{2}\mathrm{dn}^2(u, c) + \frac{\partial}{\partial c}\mathrm{dn}^2(u, c) - \frac{1}{4}\left(\frac{\partial}{\partial x}\mathrm{dn}^2(ux, c)\right)_{x=1} + O(c')$$
(A10)

Note that the first term in Eq. (A10) diverges in the limit $c \to 1$. Now we use the expansion of $\mathrm{dn}^2(u, c)$ for $c' \ll 1$,[16]

$$\mathrm{dn}^2(u, c) \simeq \mathrm{sech}^2 u + \frac{c'}{2}(1 - \mathrm{sech}^2 u + u \tanh u\, \mathrm{sech}^2 u) \quad \text{(A11)}$$

Then, we see that the terms in Eq. (A10), except for the first term, cancel each other. Furthermore, the first term in Eq. (A10) cancels with the first term in the right-hand side of Eq. (A1) [cf. Eq. (A8)].

Finally, we obtain

$$\frac{\partial}{\partial c}\mathrm{dn}^2(u, c) = -\frac{1}{2} + \frac{\pi}{2}\Delta l \sum_{n=-\infty}^{+\infty}{}' \frac{l_n^2 \coth[\pi l_n(1 + c'/4)]}{\sinh[\pi l_n(1 + c'/4)]}e^{2il_n u} + O\left(\frac{1}{\ln c'}\right)$$
(A12)

where we have used the expansion Eq. (A5b).

Substituting the Fourier expansion of $\mathrm{dn}^2(u, c_a)$, Eq. (21) and Eq. (A12), into Eq. (32), we have an expansion of $\mathrm{dn}^2(u, c)$. This expansion coincides with Eq. (33) to first order of $(c' - c_a')$, and we have the desired result.

REFERENCES

1. V. I. Arnol'd, *Usp. Mat. Nauk* **18**, 9 (1963) [*Russ. Math. Surv.* **18**, 85 (1963)].
2. R. Balescu, *Statistical Mechanics of Charged Particles* (Interscience, New York, 1963).
3. R. Balescu, *Equilibrium and Non-Equilibrium Statistical Mechanics* (Wiley, New York, 1975).
4. A. B. Channon and J. L. Lebowitz, In *Annals of the New York Academy of Sciences* **357**, 108 (1980).
5. B. V. Chirikov, *Phys. Rep.* **52**, 263 (1979).
6. J. Ford, In *Fundamental Problems in Statistical Mechanics III*, E. D. G. Cohen, ed. (North-Holland, Amsterdam, 1975), p. 215.
7. H. G. Helleman, In *Fundamental Problems in Statistical Mechanics*, E. D. G. Cohen, ed. (North-Holland, Amsterdam, 1980), p. 165.
8. M. Hénon and C. Heiles, *Astron. J.* **69**, 73 (1964).
9. P. J. Holmes, *SIAM Appl. Math.* **18**, 65 (1980).
10. C. F. Karney, Report *PPPL*-1938, 1982, 1.
11. A. N. Kolmogorov, *Dokl. Akad. Nauk.* **98**, 527 (1954) (English translation, Los Alamos Scientific Laboratory Translation LA-TR-71-67).
12. V. K. Melnikov, *Moscow Math. Soc.* **12**, 1 (1963).
13. B. Misra and I. Prigogine, In *Long-Time Prediction in Dynamics*, C. W. Horton, L. E. Reichl, and A. G. Szebehely, eds. (Wiley, New York, 1983).
14. J. Moser, *Nach. Akad. Wiss. Gottingen II Math. Phys. Kl.* **1**, 1 (1962).

15. J. Moser, *Stable and Random Motions in Dynamical Systems* (Princeton University Press, Princeton, New Jersey, 1973).
16. E. H. Neville, *Jacobian Elliptic Functions* (Clarendon Press, Oxford, 1951), Chap. XV.
17. T. Y. Petrosky and W. C. Schieve, *J. Stat. Phys.* **28**, 711 (1982).
18. I. Prigogine, *Non-Equilibrium Statistical Mechanics* (Interscience, New York, 1962).
19. H. E. Ranch and A. Liebowitz, *Elliptic Functions, Theta Functions, and Riemann Surfaces* (Williams & Wilkins, Baltimore, Maryland, 1973).
20. P. Resibois, In *Physics of Many-Particle Systems*, E. Meeron, ed. (Gordon and Breach, New York, 1966).
21. E. T. Whittaker and G. N. Watson, *A Course of Modern Analysis* (Cambridge University Press, Cambridge, 1946).

COLLISIONAL DIFFUSION IN A TORUS WITH IMPERFECT MAGNETIC SURFACES

R. B. White

Plasma Physics Laboratory, Princeton University, Princeton, New Jersey

Abstract. *A Hamiltonian formulation of the guiding center drift equations is used to investigate the modification of neoclassical diffusion for low collisionality in a toroidal magnetic field with partially destroyed magnetic surfaces. The magnetic field is assumed to be given by the small perturbation of an axisymmetric system. The results are applicable to particle diffusion in realistic confinement systems, midway between axisymmetric and purely stochastic ones. Significant enhancement of electron diffusion over neoclassical rates is found. This increase can be accounted for by the contributions due to the first few island chains in the Fibonacci sequence generated by the zero-order island and by associated stochastic domains.*

1. INTRODUCTION

In a recent series of papers[1-3] a Hamiltonian formulation of the guiding center drift equations has been developed that is applicable to toroidal systems in which the magnetic surfaces are partially or wholly destroyed. These equations were used to investigate numerically the modification of neoclassical diffusion as a function of the number and density of magnetic islands perturbing an initially axisymmetric equilibrium. In the limit of large perturbation and magnetic-island overlap, the field becomes stochastic and the random-phase approximation can be used to obtain analytic expressions for the diffusion, which depends on ratios of collision lengths and correlation lengths.[4,5] Of more interest is the case of small nonoverlapping magnetic islands, which probably represents more closely the situation existing in actual confinement devices. Unfortunately, for very low collisionality and small gyro radius such a system is difficult to investigate through Monte Carlo simulation because of the very small diffusion rates encountered. For typical fusion-oriented confinement parameters a significant modification of neoclassical diffusion is found for electrons, but not for ions.

A simple estimate of neoclassical diffusion in the collisionless regime leads to the banana diffusion rate,[6,7] $D_{NC} \sim \Delta_b^2 \nu_{eff}$, where Δ_b is the banana width and ν_{eff} is the effective collision frequency. For an equal-energy electron–ion plasma the ratio of the two rates is given by $(m_e/m_i)^{1/2}$. In a steady-state axisymmetric system momentum conservation implies that the rates are in fact equal and close to the electron value,[8] and in a nonaxisymmetric system charge neutrality implies that a radially inward electric field develops and ensures ambipolar diffusion on the average. What is observed experimentally is that the ion rates are approximately neoclassical and the electron rates approximately two orders of magnitude $[\sim(m_i/m_e)^{1/2}]$ anomalously large. One possible explanation is that field perturbations preferentially act on electrons and allow them to diffuse at the ion neoclassical rate. In this work the mechanisms whereby the diffusion rates are modified below stochastic threshold are examined, and a means of numerical estimation of diffusion rates that does not involve Monte Carlo simulation is described. The method involves an approximate division of the torus into shells in which the magnetic surfaces are topologically toroidal, island chain, or stochastic in nature. Particle diffusion is then estimated in each shell, and the results are used to estimate an average diffusion rate for the configuration. Below stochastic threshold results are in reasonable agreement with Monte Carlo simulations. In Section 2 the Hamiltonian formulation of the guiding center drift equations is reviewed. In Section 3 the magnetic field is divided into toroidal bands of different field structure. In Section 4 the modification of the neoclassical rate due to field stochasticity and magnetic islands is calculated, and used to model the diffusion in a general system.

2. GUIDING CENTER DRIFT EQUATIONS

It is particularly convenient to use magnetic coordinates for confinement studies. Besides defining the surface across which diffusion is relevant, these coordinates separate fast and slow time scales related to motion along and across the field. They are also intimately related to the canonical variables for the guiding center motion. Of course a stochastic field has no such coordinates, but the systems of interest for confinement studies are necessarily only small perturbations of systems with good magnetic surfaces. A nearby field **B**, with well-defined magnetic surfaces, is therefore assumed to exist.

Two representations for **B**, the familiar Clebsch representation and a covariant representation, are used:

$$\mathbf{B} = \nabla \psi \times \nabla \theta_0 = \lambda(\nabla \chi + \beta^* \nabla \psi) \quad (1)$$

The variable ψ labels the flux surfaces, and θ_0 is a measure of the angle around a flux surface across the field. The quantity λ is the permeability, defined by the condition $\nabla \psi \cdot \nabla \times (\mathbf{B}/\lambda) = 0$. For scalar pressure equilibria $\lambda = 1$, and Grad[9] has shown that for tensor pressure equilibria $1/\lambda = 1 + (P_\perp - P_\parallel)/B^2$. The quantity β^* is related to the parallel current and pressure. For a scalar pressure equilibrium, the pressure balance $\mathbf{J} \times \mathbf{B} = \nabla P$ gives $P = P(\psi)$ with

$$\frac{\partial P}{\partial \psi} = B^2 \frac{\partial \beta^*}{\partial \chi} \quad (2)$$

Furthermore, the parallel current is given by

$$\mathbf{B} \cdot \mathbf{J} = B^2 \frac{\partial \beta^*}{\partial \theta_0} \quad (3)$$

The variable χ measures the distance along a field line.

Given any explicit representation of **B** in toroidal coordinates r, θ, and ϕ, the magnetic coordinates can be found by comparison with Eq. (1). Consider a torus of aspect ratio $\varepsilon^{-1} = R/a$ with R and a the major and minor radii. The simplest nontrivial toroidal system is given by the truncation to first order ε of a zero-pressure equilibrium[10]

$$\mathbf{B} = \frac{r}{q(r)} \nabla \phi \times \nabla r + R \nabla \phi \quad (4)$$

where **B** is normalized to its value on axis. The coordinate r labels the magnetic surface, and it is convenient to take $\psi = r^2/2$, where r is normalized to the minor radius. The representation Eq. (1) then gives the field described by Eq. (4) to first order in ε with

$$\theta_0 = \theta - \frac{\phi}{q} - \varepsilon r \sin \theta \quad (5)$$

$$\chi = \frac{\phi}{\varepsilon} + \frac{\varepsilon r^2 \theta}{q} \quad (6)$$

$$\beta^* = -\frac{\varepsilon \theta}{r} \frac{d}{dr}\left(\frac{r^2}{q}\right) \quad (7)$$

These expressions are readily evaluated to higher order in ε using higher-order expressions for **B**.

Represent the exact magnetic field of interest through $\mathbf{B}_e = \mathbf{B} + \delta \mathbf{B}$, where **B** is described by Eq. (1) $\delta \mathbf{B}$ represents the effect of magnetic-field coils, plasma modes, etc. As has been shown,[11] one component of **B**, namely $\delta \mathbf{B} \cdot \nabla \psi$, dominates the structure of the perturbed surfaces. Other components of $\delta \mathbf{B}$ contribute only nonresonant distortions of the equilibrium and are unimportant. Represent this component exactly through

$$\delta \mathbf{B} = \nabla \times \alpha \mathbf{B} \tag{8}$$

with α being a general function of position. If α is represented through its Fourier series

$$\alpha = \sum_{m,n} \alpha_{nm}(r) \sin(n\phi - m\theta - \delta_{mn}) \tag{9}$$

we find that α changes the topology of $\mathbf{B}$ at rational surfaces $q = m/n$, producing islands of width

$$\Delta \psi = 4 \left| \frac{q^2 \alpha_{nm}}{\varepsilon (dq/d\psi)} \right|^{1/2} \tag{10}$$

The derivation of the Hamiltonian formulation of the guiding center drift equations has been given elsewhere.[1] They have the form

$$\dot{\chi} = \frac{\partial H}{\partial \rho_c}, \quad \dot{\rho}_c = -\frac{\partial H}{\partial \chi} \tag{11}$$

$$\dot{\theta}_c = \frac{\partial H}{\partial \psi}, \quad \dot{\psi} = -\frac{\partial H}{\partial \theta_c} \tag{12}$$

where the Hamiltonian is

$$H = \tfrac{1}{2} B^2 \rho_\|^2 + \mu B + \Phi \tag{13}$$

and

$$\rho_\| = \frac{V_\|}{\omega_c} = \frac{(2W - 2\mu B - 2\Phi)^{1/2}}{B} \tag{14}$$

is the "parallel gyro radius," where W is the kinetic energy, Φ is the electric potential, and μ is the magnetic moment, $\mu = \tfrac{1}{2} v_\perp^2 / B$. The canonical variables are χ, ρ_c, θ_c, and ψ with $\rho_c = \rho_\| + \alpha$ and $\theta_c = \theta_0 - \beta^* \rho_c$. Here, and in the following, lengths are given in terms of the minor radius and times in terms of the on-axis gyro frequency $\omega_0 = eB_0/mc$. In these units energy is given in terms of the basic unit $W_0 = ma^2 \omega_0^2$.

From the form of the Hamiltonian it is clear that only the magnitude of B enters the formalism. Given $B(\theta_0, \chi, \psi)$, $\beta^*(\theta_0, \chi, \psi)$, and $\alpha(\theta_0, \chi, \psi)$ the motion is completely prescribed. Realization of the trajectories in toroidal coordinates requires also Eqs. (5)–(7) or a similar set.

In the present work we will make use of the first-order model, Eq. (4), which gives

$$B = 1 - \varepsilon r \cos \theta \tag{15}$$

with $\theta = \theta_0 + \varepsilon \chi / q(\psi)$ and $\chi = \phi/\varepsilon$. Note that Eq. (5) for θ can be taken to zero order in ε since the $O(\varepsilon)$ term changes B to order ε^2. Furthermore, β^* can be neglected since it produces a correction to B of order $\varepsilon^2\rho_\|$. The equations of motion, which exactly conserve the Hamilitonian constructed of this model field B, then take the form

$$\dot{\chi} = \rho_\| B^2 \tag{16}$$

$$\dot{\rho}_c = -(\rho_\|^2 B + \mu)\frac{\partial B}{\partial \chi} - \frac{\partial \Phi}{\partial \chi} + \rho_\| B^2 \frac{\partial \alpha}{\partial \chi} \tag{17}$$

$$\dot{\theta}_0 = (\rho_\|^2 B + \mu)\frac{\partial B}{\partial \psi} - \frac{\partial \Phi}{\partial \psi} - \rho_\| B^2 \frac{\partial \alpha}{\partial \psi} \tag{18}$$

$$\dot{\psi} = -(\rho_\|^2 B + \mu)\frac{\partial B}{\partial \theta_0} - \frac{\partial \Phi}{\partial \theta_0} + \rho_\| B^2 \frac{\partial \alpha}{\partial \theta_0} \tag{19}$$

To study particle diffusion we introduce energy-conserving pitch-angle scattering. The pitch $\lambda = V_\|/V$ is changed every time step by

$$\lambda' = \lambda(1 - \nu\tau) \pm [(1 - \lambda^2)\nu\tau]^{1/2} \tag{20}$$

where ν is the effective collision frequency in units of ω_0, τ is the time step with $\nu\tau \ll 1$, and the ($\pm$) implies a random sign. This can be shown to give an equivalent Lorentz collision operator.[12]

In our units the neoclassical banana diffusion rate[6,7] is

$$D_{NC} = (2\varepsilon r)^{1/2}(0.7)q^2\frac{W\nu}{\varepsilon^2} \tag{21}$$

where this refers to diffusion in the variable ψ. The units of time refer to the gyro period of the species. Thus, for two species of equal energy $D_{NC} \sim W\nu$ per cycle or $W\nu\omega_0 \sec^{-1}$, and scaling with mass is $W \sim m$, $\nu \sim m^{1/2}$, and $\omega_0 \sim 1/m$.

3. THE MAGNETIC FIELD

Owing to work by Kolmogorov,[13] Arnold,[14] and Moser[15] (KAM), it is known that for sufficiently small perturbation $\delta \mathbf{B}$, the nature of the perturbed magnetic field $\mathbf{B} + \delta \mathbf{B}$ can be approximately described in terms of topologically toroidal surfaces in all but a finite volume proportional to $\delta \mathbf{B}$. Consider a magnetic perturbation consisting of amplitudes α_{nm} for $(n, m) = (n_1, m_1)$ and (n_2, m_2). For very small amplitudes the magnetic surfaces consist of island chains at the two surfaces $q_1 = m_1/n_1$ and $q_2 = m_2/n_2$ with widths given by Eq. (10). We refer to these as the zero-order islands. In the approximation of infinite separation the space between these island chains consists of surfaces that are topologically toroidal. The complexity that develops in this region as the separation distance is decreased or the perturbation amplitudes are increased has been studied by many, and exhibits universal behavior, independent of the nature of the map induced by the field.[16-19] At all rational values of q between q_1 and q_2 there exist chains of magnetic islands. However, the width of these islands decreases exponentially as one progresses through the Fibonacci series[20] generated from q_1, q_2. There is one term of first order in this series,

$$q = \frac{m_1 + m_2}{n_1 + n_2} \qquad (22)$$

and successive terms in the series are given by combining lower-order terms in a similar manner. As a concrete example to illustrate these calculations consider the field of Ref. 2, with a q profile linear in ψ, $1 < q < 4$, and a zero-order perturbation with $\alpha_{nm}(\psi_r) = s$ for $m/n = 7/4$, $8/4$, $9/4$, with $q(\psi_r) = m/n$. A Poincaré plot of the region from $q = 8/4$ to $q = 9/4$ is shown in Fig. 1 for $s = 10^{-5}$, along with the locations of the island chains corresponding to the first three orders in the series. The exponential dependence of the island widths on the Fibonacci order is shown in Fig. 2.

Structure in particle orbits produces diffusion through the particle collisions. A collision changes the detailed nature of the particle orbit instantaneously, thus producing a step across flux surfaces the size of the orbit excursion away from the flux surface. Even with perfect flux surfaces the particles diffuse at a rate determined by some width Δ given, for example,

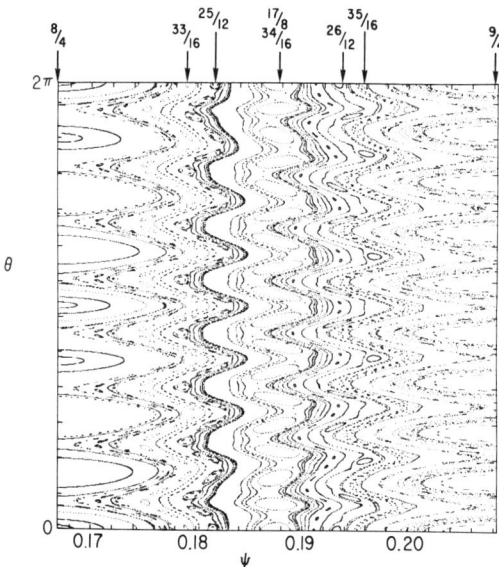

Figure 1. A poincaré plot of the magnetic field for a perturbation consisting of three modes with $n = 4$ and $m = 7, 8, 9$, with amplitude $\alpha = 10^{-5}$. Shown is the domain between $q = 8/4$ and $9/4$. Also visible are the higher-order islands associated with the Fibonacci series given by the $8/4$ and $9/4$ island chains.

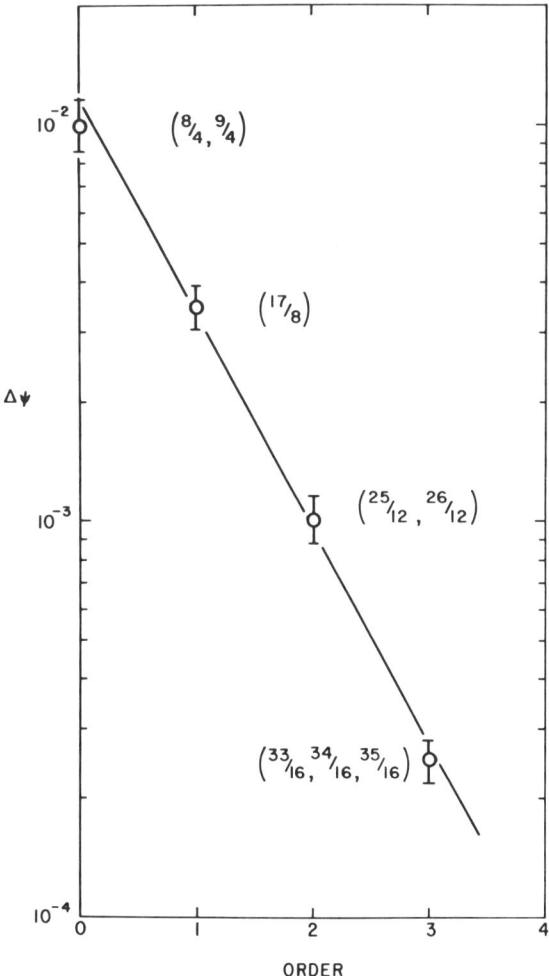

Figure 2. The island widths as a function of Fibonacci order, showing the exponential decrease.

by the banana width or the gyro radius, depending on the collision rate. Magnetic-field structure on a scale smaller than Δ is thus irrelevant, not because the particles do not sense such structure, but because diffusion due to it is small compared to that given by Δ. Ignoring structure smaller than a critical size Δ means that the exact field consists of a sequence of toroidal shells, where within each shell the field can be described as either toroidal, island, or stochastic. Thus the discrete scale associated with particle collisions simplifies the infinite complexity of the magnetic field. The exponential decrease in width means that few orders in the Fibonacci series are relevant. As the stochastic threshold for the field is approached, the island shells become more and more dense and finally break up into a stochastic sea surrounding the remaining islands. At this point the approximation of the field as topologically toroidal bands is no longer valid.

An island chain at the rational surface $q = m/n$ has associated with it a periodic orbit of length m, made up of the elliptic points of the islands in the chain, that is, m toroidal circuits are required before returning to the initial elliptic point. Consider the discrete map of points in the ψ, θ plane consisting of the transformation produced by one toroidal circuit. Orbits in the neighborhood of the elliptic points can be computed in the linear differential approximation. The domain of this approximation is called the tangent space. Each toroidal transit is then represented by a matrix M_κ, acting on the initial vector $(\delta\psi, \delta\theta)$. Over the full cycle of the periodic map the linear transformation is

$$M = \prod_1^m M_\kappa \tag{23}$$

Since $\nabla \cdot \mathbf{B} = 0$, this map is area preserving and hence $\det M = 1$. The eigenvalues of M depend only on its trace. Greene[18] has defined the residue

$$R = \frac{2 - \text{trace } M}{4} \tag{24}$$

Clearly in the limit of $\delta \mathbf{B} = 0$ the residue is zero. The eigenvalues of M are given in terms of the residue through

$$\lambda = 1 - 2R \pm 2[R(R-1)]^{1/2} \tag{25}$$

As δB increases from zero, the residue increases monotonically from zero. When R is between zero and one, λ is complex with magnitude one, and the tangent-space orbits, which continue over many toroidal circuits, rotate about the elliptic periodic points. Writing $\lambda = \exp(2\pi i/q_I)$, q_I is the ratio of toroidal transits divided by the number of revolutions about the periodic points. As the residue increases monotonically, q_I decreases monotonically from infinity. This behavior appears to be independent of the details of the map M.

In the case of analytically given discrete maps, the residue associated with an island chain can be calculated analytically, and the bifurcation of various-order periodic orbits studied as a function of perturbation amplitude. This permits the calculation of the extent of stochastic domains. In the present case the matrix M must be obtained by integration of the differential equations (16)–(19), and even the location of the periodic elliptic points is generally beyond analytic representation. However, the different domains—toroidal, island, and stochastic—can be differentiated numerically. Consider two nearby points in the ψ, θ plane with separation

$$\boldsymbol{\delta} = \mathbf{r}_1 - \mathbf{r}_2 = \delta_0(\cos\theta_I, \sin\theta_I) \tag{26}$$

Advance the points together toroidally. If they are within an island, they rotate about each other on the average, with $\Delta\theta_I = \Delta\phi/q_I$. If they are within a stochastic domain, on the average they separate exponentially with $\delta = \delta_0 \exp(hR\Delta\phi)$, with h the Kolmogorov entropy. Finally, if good magnetic surfaces exist on the scale of the separation of the points, they will not rotate, and will separate only linearly due to the difference in shear at the two surfaces. The Kolmogorov entropy is a statistical quantity and must be averaged over various initial conditions.[21] Structure on a scale smaller than Δ is not relevant, and thus it is convenient to average h over a range of $\Delta\psi$ of this magnitude. The island rotational transform q_I is, on the other hand, well defined locally. In Fig. 3 is shown the resulting numerical determination of the island internal q_I for the field of Fig. 1. The method suffices to pick out the first four orders in the Fibonacci series. An accurate determination of q_I is made by advancing in ϕ until the trajectory rotates completely around the periodic point. The largest value of q_I that can be observed is restricted by the number of toroidal circuits made. In Fig. 4 is shown the determination of the Kolmogorov entropy using the same procedure. It is found that significant stochastic domains exist in the vicinity of the separatrices of the largest islands, that is, the 8/4 and 9/4 chains, as well as the 17/8. For these determinations an initial separation of $\delta_0 = 10^{-7}$ was used, a small value being necessary to allow exponentiation within a restricted domain. For q_I the results are insensitive to the separation as long

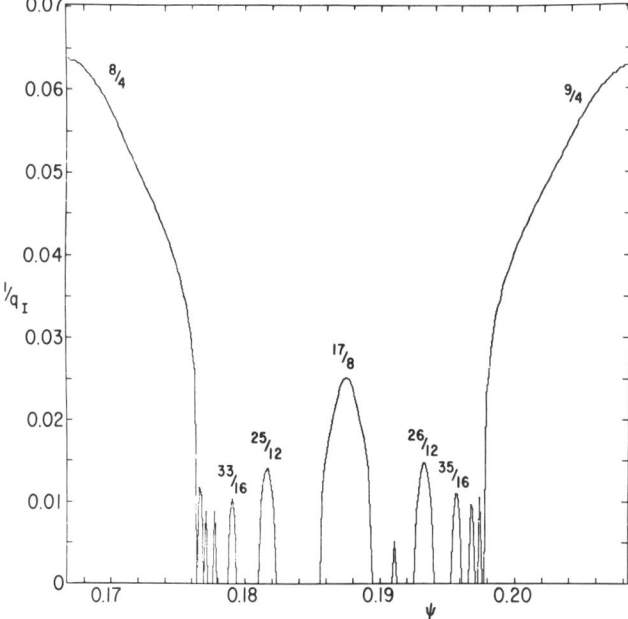

Figure 3. The rotational transform of the islands of Fig. 1, determined by following the trajectories of two closely spaced points.

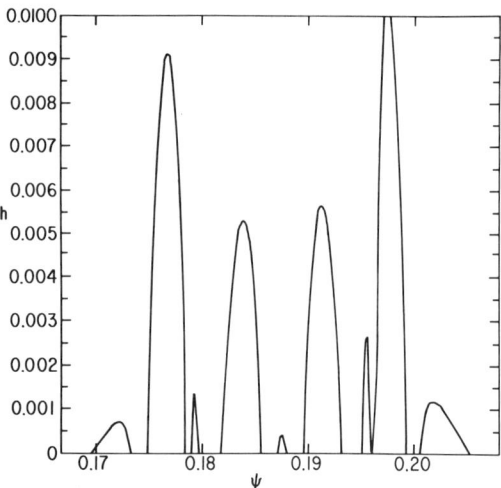

Figure 4. The Kolmogorov entropy for the field shown in Fig. 1. The values have been averaged over a range of $\nabla\psi = 3 \times 10^{-4}$, corresponding to an electron banana width. Values below $\sim 8 \times 10^{-3}$ are not significant.

as δ_0 is small compared to Δ, the smallest structure relevant. Owing to the finite number of toroidal circuits used to determine h, there is a minimum value, given by nonstochastic orbit excursions, below which h is not significant.

The universality of the structure between two nearby island chains[16-19] probably means that the distribution of island and stochastic domains shown in Figs. 3 and 4 is universal, and could be obtained for an arbitrary system using a simple map. In this case the island and stochastic domains could be described for any system using only the residues of the largest island chains.

4. DIFFUSION DUE TO STOCHASTICITY AND ISLANDS

Determinations of low-collisionality particle confinement using Monte Carlo techniques are prohibitive for large-scale systems involving very many modes. Because the diffusion rate of the electrons, under typical fusion parameters, can vary by three or four orders of magnitude with small changes in the field amplitudes, it is useful to have a means of numerically estimating the diffusion rate. Modifications of particle diffusion rates due to stochasticity and magnetic islands can be calculated to lowest order in gyro radius.

The contribution to radial diffusion due to stochasticity of the magnetic field is estimated by making a random-phase approximation for the effect of the various modes. Here we review the method by expressing the equations of motion in terms of a discrete map. Examine the drift orbit equations (16)–(19) to zero order in ρ, ignoring corrections of order ε and the ψ dependence of α. Write $\Delta\theta = \theta - n\phi/m$ and expand q in ψ around ψ_r, with $q(\psi_r) = m/n$, giving

$$\frac{d(\Delta\theta)}{d\phi} = -\frac{n^2}{m^2}q'(\Delta\psi) \tag{27}$$

$$\frac{d(\Delta\psi)}{d\phi} = -\frac{m}{\varepsilon}a_{nm}\cos(m\Delta\theta) \tag{28}$$

Now replace these differential equations with discrete time-step equations, stepping through $\Delta\phi = 2\pi/n$, that is, $\Delta t = 2\pi/(n\varepsilon\rho_\parallel)$. Substituting $x = -2\pi n/mq(\Delta\psi)$, $y = m\Delta\theta$ gives the equivalent area-preserving discrete map

$$y' = y + x \tag{29}$$

$$x' = x + k\cos y' \tag{30}$$

with $k = (2\pi)^2 \alpha_{nm} q/\varepsilon$. This has the form of the Taylor–Chirikov map.[16] In the limit of large k there is stochastic motion in x. The magnitude can be calculated by using the random-phase approximation, that is, $(\delta x)^2 = k^2 \cos^2 y \approx k^2/2$, giving for the diffusion rate in ψ, $D_s = (\delta\psi)^2/2t$,

$$D_s = \pi\left(\frac{m}{n}\right)^2 n\alpha_{nm}^2 |\rho_\parallel|/\varepsilon \tag{31}$$

For several modes this expression must be summed over m and n, and for a distribution of particles it must be averaged over the directions of **V**. For an isotropic velocity distribution the average is $\langle|\rho_\parallel|\rangle = \sqrt{2W}/\pi$.

As discussed in Section 3 the effect of magnetic islands is significant only if the island size is large compared to the magnitude of the orbit excursion from the magnetic surface, Δ. Thus to estimate the effect of the islands the particle can be considered to be following the field line. Provided the collision time is long compared to the time it takes for the particle to circumnavigate the island elliptic point, the diffusion rate due to the island can be estimated as

$$D_I = \frac{W^2 \nu}{2} f \tag{32}$$

with W being the island width and f being the fraction of particles with sufficiently long collision time. The time required to move around an island is $t_I = \pi q_I/\varepsilon\rho_\parallel$ and thus the fraction of particles f is given by all those with $|\rho_\parallel| > \pi\nu q_I/\varepsilon$. Approximating the distribution as isotropic we have $f = 1 - \pi\nu q_I/\varepsilon(2W)^{1/2}$ provided this is positive, and zero otherwise.

These results can be used to estimate the diffusion rate over the whole domain. In Fig. 5 is shown the local diffusion rate as a function of ψ, normalized to the neoclassical rate, for 1-keV electrons in the field of Fig. 1.

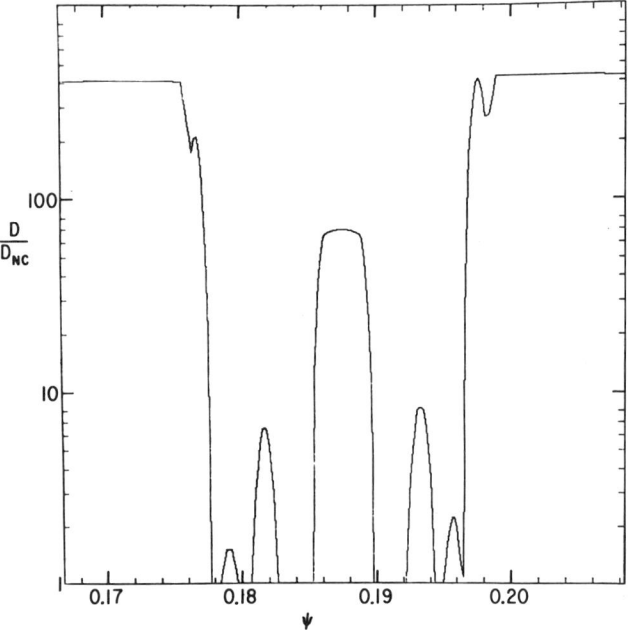

Figure 5. The local electron diffusion rate, for the field of Fig. 1, normalized to the neoclassical value, obtained using island and stochastic values as given in Section 4. The average rate for the whole domain is $D/D_{NC} \approx 4$.

The electrons have banana width $\Delta\psi \approx 3 \times 10^{-4}$, small compared to the islands of the first three Fibonacci orders. For this energy and collisionality the major island chains and stochastic domains have effectively infinite diffusion rates, and the higher-order Fibonacci islands also contribute significantly. The situation is considerably different for 1-keV deuterium ions, with banana width $\nabla\psi \approx 2 \times 10^{-2}$. Even the large islands have small effect on the diffusion, and higher-order islands and stochasticity are irrelevant.

Assuming steady-state diffusion the local density gradient is inversely proportional to the local diffusion rate, and an average diffusion rate for the whole domain is given by

$$\frac{1}{\langle D \rangle} = \int d\psi \frac{1}{D(\psi)} \tag{33}$$

This gives for 1-keV electrons in the field of Fig. 1, $\langle D \rangle = 4D_{NC}$. Values of $\langle D \rangle$ obtained in this manner are shown on Fig. 6. Good agreement with Monte Carlo determinations is obtained below stochastic threshold, where the main assumption involved in the calculation, that is, that the different domains form distinct toroidal bands, is well satisfied. This property also

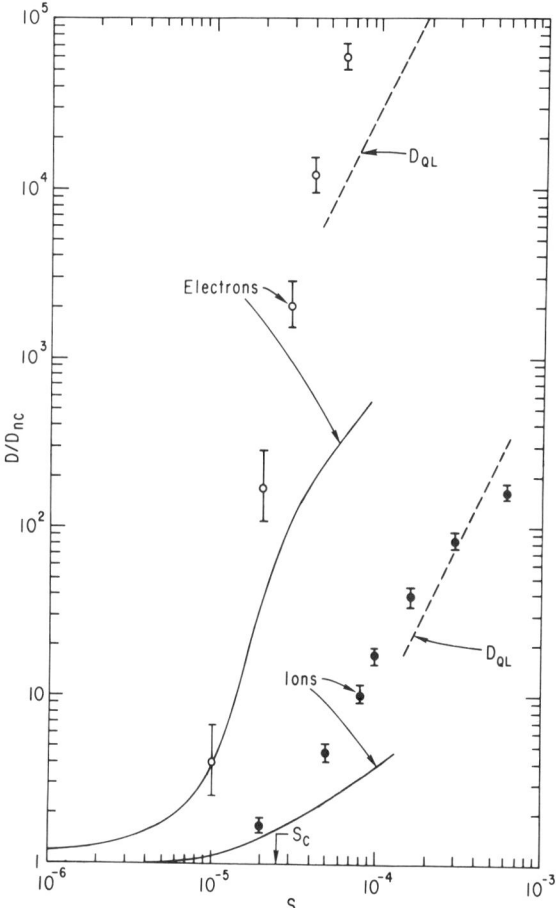

Figure 6. Modification of neoclassical banana diffusion due to magnetic perturbations. The perturbed field $\delta \mathbf{B}$ consisted of three modes with $n = 4$ and $m = 7, 8, 9$, and amplitude $\delta B/B = S$. The points are results of Monte Carlo simulations from Ref. 3, and the curves are from the numerical estimates, discussed in Section 4. Here $W = 3.5 \times 10^{-9}$ and $\nu = 2 \times 10^{-8}$ for the electrons, and $W = 1.25 \times 10^{-5}$ and $\nu = 10^{-6}$ for the ions.

makes the numerical estimation of the different domains insensitive to the value of θ at which the Kolmogorov entropy and islands are calculated. Above stochastic threshold, however, the Poincaré plot takes on the form of a stochastic sea containing isolated islands. In this configuration the diffusion rate across the islands is slower than that in the stochastic sea, but particles can easily diffuse around them, and thus their effect is overestimated by our analysis.

Another point of interest is the effect of magnetic-field fluctuations. Fluctuations have an effect similar to collisions, since they can move particle orbits radially along distances given by the field structure and in times given by the fluctuation frequency. Thus, it should be possible to include the effect of rapidly fluctuating fields through an effective collision frequency.

A confined plasma will quickly develop an electrostatic potential, which makes the diffusion ambipolar, and the correct particle diffusion rate must be determined by the solution of this self-consistent problem. However, the potential will have a strong effect on the low-energy particles responsible for heat conduction. Thus, the results obtained here, neglecting electrostatic fields, are of interest in the analysis of the anomalously large electron thermal conductivity.

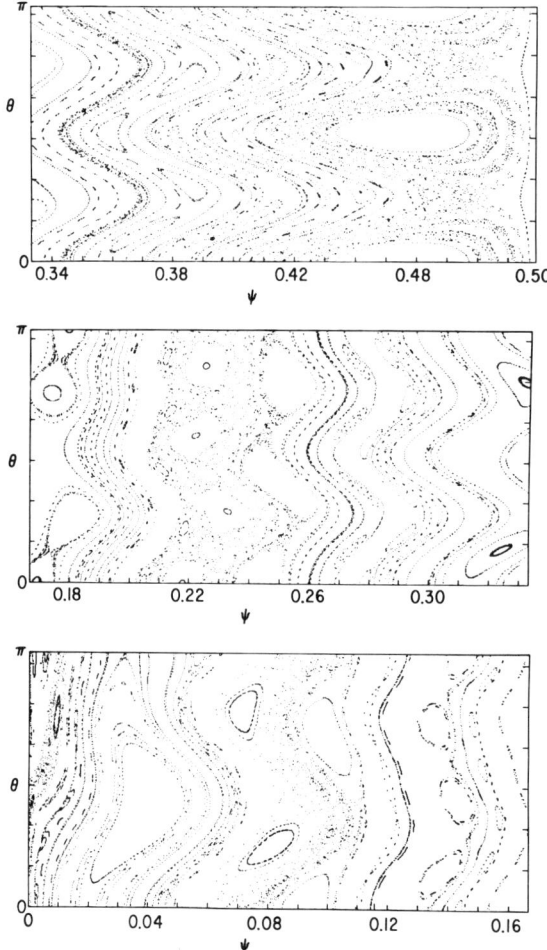

Figure 7. The Poincaré plot resulting from the magnetic-field perturbations present during large-amplitude sawtooth oscillations.

Whether or not the anomalous energy confinement times observed in tokamaks can be described by the effect of magnetic fluctuations cannot be answered without a detailed knowledge of the magnetic field fluctuation spectrum. As seen in Fig. 6 the electron diffusion rate is very sensitive to the proximity of the field to stochastic threshold. To date the only reliably known perturbation spectra are those resulting from long-wavelength MHD modes. An example of the magnetic field produced by large-amplitude sawtooth oscillations is shown in Fig. 7. There are 30 modes present, with amplitudes α_{nm} ranging from 10^{-7} to 10^{-4}, with average amplitude 1.5×10^{-5}. The harmonic structure is $1 \le n \le 4$ and $1 \le m \le 12$. There is no reason to suppose such long-wavelength modes should be responsible for the large electron thermal conductivity; this case is used merely as an example to illustrate the method. Determination of the electron diffusion rate for 1-keV electrons over the whole field required 20 min of CRAY computer time. The average was $\langle D \rangle = 2 D_{NC}$. This result would, of course, be further increased by the presence of any other magnetic fluctuations. Estimation of diffusion rates in large systems with many modes such as this using Monte Carlo codes is beyond present computing capabilities.

In conclusion, the modification of neoclassical diffusion rates below stochastic threshold observed with Monte Carlo simulation is entirely accounted for by the contributions from the islands related to the first few terms in the Fibonacci series derived from the initial perturbation spectrum, and the stochastic bands associated with the separatrices of the largest islands. These islands, and the stochastic bands, can be found numerically with a minimal amount of computing, and the resulting estimate of diffusion rates agrees well with Monte Carlo simulations. A quantitative comparison with experimentally observed electron-energy confinement times must await more precise knowledge of the fluctuation spectrum existing in magnetic confinement devices.

ACKNOWLEDGMENTS

This work was begun while on vacation in Cividale del Friuli. The author is indebted to the International Centre for Theoretical Physics, Trieste, for the use of the library. Many useful conversations were had with John Greene, Charles Karney, Robert McKay, and Ted Stringer. This work was supported by U.S. Department of Energy Contract No. DE-AC02-76-CHO-3073.

REFERENCES

1. R. B. White, A. H. Boozer, and R. Hay, *Phys. Fluids* **25**, 575 (1982).
2. A. H. Boozer and R. B. White, *Phys. Rev. Lett.* **49**, 786 (1982).

3. R. B. White, A. H. Boozer, R. J. Goldston, R. Hay, J. Albert, and C. F. F. Karney, in *Proceedings of the Ninth International Conference on Plasma Physics and Controlled Nuclear Fusion Research*, Baltimore, MD (1982), paper IAEA-CN-41/T-3.
4. A. B. Rechester and M. N. Rosenbluth, *Phys. Rev. Lett.* **40**, 38 (1978).
5. J. A. Krommes, R. G. Kleva, and C. Oberman, *J. Plasma Phys.* **30**, 11 (1983).
6. M. N. Rosenbluth, R. D. Hazeltine, and F. L. Hinton, *Phys. Fluids* **15**, 116 (1972).
7. F. L. Hinton and R. D. Hazeltine, *Rev. Mod. Phys.* **48**, 239 (1976).
8. P. H. Rutherfold, *Phys. Fluids* **13**, 482 (1970).
9. H. Grad, *Proc. Symp. Appl. Math.* **18**, 162 (1967).
10. J. M. Greene, J. L. Johnson, and K. E. Weimer, *Phys. Fluids* **14**, 67 (1971).
11. M. N. Rosenbluth, R. Sagdeev, J. B. Taylor, and G. M. Zaslavsky, *Nucl. Fusion* **6**, 297 (1966).
12. A. H. Boozer and G. Kuo-Petravic, *Phys. Fluids* **24**, 851 (1981).
13. A. N. Kolmogorov, in *Proceedings of the International Congress of Mathematicians*, Amsterdam (North Holland, Amsterdam, 1957), Vol. 1, p. 315.
14. V. I. Arnol'd, *Rus. Math. Sur.* **18**, 9 (1963).
15. J. Moser, Nachr. *Akad. Wiss Gotinen II Math. Phys. Kl* **1** (1962).
16. B. V. Chirikov, *Phys. Rep.* **52**, 263 (1979).
17. M. J. Feigenbaum, *J. Stat. Phys.* **21**, 669 (1979).
18. J. M. Greene, *J. Math. Phys. (N.Y.)* **20**, 1183 (1979).
19. R. S. MacKay, Ph.D. thesis (Princeton University, 1982).
20. Fibonacci, Liber Abaci, Pisa (1202) in *Opuscoli di Leonardo Pisano*, (Baldasarne Boncompagni, Firenze, 1956).
21. A. B. Rechester, M. N. Rosenbluth, and R. B. White, *Phys. Rev.* **A23**, 2664 (1981).

APPLICATIONS OF FRACTAL ANALYSIS TO PHASE TRANSITIONS AND OTHER PHENOMENA

Masuo Suzuki

Department of Physics, University of Tokyo, Tokyo, Japan

Abstract. A geometrical interpretation of critical exponents is given on the basis of the concept of fractals. As an interesting physical application of fractals, the Levy process, namely fractional Brownian motion, is discussed by introducing a simple model, namely, string animals.

FRACTALS IN PHASE TRANSITION

Following Mandelbrot,[3] we introduce a fractional dimensionality D by the following similarity law:

$$V_d \propto \left(\frac{1}{r}\right)^D \qquad (1)$$

where V_d denotes the volume of the relevant fractal geometrical object in the ordinary topological dimensionality d and r is the scale of length. Here we assume that Eq. (1) holds with a constrant D for any limit of small r.

For convenience we give here a formal description of fractals.[6] We consider a series of figures $\mathscr{F}_0, \mathscr{F}_1, \ldots, \mathscr{F}_n, \ldots$. We assume that all these figures are *similar* in the sense that the nth figure $\mathscr{F}_n$ is constructed from $\mathscr{F}_{n-1}$ by the same procedure $\mathscr{T}$ as

$$\mathscr{F}_{n+1} = \mathscr{T}\mathscr{F}_n \qquad (2)$$

For example, see Fig. 1 in which a, b, and c correspond to $\mathscr{F}_0, \mathscr{F}_1$, and $\mathscr{F}_2$, respectively. The limiting figure $\mathscr{F}^*$, defined by

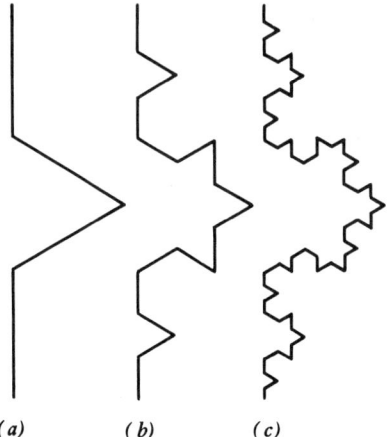

Figure 1. Koch curve: (a) $\mathscr{F}_0$, (b) $\mathscr{F}_1$, (c) $\mathscr{F}_2$.

$$\mathscr{F}^* = \lim_{n\to\infty} \mathscr{F}_n \qquad (3)$$

is the Koch curve.[3]

Now we consider an arbitrary geometrical quantity such as the total length, area, or volume associated with the figure $\mathscr{F}_n$ in an appropriate unit $b^{(n)}$ and we write it as $Q(\mathscr{F}_n; b^{(n)})$. Here, the unit $b^{(n)}$ is taken to be

$$b^{(n)} = b^{(0)}\left(\frac{1}{b}\right)^n \quad \text{for } b > 1 \qquad (4)$$

In the case of the Koch curve, we have $b = 3$. Then the fractal dimensionality d_Q of this quantity Q is defined by

$$Q(\mathscr{F}_{n+1}, b^{(n+1)}) = b^{d_Q} Q(\mathscr{F}_n; b^{(n)}) \qquad (5)$$

A critical exponent in phase transition is expressed by the fractal dimensionality of resultant dominant clusters at the critical point,[6] as in Fig. 2.

In general, the critical exponent is related to the size dependence of the relevant physical quantity at the critical point due to the finite-size scaling theory.[1,5] For example, the magnetization per unit volume of a ferromagnet with a finite size L takes the following scaling form:

$$m(L) \simeq L^{-\beta/\nu} m(h\varepsilon^{-\beta\delta}, \varepsilon L^{1/\nu}) \qquad (6)$$

where $\varepsilon = (T - T_c)/T_c$. Therefore, the total magnetization $M(L)$ at the critical point has the following size dependence:

$$M(L) \sim L^{d-\beta/\nu} \qquad (7)$$

as is well known.[2,6] On the other hand, the total susceptibility of the relevant finite system, $\chi(L)$, takes the following size dependence,

$$\chi_0 \sim L^d (L^{-1/\nu})^{-\gamma} \sim L^{d+\gamma/\nu} \qquad (8)$$

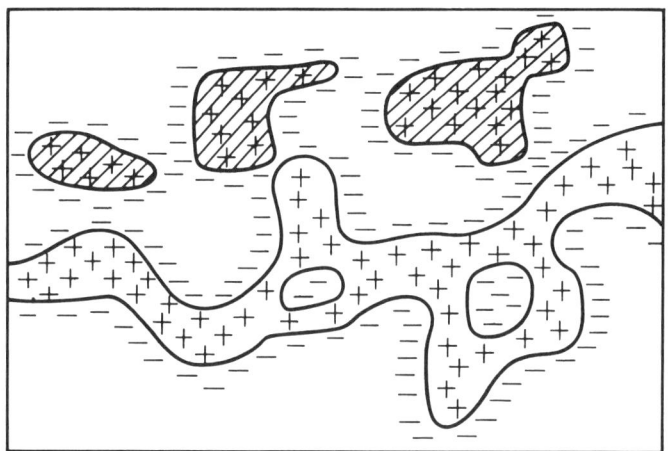

Figure 2. Resultant dominant clusters that are left by eliminating smaller clusters until the number of eliminated spins becomes $N/2$, N being the number of the total spins.

for the ordinary exponents ν and γ defined by the singularities of the correlation function ξ and susceptibility χ_0 with respect to $\varepsilon \equiv (T - T_c)/T_c$ as

$$\xi \sim \varepsilon^{-\nu} \quad \text{and} \quad \chi_0 \sim \varepsilon^{-\gamma} \tag{9}$$

The finite-size scaling theory insists that the temperature difference ε is scaled as $L^{-1/\nu}$ for a finite system. Consequently, relation (8) is derived from (9).

Now the purpose of our fractal analysis of critical phenomena is to derive, geometrically, scaling relations among critical exponents. For example, by the help of the concept of resultant dominant clusters shown in Fig. 2, namely, by noting that the total magnetization of the relevant finite system is governed by the resultant dominant cluster, we obtain the relation[6]

$$\chi_0 = \frac{\mu_B^2}{k_B T_c} \langle M^2 \rangle \simeq \frac{\mu_B^2}{k_B T_c} M^2(L) \tag{10}$$

at the critical point. Namely, we have

$$L^{d+\gamma/\nu} \sim (L^{d-\beta/\nu})^2 \tag{11}$$

with use of Eqs. (7) and (8). Thus, we arrive at the well-known scaling relation

$$2\beta + \gamma = d\nu \tag{12}$$

Similarly, as for the exponent of the specific heat, α, we obtain the following relation*:

---

* Equation (3.31) in Ref. 6 should read $C \sim V\varepsilon^{-\alpha} \sim VL^{\alpha/\nu} \sim L^{d+\alpha/\nu}$.

$$L^d(L^{-1/\nu})^{-\alpha} \sim (L^{1/\nu})^2 \tag{13}$$

namely,

$$2 - \alpha = d\nu \tag{14}$$

By combining Eqs. (12) and (14), we obtain the well-known scaling relation

$$\alpha + 2\beta + \gamma = 2 \tag{15}$$

In general, we consider a symmetry-breaking quantity Q_j and its conjugate field h_j. Now we define the following fractal dimensionalities $d(\chi_j)$ and $d(Q_j)$ as

$$\chi_j \sim L^{d(\chi_j)} \quad \text{and} \quad \Delta Q_j \sim L^{d(Q_j)} \tag{16}$$

Then, as before, our geometrical arguments lead to the following general relation:

$$d(\chi_j) = 2d(Q_j) \tag{17}$$

through the relation

$$\chi_j \simeq \langle(\Delta Q_j)^2\rangle \simeq \{\Delta Q_j(L)\}^2 \sim L^{2d(Q_j)} \tag{18}$$

This implies the following general scaling relation

$$\psi_j = d\nu - 2\phi_j \tag{19}$$

where ψ_j and ϕ_j are defined by

$$\Delta Q_j \sim L^d \varepsilon^{\phi_j} \quad \text{and} \quad \chi_j \sim L^d \varepsilon^{-\psi_j} \tag{20}$$

Thus, our geometrical interpretation of critical phenomena yields a very simple intuitive derivation of all the known scaling relations as well as new possible scaling relations for new possible symmetry-breaking quantities.

FRACTIONAL BROWNIAN MOTION IN PHYSICS

In the present section, we discuss the fractal scaling exponent with respect to time t. We assume that a physical quantity $Q(t)$ at time t takes the following asymptotic form:

$$Q(t) \sim t^\psi \tag{21}$$

for large t. Here, ψ is called a fractal scaling exponent, because Eq. (21) is scale invariant for the following scaling transformation:

$$t' = bt \quad \text{and} \quad Q' = b^\psi Q \tag{22}$$

in the limit of large t.

The most typical example is the Wiener process, whose correlation is defined by

$$\langle W(t)W(t')\rangle = 2D \min(t, t') \tag{23}$$

In particular, the fluctuation $\langle W^2(t) \rangle$ takes the form

$$\langle W^2(t) \rangle = 2Dt^\psi; \quad \psi = 1 \tag{24}$$

Although ψ itself is not fractional in the Wiener process, its sample path is fractal, because the infinitesimal increment ΔW is proportional to the root of the time increment Δt as

$$\Delta W \sim (\Delta t)^{1/2} \tag{25}$$

This relation is extended[3] as

$$\Delta B \sim (\Delta t)^H \tag{26}$$

Correspondingly, the fluctuation of the fractional Brownian motion $B(t)$ is given by

$$\langle B^2(t) \rangle \sim t^{2H} \tag{27}$$

In general, the exponent $2H$ is fractional, as will be seen later.

The ordinary Levy process[3] corresponds to the case $H < 1/2$ (or $\psi < 1$), while the opposite case $H > 1/2$ (or $\psi > 1$) is realized in many interesting physical phenomena such as clumps in turbulent plasma ($\psi = 3$) and relative diffusion in turbulent velocity fields ($\psi \simeq 3$, i.e., Richardson's law).[4,7]

In the present chapter, we explain a simple physical model, so-called string animals introduced by the present author,[8] as is shown in Fig. 3.

This string animal is equivalent[8] to a dynamical process of an interface in the two-dimensional Ising model, which is shown in Fig. 4.

More explicitly, the above model is described by the kinetic Ising model with energy conservation. For more details, see a paper by the present author.[8] The fluctuation of the coordinate of the jth piece denoted by x_j in Fig. 3 takes the following asymptotic form:

$$\langle x_j^2(t) \rangle \sim t^\psi \tag{28}$$

for large t in our model. Our Monte Carlo simulation[9] gives such an estimate of ψ as

$$\psi = 0.65 \sim 0.67 \simeq \tfrac{2}{3} \tag{29}$$

in two dimensions, on the basis of the finite-size scaling theory[1,5] namely,

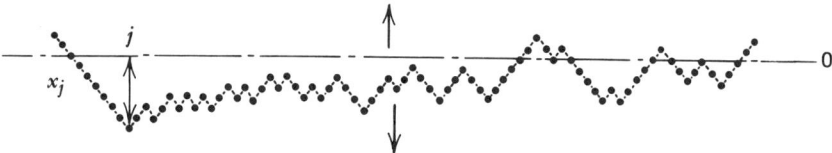

Figure 3. A string animal[8] moving upward and downward in two dimensions: only corner points can move at each moment.

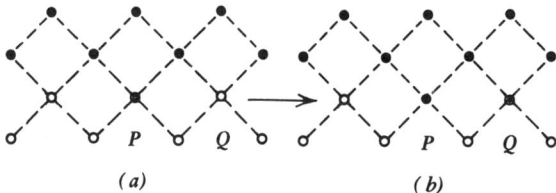

Figure 4. Dynamical elementary process of the temporal change of an interface: A white piece P can change into a black one without changing energy and so does a white piece Q.[8]

$$\langle x_j^2(t) \rangle \simeq D_n t; \quad D_n \simeq \frac{c}{n^\phi} \tag{30}$$

for a finite geometrical string animal. This size-dependence of the diffusion constant yields the value of ψ through the scaling relation[8]

$$\psi = \frac{2}{\phi + 2} \tag{31}$$

From our simulation[9] we get $\phi \simeq 1.0$–1.1, which gives Eqs. (29) through (31).

This is a typical fractional Brownian motion due to a topological (or geometrical) restriction ($\psi < 1$).

The above model is easily extended to three-dimensional systems and other complicated physical systems with geometrical restrictions.

These studies will be reported in future.

ACKNOWLEDGMENTS

The author would like to thank Professor R. Kubo for encouraging discussions and also to thank Dr. F. Sasagawa for his collaboration with Monte Carlo simulations. This study is partially financed by the Research Fund of the Ministry of Education.

REFERENCES

1. M. E. Fisher and M. N. Barber, "Scaling Theory for Finite-Size Effects in the Critical Region," *Phys. Rev. Lett.* **28:**, 1516–1519. (1972).
2. Y. Gefen, A. Aharony, B. B. Mandelbrot, and S. Kirkpatrick, "Solvable Fractal Family, and Its Possible Relation to the Backbone at Percolation," *Phys. Rev. Lett.* **47**, 1771–1774 (1981).
3. B. B. Mandelbrot, *The Fractal Geometry of Nature* (W. H. Freeman and Company, San Franscisco, 1982).
4. H. Mori, this volume.

5. M. Suzuki "Static and Dynamic Finite-Size Scaling Theory Based on the Renormalization Group Approach," *Prog. Theor. Phys.* **58**. 1142–1150 (1977).
6. M. Suzuki, "Phase Transition and Fractals," *Prog. Theor. Phys.* **69**, 65–76 (1983).
7. M. Suzuki, this volume.
8. M. Suzuki, "Dynamics of Topological Disorder—Brownian Motion with Geometrical Restriction, in *Topological Disorder in Condensed Matter*, (Springer-Verlag, Berlin, 1983).
9. M. Suzuki, and F. Sasagawa, (unpublished).

Part 4

TURBULENCE THEORY

TIME EVOLUTION OF VORTICITY FIELD AND TURBULENT DIFFUSION

Hazime Mori
Department of Physics, Kyushu University, Fukuoka, Japan

Abstract. *We shall discuss the following two fundamental problems of fully-developed turbulence: (1) How the turbulent energy is distributed among various length scales. (2) How the mean square of the relative distance of a pair of probe particles evolves in time (i.e., the relative diffusion).*

STEADY TURBULENCE

Let us first consider a steady turbulence whose small-scale fluctuations are statistically homogeneous and isotropic. The mean turbulent energy per unit mass can be decomposed into wavenumber components as

$$\frac{\langle u^2 \rangle}{2} = \int_0^\infty E(k)\, dk \qquad (1)$$

where **u** is the turbulent fluctuation of the fluid velocity and k is the wavenumber. It is well known that the energy-spectrum density $E(k)$ takes the form

$$E(k) = Ck^{-\delta_0}(kL)^{\mu\delta_1} \sim k^{-\delta} \qquad (2)$$

in the normal cascade range $L^{-1} < k < l_d^{-1}$, where $\delta = \delta_0 + \mu\delta_1$ and L is the length scale of the largest eddies of the normal cascade range, C is a positive constant independent of L, l_d is the length scale of the dissipation range, and μ is the intermittency exponent. As is well known, we have $\delta_0 = 5/3$ (Kolmogorov's exponent) and $\delta_1 = 1/3$ for the three-dimensional energy cascade. For other normal cascades, δ takes different values, as shown in Table 1, where the first line is the $3d$ energy cascade, the second is the $2d$ enstrophy cascade, the third is the $3d$ temperature-fluctuation cascade of the Bénard convection, and the fourth is the $3d$ helicity cascade.

Table 1. Characteristic Exponents of Normal Cascades[a,b]

Cascade	d	δ	a	μ	a	$\alpha \equiv a - \mu$
$\langle u^2 \rangle/2$	3	$(5+\mu)/3$	$(2+\mu)/3$	0.341 ≃1/3	0.780 7/9	0.439 4/9
$\langle \omega^2 \rangle/2$	2	$(9+\mu)/3$	$\mu/3$	0	0	0
$\langle \theta^2 \rangle/2$	3	$(11+3\mu)/5$	$(2+\mu)/5$	0.261 ≃1/4	0.452 9/20	0.191 1/5
$\langle \mathbf{u} \cdot \boldsymbol{\omega} \rangle/2$	3	$(7+\mu)/3$	$(1+\mu)/3$	0.203 ≃1/5	0.401 2/5	0.198 1/5

[a] $\boldsymbol{\omega} = \operatorname{rot} \mathbf{u}$ denotes the vorticity, and $\langle \theta^2 \rangle$ the temperature fluctuation in the Bénard convection.
[b] For each cascade, two values of μ and the corresponding results are shown.

These partition laws of energy are quite different from the equipartition law of energy for thermal equilibrium, which leads to $E(k) \sim k^{d-1}$, d being the dimensionality.

Exponent δ_0 can be determined by dimensional analysis. The intermittency exponent μ, however, is a statistical quantity that is determined by the fluctuation of the transfer rate of cascade quantity. Though $\mu\delta_1$ gives at most a 10% correction to δ, it will turn out that the intermittency gives an important contribution to the relative diffusion.

The most important quantity of eddy dynamics is the mean vorticity amplitude $\omega(l)$ of eddies of length scale $l\ (=1/k)$, which takes the form

$$\omega(l) = \frac{v(l)}{l} \sim l^{-a} \tag{3}$$

where $v(l)$ is the mean turnover velocity of eddies of size l. The exponent a is related to δ by

$$a = \frac{(3 - \delta + \mu)}{2} \geqq 0 \tag{4}$$

as shown in a previous paper.[5] The intermittency exponent μ is determined in terms of this a. Namely, it can be shown that the Fujisaka–Mori variational principle[1] of maximizing an information entropy of intermittency leads to

$$\mu + a = \log_2(2^{1+a_0} - 1) \tag{5}$$

with $a_0 \equiv a(\mu = 0)$.[6] This determines μ explicitly, as shown in Table 1. Thus we obtain $\mu \doteq 0.341 \simeq 1/3$ for the $3d$ energy cascade and $\mu = 0$ for the $2d$ enstrophy cascade. These agree with experiments very well.[3,6,9]

The relative diffusion is caused by the eddy elongation of those pairs of eddies whose eddy diameters are not larger than the root-mean-square separation $L_*(t)$ of the pair particles so as to contain the pair particles separately. This is formulated to give

$$\frac{d}{dt}L_*^2 = \begin{cases} \tilde{A}L_*^2 \omega(L_*)(L_*/L)^\mu, & L > L_* \gg l_d & (6a) \\ \tilde{A}L^2 \omega(L), & L_* > L & (6b) \end{cases}$$

where $\tilde{A}$ is a constant number of order unity.[5] Using Eq. (3) for $\omega(L_*)$ of Eq. (6a) we obtain

$$dL_*^2/dt = AL_*^{2-\alpha} \quad (\alpha \equiv a - \mu \geq 0) \tag{7}$$

for $L > L_*(t) \gg l_d$, where $A = \tilde{A}L^\alpha \omega(L)$. This is integrated to give

$$L_*^2(t) \sim \begin{cases} \exp(At), & \text{if } \alpha = 0 & (8a) \\ t^{2/\alpha} & \text{if } \alpha > 0 & (8b) \end{cases}$$

where $L > L_*(t) \gg L_*(0)$ has been assumed. α takes different values as shown in Table 1. The diffusion law [Eq. (8)] is summarized in Table 2 for the four normal cascades. $L_*(t)$ reaches L at a time t_c. After this crossover time t_c, $L_*(t)$ obeys Eq. (6b), which is integrated to give the linear law

$$L_*^2(t) = L^2 + K(t - t_c) \tag{9}$$

with $K \equiv \tilde{A}L^2\omega(L)$.

For the $3d$ energy cascade we thus obtain[9]

$$2 - \alpha \doteq 1.56, \quad \psi = 2/\alpha \doteq 4.56 \tag{10}$$

Table 2. Diffusion Exponent ψ of the Relative Diffusion

	d	$L_*^2(t) \sim t^\psi$	ψ	
Steady $(p=0)$	2	$e^{At} \quad (\alpha = 0)$	∞	$\langle \omega^2 \rangle/2$
$L_*(t) < L$	3	$t^{2/\alpha} \quad (\infty > \psi = 2/\alpha > 2)$	10.5 10.1 4.56	$\langle \theta^2 \rangle/2$ $\langle \mathbf{u} \cdot \boldsymbol{\omega} \rangle/2$ $\langle u^2 \rangle/2$
$L_*(t) \gg L$		t	1	Purely random
Decaying $(p \neq 0)$	2	$(1+pt)^{A_0/p} \quad (\alpha = 0)$		
$L_*(t) < L_t$	3	$[(1+pt)^{1-x} - 1]^{2/\alpha} \quad (x \neq 1)$ $[\ln(1+pt)]^{2/\alpha} \quad (x = 1)$		
$L_*(t) \gg L_t$		$t^{1-z} \quad (1 > \psi = 1 - z > 0)$ $\ln t \quad (z = 1)$	0.80 0.57 0	$z = 1/5$ $z = 3/7$

If one neglects the intermittency, then one gets $2 - \alpha = 4/3$, $\psi = 3$, which agree with Richardson's 4/3 law and t^3 law. Recently, however, Hentschel and Procaccia[2] reanalyzed Richardson's data and obtained $dl_*^2/dt \sim L_*^{1.57}$. This agrees with Eq. (10) very well. For the $2d$ enstropy cascade we have the exponential law (8a), which agrees with a constant-level balloon experiment in the stratosphere on a large geophysical scale.

Thus the diffusion exponent $\psi = 2/\alpha$ takes different values, depending on the cascade process. The exponential law [Eq. (8a)] corresponds to $\psi \to \infty$. In the cascade regime $L_*(t) < L$, the diffusion law deviates from the linear law, Eq. (9). This deviation is due to the coherence of the vortex motion and the eddy elongation of vortices.

DECAYING TURBULENCE

Next let us consider a decaying turbulence, such as a turbulent flow behind a grid in a large wind tunnel. It is well known that at high Reynolds numbers a similarity region that is statistically homogeneous and isotropic is established in a short time. We set the origin of time after this transient period. Then the energy-spectrum density at time $t \geq 0$ is obtained by replacing C and L of Eq. (2) by slowly varying functions of time C_t and L_t, respectively:

$$E_t(k) = C_t k^{-\delta_0} (kL_t)^{-\mu\delta_1} \tag{11}$$

for the normal cascade range $L_t^{-1} < k < l_{dt}^{-1}$, where L_t and l_{dt} are the length scale of the largest eddies of the normal cascade range and the length scale of the dissipation range at time t, respectively. It is convenient to define a decay exponent z by the Reynolds number at time t

$$R_t = \frac{R_0}{(1 + pt)^z} \tag{12}$$

where p is a decay constant. z takes different values, depending on the large-scale structure of the initial energy-spectrum density $E_0(k)$. Let σ be its exponent; $E_0(k) \sim k^\sigma$ ($kL_0 \ll 1$). Then

$$z = \frac{\sigma - 1}{\sigma + 3} \quad (1 \geq z > 0) \tag{13}$$

Typical values of z are $z = 1/5$ ($\sigma = 2$), $z = 3/7$ ($\sigma = 4$), and $z = 1$ ($\sigma \to \infty$). In terms of this decay exponent, we have

$$C_t = \frac{C_0}{(1 + pt)^w} \tag{14}$$

$$L_t = L_0(1 + pt)^{(1-z)/2} \tag{15}$$

where $w \equiv 2 - (3 - \delta_0)(1 - z)/2 > 0$.

The mean vorticity amplitude of eddies of size l at time t takes the form

$$\omega_t(l) \sim \frac{l^{-a}}{(1+pt)^y} \tag{16}$$

where $y \equiv 1 - a(1-z)/2 \geqq (2-a)/2$. The decay constant p takes the form

$$p = \tilde{p}\omega_0(L_0) \tag{17}$$

with a positive number $\tilde{p}$ of order unity.

The diffusion law [Eq. (6)] can be extended to the decaying turbulence by replacing $\omega(l)$, L, and l_d by $\omega_t(l)$, L_t, and l_{dt}, respectively:

$$\frac{d}{dt}L_*^2 = \begin{cases} \tilde{A}L_*^2\omega_t(L_*)(L_*/L_t)^\mu, & L_t > L_* \gg l_{dt} \\ \tilde{A}L_t^2\omega_t(L_t), & L_* > L_t \end{cases} \tag{18}$$

Using Eq. (16) for $\omega_t(l)$ we obtain

$$\frac{d}{dt}L_*^2 = \begin{cases} A_0 L_*^{2-\alpha}/(1+pt)^x, & L_t > L_* \gg l_{dt} \tag{19a} \\ K_0/(1+pt)^z, & L_* > L_t \tag{19b} \end{cases}$$

with

$$x = 1 - \alpha(1-z)/2 \quad (1 \geqq x \geqq y) \tag{20}$$

where $A_0 \equiv \tilde{A}L_0^\alpha \omega_0(L_0)$, $K_0 \equiv \tilde{A}L_0^2\omega_0(L_0)$. If $\alpha = 0$, then $x = 1$, and Eq. (19a) leads to $L_*^2(t) \sim (1+pt)^{A_0/p}$. This is a generalization of Eq. (8a). If $\alpha \neq 0$, then Eq. (19a) leads to a different form. Equation (19b) can also be integrated easily. Thus we obtain the relative diffusion summarized in the second part of Table 2.

Thus the diffusion exponent ψ takes different values, ranging from 0 to ∞. This is in strong contrast to the Brownian motion of particles in an equilibrium fluid which always leads to $\psi = 1$. The deviations of ψ from unity represent ordered motions of vortices of various length scales.

REFERENCES

1. H. Fujisaka and H. Mori, "A Maximum Principle for Determining the Intermittency Exponent μ," *Prog. Theor. Phys.* **62**, 54–60 (1979).
2. H. G. E. Hentschel and I. Procaccia, "The Fractal Nature of Turbulence as Manifested in Turbulent Diffusion" (preprint) (1979).
3. A. S. Monin and A. M. Yaglom, *Statistical Fluid Mechanics* (MIT Press, Cambridge, MA, 1975), Chap. 25.
4. P. Morel and M. Larcheveque, "Relative Dispersion of Constant-Level Balloons in the 200-mb General Circulation," *J. Atmos. Sci.* **31**, 2189–2196 (1974).
5. H. Mori, "Entropy Cascade and Relative Diffusion in Two-dimensional Fully-Developed Turbulence," *Prog. Theor. Phys.* **69**, 756–772 (1983).
6. H. Mori, "Intermittency and Relative Diffusion in Fully-Developed Turbulence," *Prog. Theor. Phys.* **70** (1983).

7. H. Mori and F. Takayoshi, "Vortex Stretching and Relative Diffusion in Grid Turbulence," *Prog. Theor. Phys.* **69** (1983).
8. M. Nelkin, "Do the Dissipation Fluctuations in High Reynolds Number Turbulence Define a Universal Exponent?" *Phys. Fluids* **24**, 556 (1981).
9. F. Takayoshi and H. Mori, "Diffusion of Particles in Fully-Developed Turbulence," *Prog. Theor. Phys.* **68**, 439–447 (1982).

TOPICS IN THE THEORY OF STATISTICAL CLOSURE APPROXIMATIONS FOR PLASMA PHYSICS

John A. Krommes

Princeton University, Plasma Physics Laboratory, Princeton, New Jersey

Abstract. Recent advances and directions in the theory of statistical descriptions for plasmas are discussed.

1. INTRODUCTION: STATISTICAL CLOSURES AND NONLINEAR DYNAMICS

In this chapter I wish to mention some of the recent advances in the theory of statistical closures for plasma physics, the relationship of such approximations to the more–general recent results in nonlinear dynamics, and directions for the future. Space limitations require that I be brief and very qualitative; however, I have provided a broad spectrum (hopefully not too stochastic!) of references. (Often, these are not to the original work, but rather to more recent articles of review nature; even these are only representative, not exhaustive.)

In the study on nonlinear dynamical systems, plasma-related or otherwise, a hierarchy of approaches may be discerned and classified by their degree of rigor. At the most fundamental level, which I shall call Level 0, there are the studies which—at least in principle, and often in practice—can lead to mathematically rigorous results. Typically, these give information about global, topological changes of the orbits as a parameter is varied; examples include the Kolmogorov–Arnol'd–Moser theorem,[1] Greene's stochasticity criterion[2] and MacKay's extensions thereof,[3] and Feigenbaum's universality theory.[4]

The recent advances in these areas are exciting indeed (for an excellent review, see Ref. 5), and further progress is to be expected. Although the applications to plasmas are not so well worked out, nonlinear dynamics has definitely had its impact. When formulating a theory of plasma turbulence, one need no longer wonder what such vague assumptions as the random–phase approximation or a "sufficiently short" correlation time really mean. To some extent at least, the limits of validity of such assumptions have been quantified,[6,7] and physical understanding of the underlying particle dynamics has improved dramatically.

However, the fundamental studies have one important general deficiency: at present, they have only very limited capabilities for quantitative prediction of macroscopic dynamical quantities such as time–correlation functions (and, therefore, transport coefficients). Of course, exactly soluble models are known, and the general situation is likely to improve. However, at present it seems that for the extremely complicated nonlinear systems of interest in applications-oriented plasma physics (such as the Klimontovich–Poisson system or the Hasegawa–Mima equation[8]) one may learn from the fundamental studies mostly qualitative information, such as in what parameter ranges one is likely to expect intrinsic stochasticity, or general predictions about the time-asymptotic nonlinear behavior. However, one must turn elsewhere for a detailed quantitative theory of the statistical dynamics in the stochastic regime.

It is well known that statistical behavior is not incompatible with deterministic orbits—there can be *intrinsic* stochasticity. There are some subtleties associated with the choice of an appropriate random variable, but let me assume that this choice has been made. Then, I can define a theory of statistical dynamics to be one which, in its most complete form, provides an approximate description of the probability distribution function $P(z)$ of the underlying random variable z. [For cases where z is a random process depending on time, one must seek a probability *functional* $P\{z(t)\}$; this is also the case when the random variable is an entire *field* $N(x, v, t)$, such as in the Klimontovich equation.] Since the full distribution is generally extremely complicated (and much of its information experimentally inaccessible), one is often content with predictions of low-order cumulants,[9] such as the mean $\langle z(t) \rangle$ or two-point correlation function $\langle \delta z(t) \delta z(t') \rangle$. Even at this level, a quantitative description is not likely to be simple. This can be emphasized by considering two of the simplest equations for which a statistical dynamics can be defined and for which the exact solution is also available. The first, studied by Kraichnan[10] and Kubo,[11] is the so-called stochastic oscillator[12]

$$\frac{\partial \psi(t)}{\partial t} = -i\omega(t)\psi(t), \qquad \psi(0) = 1, \tag{1}$$

where ω is taken to be a statistically stationary, centered Gaussian random process with $\langle \delta\omega(\tau)\delta\omega(0)\rangle = \beta^2 \exp(-|\tau|/\tau_c)$. Equation (1) is, of course, a linear equation, of little interest from the point of view of *intrinsic* stochasticity. However, because ψ is random since ω is, Eq. (1) contains a product of random variables and, following Kraichnan, may be called *stochastically nonlinear*. From the point of view of statistical dynamics, it is the stochastic nonlinearity that is relevant; even this dynamically linear equation exhibits the *statistical closure problem*. (The equation of motion for the mean $\langle\psi\rangle$ contains the unknown mixed correlation $\langle\delta\omega\delta\psi\rangle$.) Furthermore, in the limit of large Kubo number $K \equiv \beta\tau_c$, the statistics of ψ are far from Gaussian, thus precluding, for example, a straightforward application of the central-limit theorem or a quasilinear perturbation theory.[12]

The second example is a true nonlinear dynamical system, and does exhibit intrinsic stochasticity. It consists of the following mapping of the interval $[0,1]$ onto itself:

$$x_{n+1} = 4x_n(1-x_n). \qquad (2)$$

This map, one special case of the so-called logistic map,[13] also exhibits the closure problem, and contains no obvious small parameters on which to build a perturbation theory. It is known to be ergodic,[13] to have positive Lyapunov exponent, and to have invariant measure (ignoring mathematical subtleties, this is just the asymptotic stationary distribution function[14]) $P(x) = [\pi^2 x(1-x)]^{-1/2}$. Again, $P(x)$ is not Gaussian. Such a result can be expected to obtain quite generally, as has been stressed by Ruelle.

Equations (1) and (2) are useful prototypes for such equations as those of Klimontovich–Poisson, Hasegawa–Mima, or Navier–Stokes because they are quadratically nonlinear in stochastic variables and, therefore, have the closure problem. Unfortunately, in the nonintegrable regimes of the latter, no exact solutions are available. In such cases, how can one proceed to find the long-time statistics? Numerical experiments have proven to be extremely useful, but it is not always easy to extract from them analytic interrelationships or parameter dependences. Analytically, several powerful schemes have been advanced (although none are, as yet, completely developed). I mention two: the projection operator approach of Mori,[15] and the functional formalism of Martin, Siggia, and Rose.[16] The former is very convenient for modeling[17] and for interpolation between short- and long-time regimes. The latter is particularly interesting because it has close quantum-mechanical analogies, it can be realized by a path-integral representation,[18] and it is in a form convenient for applying tools such as the renormalization group. In either of these approaches, one eschews from the outset consideration of microscopic details such as the stability of fixed points or the fractal dimension of the attracting manifold, and instead develops approximate closed equations for statistical averages (like correlation functions in phase space and time). These equations contain in their own right a wealth of detail and interesting physical information. In principle, they can give rise to predictions that are just as exact and fundamental as any of the other Level 0 studies. However, in practice, the nature of the closure approximations is not always well understood, and they tend to be viewed with some (much?) skepticism. For this reason, and because the statistical theories discuss *only* averages, nothing more microscopic, I will classify the general theories of statistical closure at Level 1. (Clearly, the boundaries between levels are not sharply defined.)

Can one link the statistical closure formalisms with the recent results of nonlinear dynamics? For general equations with arbitrary nonlinearities, this is difficult and perhaps impossible. It may be that entirely new statistical methods are required, perhaps based on variational principles for the invariant measures.[19] However, some possible connections are beginning to emerge. For example, it has long been argued intuitively that the positive Lyapunov exponents[20] should have something to do with the decay of time-correlation functions.[21] Although a precise theory is lacking and may not exist in general, Grebogi and Kaufman[22] have demonstrated numerically that the exponent associated with correlation decay in certain Hamiltonian maps has scaling similar to the Lyapunov exponent. The Lyapunov exponent is known to describe the rate of coarse-grained mixing,[23] and therefore the rate of evolution of an appropriately defined entropy. Krommes and coworkers have argued in a series of papers[24-26] that approximate solutions of the Bethe–Salpeter equation[12,16] (which emerges in the closure formalisms at Level 1) may determine the Lyapunov exponent (a fundamental dynamical quantity defined at Level 0). Such a connection appears necessary in order to reconcile heuristic arguments about particle diffusion in stochastic magnetic fields with kinetic theory.[26] Shraiman, Wayne, and Martin[27] have demonstrated perhaps the most fundamental connection to date: for the special case of a one-dimensional map, they have shown how the Lyapunov exponent is determined from the mean infinitesimal response function, which plays a fundamental role in the MSR formalism.

2. PLASMA PHYSICS AND THE DIRECT-INTERACTION APPROXIMATION

Although the difficulties of statistical closures are severe, the kinds of plasma turbulence that are typically encountered do not represent the worst possible case. This point has been emphasized by Dupree[28] in a paper remarkable for its prescience (words such as "nonintegrable" or "stochasticity criterion" do not appear). Two features of the plasma must be remarked upon. First, particles rapidly stream in at least one direction (along the magnetic field, if present). For typical wave packets, this leads to a short correlation time of the particle with the waves. This, in turn, enhances the possibility of a Markovian description (with a suitable coarse graining in time), and the possibility of near-Gaussian statistics. Second, the fluctuation level in a plasma is generally not arbitrarily large. Unlike the usual way in which turbulence is generated in neutral fluids, plasmas are typically not forced externally. Rather, internal fluctuations grow up and balance linear excitations. This leads to a Reynolds number[28] which is typically of order unity, rather than arbitrarily large. Again, this is compatible with the possibility that the plasma statistics are not far from Gaussian.

Given these two special features of the plasma, a particular closure is suggested immediately. Martin, Siggia, and Rose have shown that the so-called direct-interaction approximation (DIA) plays a preferred role in their formalism—it emerges at the lowest-order theory in a renormalized expansion (in a certain well-defined sense) about Gaussian statistics. Not only is it compatible with quasilinear theory and weak–turbulence theory (which are sometimes appropriate when the correlation time is small), it also provides an appropriate, self-consistent description of the "resonance-broadening" corrections to free-particle propagation that arise because of the intrinsic stochasticity. For plasmas, the DIA was first proposed by Orszag and Kraichnan[29] as a major, nontrivial generalization of Dupree's 1966 resonance-broadening theory,[30] which did not provide a self-consistent determination of the electric field. Orszag's pioneering work was largely ignored for many years, presumably because of its complexity. Recently, however, the DIA has been explored in depth analytically.[12,31–33] In particular, well understood are its relations[33] to the various simpler, not-as-systematic closures that have been proposed, such as resonance-broadening theory in its non-energy-conserving[30] and energy-conserving[34] versions, or the Hirshman–Molvig theory.[35]

Before proceeding with a further discussion of the DIA, let me define one more level (Level 2). Here live the applications of the general closure formalisms developed at Level 1 to the very complicated geometries and constraints typical of real physical systems (such as the tokamak). Practical difficulties preclude direct application of the most fundamental closures; rather, the Level 1 schemes must be severely (sometimes brutally) mutilated. It is difficult to give a general discussion of such results, and I will not attempt it here. At best, they capture the essence of the dominant nonlinear physics and, essentially, provide a systematic way of performing dimensional analysis; at worst, they foster a false sense of security and, therefore, are dangerously misleading.

In the remainder of this paper I will discuss two specific topics related to statistical closure approximations motivated by plasma dynamics in hot tokamaks, but of relevance more generally. Related to Level 1, I will discuss the feasibility of numerical computations and tests of the direct-interaction approximation applied to the nonlinear gyrokinetic equation. Related to Level 2, I will briefly discuss certain properties of the solutions of a certain Markovian approximation to the fluid limit of the above DIA.

Unfortunately, since the work which I will mention is still in progress, the following remarks must be somewhat tentative. Nevertheless, some general conclusions can be drawn.

3. COMPUTATIONAL FEASIBILITY OF THE GYROKINETIC DIRECT-INTERACTION APPROXIMATION

The direct-interaction approximation has played a preferred role in the theory of statistical closures ever since its invention by Kraichnan in the late 1950s.[36] (A recent review oriented toward plasma physics, with references to earlier reviews and fundamental papers, has been given by Krommes.[12]) From the specific point of view of the Martin–Siggia–Rose theory, the DIA can be described as the lowest-order self-consistent renormalization. More generally, of course, such a characterization is not unique. For example, Kraichnan[37] has provided a mixed Eulerian–Lagrangian framework from which other closures emerge on the same nominal footing as the DIA. However, the DIA is apparently very useful and well balanced. It has an immediate and correct reduction to weak-turbulence theory[31,32]; it incorporates statistical effects of the turbulent fields on the particle orbits, as well as the important *self-consistent* reactions of the particles back on the collective fields. These back reactions are required in order that energy be conserved.[12,33]

The nonlinear dynamics of plasma are very rich, and this manifests itself in the complexity of such closures as the DIA. In particular, the *self-consistent* nature of the wave-particle interaction in plasmas is very incompletely understood within the statistical framework (and otherwise!). Analytic studies of conservation laws and special models have provided some insight, but progress has been hindered by the lack of detailed (numerical) solutions of the plasma dynamical equations with which to compare predictions of the closure approximations. Therefore, it is appropriate to consider the possibility of numerically solving the DIA and comparing its predictions with statistical predictions distilled from the exact nonlinear dynamics.

[It is important to stress that such a program leads to useful *qualitative* insights, not to just the comparison of (large) tables of numbers. Statistical closures such as the DIA partition the effects of the nonlinear interactions into distinct classes of terms: in one intuitive classification, into "diffusion" and "polarization" terms[38,39]; in another, into "coherent" and "incoherent" terms.[40] These terms can be turned off one by one in order to study their relative importance. This important luxury is afforded only by the statistical approach. In contrast, in direct numerical integrations of the dynamical equations there may be just one dominant nonlinearity—for example, the $\mathbf{E} \times \mathbf{B}$ term; turning this off reduces the problem to the (relatively) trivial linear theory.]

For fluids, the DIA has been compared in some depth to numerical simulations.[41,42] The beginnings of such a program for plasma were reported at this meeting last year: the DIA was solved[43] for the very simple (but nontrivial) case of three interacting fluid fluctuations characteristic of Hasegawa–Mima dynamics, but with growth or damping.[44,45] Such a system supports stochastic solutions, including strange attractors. Excellent agreement was found between saturation levels in a Gaussian ensemble and the predictions of the DIA. The key to this comparison lay in the simplicity and speed with which exact solutions of the dynamical equations could be generated numerically. For more-interesting physical problems, such solutions have not been available, and are nontrivial to obtain. Recently, however, numerical studies have been made[46] of a gyrokinetic model of plasma[47] of interest both fundamentally and (to a much lesser extent) practically. It is desirable to study the predictions of the DIA for this more complete and interesting problem, especially since the study allows for nonlinear kinetic effects. Here, however, there arise significant questions of numerical feasibility.

Because of the complexity of the self-consistent interactions at the kinetic level, as well as the geometrical nature of the relevant (drift) fluctuations, the gyrokinetic DIA is very complicated. Consider first the nature of the fluctuations. It is well known that drift fluctuations are three dimensional, with a density gradient in the x direction, a strong magnetic field predominantly in the z direction, with typical poloidal ($\hat{z} \times \hat{x}$) wavenumbers satisfying $k_y \rho_i \sim 1$, and with parallel wavenumbers satisfying $k_\parallel / k_\perp \ll 1$. Furthermore, at least the parallel velocity $v_\parallel$ must be retained in order to describe Landau resonances. (Lee and others have argued that the perpendicular velocities can be approximately averaged over a Maxwellian.) Furthermore, electrons and ions play distinct roles in the modal dynamics. Therefore, the drift wave problem has a very low degree of symmetry, and requires many variables for its description. As if this were not enough, note that the DIA advances two-point correlations in time, which in principle doubles the number of variables. (Two variables can be saved by using translational invariance in the y and z directions.)

To these formidable requirements, one must compare available computer resources, which I take to be the "D" machine (Cray Research model 1s) at the National Magnetic Fusion Energy Computer Center at Livermore, CA. This is a vector machine with a 12.5 ns clock. The operating system contraints are: user-accessible memory, 1.2×10^6 words (64 bits/word); maximum length of a disk file, $\simeq 12 \times 10^6$ words. Optimized machine language subroutines are available that expedite vector operations; in that mode, approximately one multiplication can be performed each clock cycle. I conservatively assume that the maximum feasible time per run is approximately 1 hr, or 3×10^{11} cycles.

Because of limited space, I shall write here neither the gyrokinetic equation[47,48] nor the full details of the gyrokinetic DIA. However, let me make some general remarks about the structure of the latter. The general variable dependence of the two-point correlation function C of the fluctuating distribution function is $C(x, y - y', z - z', v, s, t; x', v', s', t')$. In the DIA, this function will evolve according to

$$0 = \frac{\partial}{\partial t} C_k(x, v, s, t; x', v', s', t') + \text{(linear terms)}$$

$$+ \int_0^t dt'' \int dx'' \int dv'' \sum_{s''} \sum_{\mathbf{k+p+q}=0} (R_\mathbf{p} C_\mathbf{q})(x, v, s, t; x'', v'', s'', t'')$$
$$\times C_k(x'', v'', s'', t''; x', v', s', t') + \text{(one similar nonlinear term)}, \quad (3)$$

where R is the mean infinitesimal response function and where $\mathbf{k}$ is the Fourier variable conjugate to $(y - y', z - z')$. The convolution nature of Eq. (3) is severely restrictive. Elementary estimates lead one to conclude immediately that if kinetic effects (velocity space) are to be described, then full two-point effects in x cannot be. Fortunately, Lee's initial studies have been done in a slab homogeneous in x, and use a two-dimensional wavevector (k_x, k_y). (The parallel wavenumber is computed according to $k_\parallel = \alpha k_y$, with $\alpha \ll 1$.) For these studies, one may then drop the x and x' variables.

Next, consider the time dependence. If the time convolution were to be integrated faithfully for N_T steps and all information held in memory, the number of operations would scale as N_T^3 and the number of memory locations would be proportional to N_T^2. Both scalings turn out to be prohibitive. Since memory provides an absolute restriction, one must adopt drastic measures, which are to hold in memory only enough information to move from one time step to the next (via a second-order predictor–corrector scheme). A disk file may be used as a dynamic subsidiary storage medium. History information for the time convolutions must be read in at each time step; updated information must be written out. Efficient real-time speed can be obtained by overlapping these input/output operations with computations. Although this scheme is very simple in principle, its implementation in practice is complex.

The prohibitive scaling of run time with N_T^3 can be reduced by noting that, although in principle one must integrate over all difference times $\tau \equiv t - t' < t$, in practice correlations will have substantially decayed in a correlation time τ_c; one can restrict the integrations in difference time to $\tau < \tau_c$. (Some trial and error may be necessary to determine an appropriate value of τ_c.) The possibility of restricting the convolutions to a finite range was already recognized in early numerical work on the DIA[49]; in the present problem, it is absolutely essential to do so.

Now consider the species dependence. Here again, memory provides a limitation. The correlation matrix for two species requires four times as much memory as does one species. I choose to retain kinetic effects for the electrons only; I make the fluid approximation for the ions. Unfortunately, this precludes study of certain saturation mechanisms, such as enhanced ion Landau damping due to fluctuation-induced resonance broadening.[34] However, this mechanism does not seem to play an important role in Lee's studies, and much physics, including fluid ion nonlinearities, is retained in the final approximation.

In addition to an equation of the form (3), the DIA provides a second equation for the mean infinitesimal response function. In the interest of saving factors of two in both time and memory, I will make one further approximation, which is to replace the two-time dependence of $C(\tau)$ with that of $R(\tau)$. That is, I will make the so-called self-consistent-field approximation.[50,51]

Memory requirements for the resulting approximation are proportional to $(2 \times 3) N_v^2 N_k N_\tau$, where the factor of 2 accounts for complex numbers, the factor of 3 is the number of input/output and computation buffers, N_v is the number of velocity grid points, N_k is the number of modes, and N_τ is the number of difference times. A fully vectorized code implementing the closure scheme has been written. Timing tests indicate that nonlinear saturation can be achieved in a reasonable amount of computer time. Linear theory has been recovered. Tests of the nonlinear terms continue.

Considerations similar to the above lead one to conclude that if one is willing to give up the kinetic description (i.e., to work at the fluid level), many more modes and/or more realistic geometry can be studied. This is a fruitful area for future investigation.

In concluding this section, the following very important point must be stressed. The original applications of the DIA were to homogeneous, isotropic fluid turbulence. There, the high degree of symmetry made it easy to conclude that it was much faster to solve numerically the DIA than to directly integrate an ensemble of realizations, then average directly. For drift wave problems, this is not so obvious. Instead, the rationale must be that studies of the DIA directly aid one's theoretical understanding of the various physical mechanisms. In view of the complexity of the physics, it is my opinion that such studies are absolutely essential.

4. MARKOVIAN APPROXIMATIONS

It is possible to generate from the DIA *reduced* approximations. For example, in a fluid theory one may model the two-time dependence of the response function by[12,41,52] $R_k(\tau) \simeq \exp(-\eta_k \tau) H(\tau)$, where $H(\tau)$ is the Heaviside function. Since this approximation no longer solves the DIA for all τ, one must determine η_k by, for example, averaging the DIA equations over all τ. When dependence on the mean time is weak, one is then led to nonlinear *algebraic* equations for the η_k's. For example, for the case of three interacting fluid fluctuations labeled by wavenumbers K, P, and Q, one is led to the system

$$\eta_K = -i\Omega_K + \frac{N_K}{\eta_P^* + \eta_Q^*} \text{ (plus two more cyclically permuted equations),}$$
(4)

where Ω_k is the complex mode frequency and where the N_k's are specified in terms of the spectral intensities (which evolve slowly according to an additional equation which is omitted here). Other procedures [52] lead to different η dependence. A detailed discussion of such Markovian closures will be given elsewhere. Here, I wish to warn that equations of the general form (4) are not necessarily as simple as they seem. By averaging over time, one may lose contact with the physical solution which emerges from $t = 0$; spurious nonphysical branches may appear. For example, Eq. (4) has seven roots, only one of which is the physical one. This implies that special care is required in any numerical scheme which implements closures such as Eq. (4).

5. CONCLUSIONS

In view of work such as described in Refs. 18, 26, or 27, it would appear that analytic connections are beginning to be made between statistical closures and the impressive body of recent advances in nonlinear dynamics. Furthermore, the state of the art in numerical computation has advanced to the point where nontrivial tests of such closures as the direct-interaction approximation can be undertaken for situations of interest to applications-oriented plasma physics. Although all of these efforts are still in their infancy, I expect progress to continue. Perhaps we are witnessing the slow transformation of nonlinear plasma physics from an art to a science. If so, this is truly exciting.

ACKNOWLEDGMENTS

This work was supported by United States Department of Energy Contract No. De-AC02-76-CHO-3073.

REFERENCES

1. See, for example, V.I. Arnol'd, *Mathematical Methods for Classical Mechanics*, translated by K. Vogtmann and A. Weinstein (Springer, New York, 1978), App. 8.
2. J. M. Greene, in *Nonlinear Dynamics* (New York Academy of Sciences, New York, 1980), p. 80, and references therein.
3. R. S. Mackay, *Renormalization in Area-Preserving Maps*, Ph.D. Thesis, Princeton University, 1982; J. M. Greene, this volume.

4. M. J. Feigenbaum, in *Nonlinear Dynamics* (New York Academy of Sciences, New York, 1980), p. 330, and references therein.
5. R. H. G. Helleman, in *Fundamental Problems in Statistical Mechanics*, E. G. D. Cohen, ed. (North-Holland, Amsterdam, 1980), Vol. V, p. 165.
6. G. M. Zaslavskii and N. N. Filonenko, Zh. Eksp. Teor. Fiz. **54**, 1590 (1968) [Sov. Phys. JETP **27**, 851 (1968)].
7. See the various articles in *Intrinsic Stochasticity in Plasmas*, G. Laval and D. Gresillon, eds. (Les Editions de Physique Courtabuef, Orsay, France, 1979).
8. A Hasegawa and K. Mima, *Phys. Fluids* **21**, 87 (1978).
9. R. Kubo, *J. Phys. Soc. Japan* **17**, 1100 (1962).
10. R. H. Kraichnan, *J. Math. Phys.* **2**, 124 (1961).
11. R. Kubo, *J. Math. Phys.* **4**, 174 (1963).
12. J. A. Krommes, "Statistical Descriptions and Plasma Physics," Princeton University, Plasma Physics Laboratory Report No. 1568 (1980); to appear in *Handbook of Plasma Physics*, Vol. II, R. N. Sudan and A. A. Galeev, eds. (North-Holland, Amsterdam).
13. R. M. May, *Nature* **261**, 459 (1976), and references therein.
14. D. Ruelle, in *Bifurcation Theory and Applications in Scientific Diciplines*, O. Gurel and O. E. Rössler, eds. (New York Academy of Sciences, New York, 1979), p. 408.
15. H. Mori, *Prog. Theor. Phys.* **33**, 423 (1965).
16. P. C. Martin, E. D. Siggia, and H. A. Rose, *Phys. Rev. A* **8**, 423 (1973).
17. B. J. Berne, in *Physical Chemistry*, D. Henderson, ed. (Academic Press, New York, 1971), Vol. 8B, p. 539.
18. R. V. Jensen, *Functional Integral Approach to Classical Statistical Dynamics*, Ph.D. Thesis, Princeton University, 1981; *J. Stat. Phys.* **25**, 183 (1981), and references therein.
19. D. Ruelle, in *Lecture Notes in Physics* (Springer, Berlin, 1978), Vol. 80, p. 341.
20. G. Benettin, L. Galgani, and J.-M. Strelcyn, *Phys. Rev. A* **14**, 2338 (1976).
21. B. V. Chirikov, Nuclear Physics Institute of the Siberian Section of the USSR Academy of Sciences Report No. 267; translated by A. T. Sanders, CERN-Trans. 71-40, Geneva, 1971.
22. C. Grebogi and A. N. Kaufman, *Phys. Rev. A* **24**, 2829 (1981).
23. I. Hamilton and P. Brumer, *Phys. Rev. A* **25**, 3457 (1982).
24. J. A. Krommes, R. G. Kleva, and C. Oberman, Princeton University, Plasma Physics Laboratory Report No. 1389 (1978).
25. J. A. Krommes, *Suppl. Prog. Theor. Phys.* **64**, 137 (1978).
26. J. A. Krommes, C. Oberman, and R. G. Kleva, Princeton University, Plasma Physics Laboratory Report No. 1915 (1982); *J. Plasma Phys.* **30**, 11 (1983).
27. B. Shraiman, C. E. Wayne, and P. C. Martin, *Phys. Rev. Lett.* **46**, 935 (1981).
28. T. H. Dupree, in *Turbulence of Fluids and Plasmas* (Polytechnic Press, Brooklyn, 1968), p. 3.
29. S. A. Orszag and R. H. Kraichnan, *Phys. Fluids* **10**, 1720 (1967).
30. T. H. Dupree, *Phys. Fluids* **9**, 1773 (1966).
31. J. A. Krommes, in *Theoretical and Computational Plasma Physics* (International Atomic Energy Agency, Vienna, 1978), p. 405.
32. D. F. Dubois and M. Espedal, *Plasma Phys.* **20**, 1209 (1978).
33. P. Similon, *A Renormalized Theory of Drift Wave Turbulence in Sheared Geometry* (Ph.D. Thesis, Princeton University, 1981).
34. T. H. Dupree and D. J. Tetreault, *Phys. Fluids* **21**, 425 (1978).

35. S. P. Hirshman and K. Molvig, *Phys. Rev. Lett.* **42**, 648 (1979).
36. R. H. Kraichnan, *J. Fluid Mech.* **5**, 497 (1959).
37. R. H. Kraichnan, *J. Fluid Mech.* **83**, 349 (1978).
38. J. A. Krommes and R. G. Kleva, *Phys. Fluids* **22**, 2168 (1979).
39. J. A. Krommes and M. Kotschenreuther, *J. Plasma Phys.* **27**, 83 (1982).
40. T. H. Dupree, *Phys. Fluids* **15**, 334 (1972).
41. D. C. Leslie, *Developments in the Theory of Turbulence* (Clarendon Press, Oxford, 1973), and references therein.
42. J. R. Herring and R. H. Kraichnan, in *Statistical Models and Turbulence*, M. Rosenblatt and C. van Atta, eds. (Springer, Berlin, 1972), p. 148.
43. J. A. Krommes, *Phys. Fluids* **25**, 1393 (1982).
44. J.-M. Wersinger, J. M. Finn, and E. Ott, *Phys. Fluids* **23**, 1142 (1980).
45. P. Terry and W. Horton, *Phys. Fluids* **25**, 491 (1982).
46. W. W. Lee, *Bull. Am. Phys. Soc.* **27**, 945 (1982).
47. W. W. Lee, *Phys. Fluids* **26**, 556 (1982).
48. D. H. E. Dubin, J. A. Krommes, C. Oberman, and W. W. Lee, *Phys. Fluids* **26**, 3524 (1983).
49. R. H. Kraichnan, *Phys. Fluids* **7**, 1030 (1964).
50. J. R. Herring, *Phys. Fluids* **8**, 2219 (1965).
51. J. A. Krommes and C. Oberman, *J. Plasma Phys.* **16**, 193 (1976).
52. A. E. Koniges and J. A. Krommes, *Bull. Am. Phys. Soc.* **27**, 947 (1982).

STATISTICAL DESCRIPTION OF DRIFT WAVE TURBULENCE

Wendell Horton, Jr.
Institute for Fusion Studies, University of Texas at Austin, Austin, Texas

Abstract. *Dissipative drift wave fluctuations are studied with the Terry-Horton nonlinear drift wave model. The $\mathbf{k}\omega$ spectral characteristics of the fluctuations are parametrized in terms of the nonlinear frequency $\omega_\mathbf{k}$ and line width $\nu_\mathbf{k}$ from computer simulations and the renormalized wave–kinetic equation. The probability distributions of the fluctuations are analyzed to assess the validity of the quasinormal approximation made in the closure of the hierarchy of correlations in statistical turbulence theory.*

1. INTRODUCTION AND PHYSICAL MODEL

The drift wave instability of the inhomogeneous magnetized plasma and the transition of the nonlinear oscillations to turbulence has been the subject of numerous experimental and theoretical studies in plasma physics. Only recently, however, has it been possible to explain the $\mathbf{k}\omega$ spectral features of the fluctuation spectrum in terms of theoretical models. In this chapter we show how the broad frequency spectra observed in drift wave turbulence can be understood in terms of a chaotic attractor in the phase space of the dissipative nonlinear dynamics. The randomness produced by the chaotic attractor as measured by the probability distributions of the fluctuating fields is shown to be sufficient to suggest that closure of the correlation hierarchy by the quasinormal approximation is a good first approximation.

  The drift wave instability arises from quasineutral collective oscillations with $\omega \sim k_y v_{de}$ in magnetized plasmas with density and temperature gradients across the magnetic field.[1] Referring to Fig. 1, which shows the mechanism of the drift wave instability, we define the gradient scale lengths by

$$\frac{1}{r_n} = \frac{-1}{N}\frac{dN}{dx}, \quad \frac{1}{r_T} = \frac{-1}{T_e}\frac{dT_e}{dx}, \quad \text{and} \quad \eta_e = \frac{r_n}{r_T} \quad (1)$$

Long-wave oscillations propagate at the electron diamagnetic drift velocity v_{de}, where $v_{de} = cT_e/eBr_n = (\rho/r_n)c_s$, where $c_s = (T_e/m_i)^{1/2}, \rho = c_s/\omega_{ci}$, and $\omega_{ci} = eB/m_i c$.

Since the drift wave oscillations $\omega \simeq k_y v_{de}$ are slow compared with the ion cyclotron gyrations ω_{ci}, the cross-field velocity $\mathbf{v}(\mathbf{x}, t)$ of the ions can be derived from the momentum balance equation by an expansion in $1/\omega_{ci}$. In the presence of an electrostatic field $\mathbf{E}(\mathbf{x}, t) = -\nabla\phi(\mathbf{x}, t)$ the ion velocity is

$$\mathbf{v}(\mathbf{x}, t) = \mathbf{v}_E + \mathbf{v}_p + \mathbf{v}_\mu \quad (2)$$

where

$$\mathbf{v}_E = c_s \rho \hat{z} \times \nabla\phi$$

$$\mathbf{v}_p = -\rho^2 \nabla_\perp \left(\frac{\partial\phi}{\partial t} + \mathbf{v}_E \cdot \nabla\phi\right)$$

$$\mathbf{v}_\mu = \mu\rho^2 \nabla^2 \nabla\phi \quad (3)$$

where in Eq. (3) the potential is in units of T_e/e and $\mu \sim \nu_{ii}$ is the cross-field viscosity coefficient due to ion–ion collisions.[2]

The drift motions of the plasma are quasineutral with the equation $\partial_t \rho_Q = -\nabla \cdot \mathbf{j} = 0$ determining the evolution of the potential $\phi(\mathbf{x}, t)$. The

(a) DISSIPATIONLESS

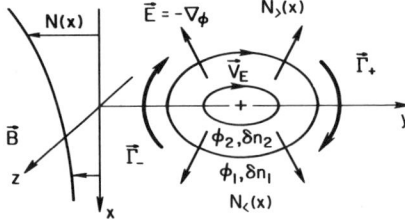

(b) DISSIPATIVE

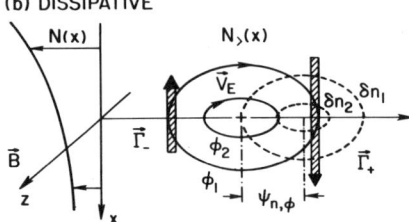

Figure 1. Geometry and mechanism of the dissipative drift wave instability.

electron $\mathbf{E} \times \mathbf{B}$ drift cancels the ion $\mathbf{E} \times \mathbf{B}$ current so that quasineutrality reduces to $\nabla \cdot [n(\mathbf{v}_p + \mathbf{v}_\mu)] = -\nabla_\parallel(j_\parallel^e/e)$ where $j_\parallel^e$ is the parallel electron current.

The divergence of the parallel electron current and the parallel electron momentum equation

$$\nabla_\parallel(j_\parallel^e/e) = \frac{\partial n}{\partial t} + \mathbf{v}_E \cdot \nabla n$$

$$j_\parallel^e = \left(\frac{ne^2}{m_e \nu_{ei}}\right)\left(E_\parallel + \frac{1}{ne}\nabla_\parallel p_e\right) \qquad (4)$$

complete the system of equations for $\nu_e > k_\parallel v_e$. The remaining details of the derivation are given in Hinton and Horton[2] for the collisional regime $\nu_e > k_\parallel v_e$ and in Terry and Horton[3,4] for the collisionless regime $\nu_e < k_\parallel v_e$. Finally, all derivatives parallel to the magnetic field are approximated by $\nabla_\parallel \sim 1/L_c$, where L_c is the characteristic length along $\mathbf{B}$ of the system under consideration.

Recently, a kinetic-theory derivation of the Terry–Horton equation, and its higher-order generalizations, is given by Dubin et al.[5] based on the drift wave ordering of $\mathbf{k}\omega$.

In terms of the natural dimensionless variables of the dynamics $t[r_n/c_s]$, $(x, y)[\rho]$, and $\phi[\rho/r_n]$ the nonlinear dynamics of the drift wave fluctuations is given by

$$(1 + \hat{\mathscr{L}})\frac{\partial \phi}{\partial t}(x, y, t) = -\frac{\partial \phi}{\partial y} - \left[\frac{\partial \phi}{\partial x}\frac{\partial}{\partial y}\hat{\mathscr{L}}\phi - \frac{\partial \phi}{\partial y}\frac{\partial}{\partial y}\hat{\mathscr{L}}\phi\right] - \mu\nabla^4\phi \qquad (5)$$

as given earlier by Terry and Horton.[3,4] The linear operator is $\hat{\mathscr{L}} = \mathscr{L}^h + \mathscr{L}^{ah}$ where $\mathscr{L}^h$ and $\mathscr{L}^{ah}$ are the Hermitian and anti-Hermitian parts of $\hat{\mathscr{L}}$, respectively. The Hermitian operator $\mathscr{L}^h$ determines the wave dispersion, and the anti-Hermitian $\mathscr{L}^{ah}$ the dissipation in the oscillations.

The exact forms of $\mathscr{L}^h$ and $\mathscr{L}^{ah}$ depend on the geometry and collisionality regime of the system. In this study we take

$$\mathscr{L} = -\nabla_\perp^2 + \delta(\tfrac{1}{2}\eta_e + \nabla_\perp^2)\frac{\partial}{\partial y} \qquad (6)$$

For higher collisionality regimes $\eta_e \to 3\eta_e$ and for the trapped electron regime $\eta_e \to -\eta_e$. From Eqs. (5) and (6) it is evident that η_e controls the dissipation at small $|\mathbf{k}|$ and μ controls the dissipation at large $|\mathbf{k}|$. For the dissipationless system $\delta = \mu = 0$ the equation becomes Hasegawa–Mima equation[6] or the Rossby-Wave equation[6-8] and has exact 2D solitary-wave solutions.[8]

For a plasma system with frozen background gradients ($r_n, r_T = \text{const}$) we may consider the local two-dimensional turbulence in a box of volume $L_x \times L_y$ and write

$$\phi(\mathbf{x}, t) = \sum_{\mathbf{k}} \phi_{\mathbf{k}}(t) \exp(i\mathbf{k} \cdot \mathbf{x}) \qquad (7)$$

with $\mathbf{k} = (2\pi n/L_x, 2\pi m/L_y)$. Truncating the $\mathbf{k}$ space to $|\mathbf{k}| \leq K$ we derive from Eqs. (5), (6), and (7) the dynamics in the truncated $\mathbf{k}$ space

$$(1+\chi_\mathbf{k})\frac{d\phi_\mathbf{k}}{dt} = -(ik_y + \mu k_\perp^4)\phi_\mathbf{k} + \sum_{\mathbf{k}_1+\mathbf{k}_2=\mathbf{k}} \frac{\mathbf{k}_1 \times \mathbf{k}_2 \cdot \hat{\mathbf{z}}}{2}(\chi_{\mathbf{k}_2} - \chi_{\mathbf{k}_1})\phi_{\mathbf{k}_1}\phi_{\mathbf{k}_2} \quad (8)$$

where

$$\chi(\mathbf{k}) = \chi_\mathbf{k}' + i\chi_\mathbf{k}'' = k_\perp^2 + i\delta k_y(\tfrac{1}{2}\eta_e - k_\perp^2).$$

The linear modes of Eq. (8) are $\omega_0 = (k_y - i\mu k_\perp^4)/(1+\chi_\mathbf{k})$, which defines the linear frequency and growth-damping rate by

$$\omega_\mathbf{k}^l = \frac{k_y(1+\chi_\mathbf{k}') - \mu k_\perp^4 \chi_\mathbf{k}''}{(1+\chi_\mathbf{k}')^2 + (\chi_\mathbf{k}'')^2} \quad \text{and} \quad \gamma_\mathbf{k}^l = -\frac{k_y \chi_\mathbf{k}'' + \mu k_\perp^4(1+\chi_\mathbf{k}')}{(1+\chi_\mathbf{k}')^2 + (\chi_\mathbf{k}'')^2} \quad (9)$$

The stability parameters of the problem are μ, δ, and η_e.

2. PHASE-SPACE DEFINITIONS AND PROPERTIES

For a truncation with N independent $\mathbf{k}_l$ vectors, $l = 1, 2, \ldots N$ (excluding $-\mathbf{k}_l$ where $\phi_{-\mathbf{k}} = \phi_\mathbf{k}^*$), there are $2N$ first-order differential equations giving a deterministic trajectory for $\mathbf{Y}^{2N}(t) = \{y_i(t)\}_{i=1}^{2N}$ in a particular realization of the system. The most-convenient definition of the phase-space coordinates $y_i(t)$ is

$$(1+\mathbf{k}_l^2)^{1/2}\phi_{\mathbf{k}_l}(t) = y_{2l-1}(t) + i\, y_{2l}(t), \quad l = 1, 2, \ldots N \quad (10)$$

For an ensemble of systems the associated probability density $\rho^{2N}(\mathbf{Y}, t)$ in Γ^{2N} evolves according to the conservation equation

$$\frac{\partial \rho^{2N}}{\partial t} + \sum_{i=1}^{2N} \dot{y}_i \frac{\partial \rho^{2N}}{\partial y_i} = -\rho^{2N} \sum_{i=1}^{2N} \frac{\partial \dot{y}_i}{\partial y_i}$$

where $\dot{y}_i = dy_i/dt$.

2.1. Phase-Space Metric and Volume Contraction

The physical energy density W in the fluctuation field as a fraction of the electron thermal energy density is given by $W/n_e T_e = \tfrac{1}{2}\int[\phi^2 + (\nabla_\perp \phi)^2]dx$, and it is easily shown from Eqs. (5) and (6) that $W = \text{const}$ for $\mu = \delta = 0$. In the dissipationless limit there is an additional invariant of the motion[9,10] called the potential enstrophy $U = \tfrac{1}{2}\int[(\nabla_\perp \phi)^2 + (\nabla_\perp^2 \phi)^2]dx$, which is also easily shown from Eqs. (5) and (6) to be constant for $\mu = \delta = 0$.

In the phase space of the truncated system the energy becomes

$$E(t) = \frac{1}{2}\sum_\mathbf{k}(1+k_\perp^2)|\phi_\mathbf{k}(t)|^2 \equiv \frac{1}{2}\sum_{i=1}^{2N} y_i^2(t) \quad (11)$$

and the potential enstrophy is

$$U(t) = \frac{1}{2}\sum_{\mathbf{k}} k_\perp^2(1+k_\perp^2)|\phi_{\mathbf{k}}|^2 \equiv \frac{1}{2}\sum_{i=1}^{2N} k_i^2 y_i^2(t) \qquad (12)$$

The microcanonical ensemble for the system has entropy $S = k_B \ln \Sigma^{2N}(E)$ where $\Sigma^{2N}(E)dE = d\Omega^{2N}(E)$ with the volume of the $2N$ sphere in phase space is $\Omega^{2N}(E) = \pi^N(2E)^N/N!$. The statistical physics and the evolution to equipartition at constant E and U for the dissipationless system are studied by Montgomery et al.[9,10] The system is ergodic with Gibbs canonical ensemble $\rho^{2N} = \exp(-\alpha W - \beta U)/Z$ predicting the time averaged $|\phi_{\mathbf{k}}|^2$.

For finite values of E the positive definite quadratic form (11) defines the rectangular Cartesian metric for the phase space. The square of the radius vector $\mathbf{Y}(t)$ to the system's phase state is twice the energy of the system.

The volume of the region R in the phase space with this metric is

$$\Omega_R^{2N} = \int_R dy_1 dy_2 \cdots dy_{2N} \qquad (13)$$

and the rate of change of the volume defined by the flow in Eq. (8) is

$$\frac{1}{\Omega^{2N}}\frac{d\Omega^{2N}}{dt} = \sum_{i=1}^{2N} \frac{\partial}{\partial y_i}\left(\frac{dy_i}{dt}\right) = 2\sum_{\mathbf{k}} \gamma_{\mathbf{k}}^l \equiv 2\gamma_T(\delta, \mu, \eta_e) \qquad (14)$$

with $\gamma_{\mathbf{k}}^l$ given by Eq. (9). Since the $\gamma_{\mathbf{k}}^l$'s are constants and functions of the stability parameters there is

$$\text{uniform rate of volume contraction} \Leftrightarrow \sum_{\mathbf{k}} \gamma_{\mathbf{k}}^l < 0$$

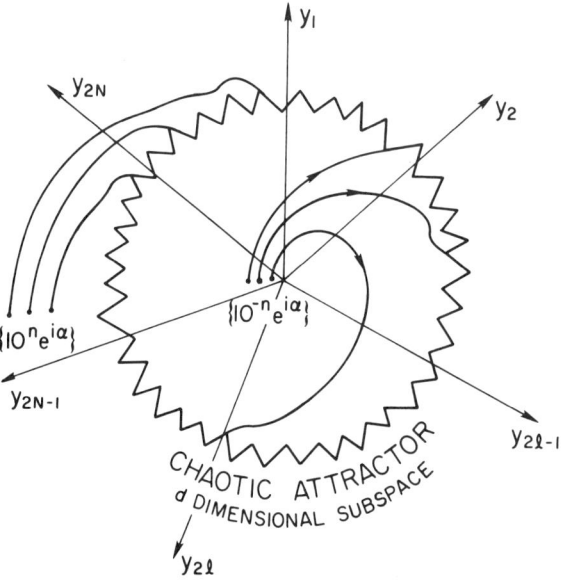

Figure 2. Schematic diagram of the chaotic attractor in the Γ^{2N} phase space.

When $\gamma_T \equiv \sum_{\mathbf{k}} \gamma_k^l < 0$ every phase volume $\Omega_R^{2N}(t)$ contracts to zero at $t \to \infty$ according to

$$\Omega_R^{2N}(t) = \Omega_R^{2N}(t_0) \exp[-2\gamma_T(t - t_0)] \tag{15}$$

Equation (15) shows that after a transient, the steady-state fluctuations occur on a lower-dimensional $d < 2N$ manifold in the phase space. The situation is shown schematically in Fig. 2. We conjecture that d is at most $\approx N$ and may be as small as the number of linearly unstable modes in the system.

2.2. Ergodic Behavior: Phase and Time Averages

Computer experiments indicate that to a good approximation the time averages of the quantities such as $E(t)$, $U(t)$, and the anomalous flux $\Gamma(t) = \sum_k k_y \chi_k'' |\phi_\mathbf{k}(t)|^2$ are equal to the ensemble average over random initial phases. The random-phase ensemble has initial data $\phi_\mathbf{k}(t_0) = |\phi_\mathbf{k}(t_0)| \exp(i\alpha_\mathbf{k})$ with $P(\alpha_\mathbf{k}) d\alpha_\mathbf{k} = (1/2\pi)^N d\alpha_1 d\alpha_2 \cdots d\alpha_N$

Owing to the phase independence of the physical observables such as energy and flux, it is useful to consider the reduced dynamics in the random-phase approximation (RPA). We define the reduced state space of N dimensions with positive definite coordinates $I_{\mathbf{k}_l}(t) = y_{2l-1}^2(t) + y_{2l}^2(t) = |\phi_\mathbf{k}(t)|^2$. The system energy is $E = \frac{1}{2} \sum_k I_k$. Near the origin of the RPA phase space the volume contaction $\sum_k \partial \dot{I}_k / \partial I_k$ is given by $-2\gamma_T$; however, for large $E = \frac{1}{2} \sum_k I_k$ the rate of change of volume depends on $\{I_l\}$ and can be positive or negative.

2.3. Divergence of Neighboring Trajectories

Computations for the Lyapunov exponents for neighboring trajectories shows that the largest Lyapunov exponent λ is positive and typically of order unity in the saturated state. Details of the calculation for a three-wave problem in a 4D phase space are given in Ref. 3 and also apply here.

The exponential divergence of neighboring orbits accounts for the intrinsic stochasticity observed in the time signals. In the presence of an attracting region of phase space, the exponential divergence of neighboring trajectories produces the well-defined fluctuating steady states with properties independent of the initial data $\mathbf{Y}(t_0)$.

3. CHARACTERISTICS OF THE FLUCTUATION SPECTRUM

With extensive computer experiments we have studied the characteristics of the fluctuation spectrum. Saturation or a state with bounded fluctuations in $E(t)$ occurs when the average growth-damping rate per mode

$$\langle \gamma \rangle = \gamma_T/N = \frac{1}{N} \sum_{i=1}^{N} \gamma_{k_i}^l$$

is slightly negative. For some $\gamma_k^l > 0$ and $\langle \gamma \rangle < 0$ the origin $\|\mathbf{Y}\| \to 0$ and $\|\mathbf{Y}\| \to \infty$ are repelling regions of phase space, and there appears a chaotic attracting region at finite $\|\mathbf{Y}\|$.

We find that the basin of attraction to the chaotic attractor is large by observing that for numerous experiments with initial data of the form

$$\{\phi_{\mathbf{k}_l}(t_0)\} = \{10^n e^{i\alpha_l}\} \quad \text{with } -3 \leq n \leq +3 \text{ and random } \alpha_l$$

appear to evolve to the same chaotic attractor as identified by the equivalence of the time-averaged $\mathbf{k}$ spectrum of E, U, and Γ.

That the condition $\langle \gamma \rangle = \gamma_T/N < 0$ is approximately a necessary and sufficient condition for saturation follows from the conservation of energy by the nonlinear coupling the tendency of the system to evolve to equipartition for $\gamma_{\mathbf{k}}^l = 0$. The equipartition tendency is checked with computer experiments by switching off $\gamma_{\mathbf{k}}^l$ for $t > t_1$ in a saturated nonlinear state. The total energy $E(t > t_1)$ is constant but the spectrum $E(\mathbf{k}, t)$ evolves to the distribution $E(\mathbf{k}, t \to \infty) = \text{const} = E(t_1)/N$ as shown in Figs. 5 and 6 of Ref. 4 for $N = 10$. The result that $E(\mathbf{k}, t)$ evolves to equipartition is also a well-known property of the wave–kinetic equation for $\partial_t I_{\mathbf{k}}$ in the RPA.

To calculate the $dE(t)/dt$ from Eq. (8) we note that for every triplet $\mathbf{k}_1 + \mathbf{k}_2 + \mathbf{k}_3 = 0$ ($\mathbf{k} = -\mathbf{k}_3$) interaction the nonlinear contribution to $\partial_t^{nl}(E_{\mathbf{k}_1} + E_{\mathbf{k}_2} + E_{\mathbf{k}_3})$ is proportional to $\phi_{\mathbf{k}_1} \phi_{\mathbf{k}_2} \phi_{\mathbf{k}_3}$ times

$$(\mathbf{k}_1 \times \mathbf{k}_2 \cdot \hat{z})(\chi_{\mathbf{k}_2} - \chi_{\mathbf{k}_1}) + (\mathbf{k}_2 \times \mathbf{k}_3 \cdot \hat{z})(\chi_{\mathbf{k}_3} - \chi_{\mathbf{k}_2}) + (\mathbf{k}_3 \times \mathbf{k}_1 \cdot \hat{z})(\chi_{\mathbf{k}_1} - \chi_{\mathbf{k}_3}) \equiv 0 \quad (16)$$

Thus, the rate of change of $E(t)$ is given by

$$\frac{dE}{dt} = \sum_{\mathbf{k}} 2\gamma_{\mathbf{k}}^l E_{\mathbf{k}}(t) = \left(2 \sum_{\mathbf{k}} \gamma_{\mathbf{k}}^l\right) E^*(t) \quad (17)$$

where the mean-value theorem is used in Eq. (17) to evaluate $E_{\mathbf{k}}(t)$ at a mean-value $E^*(t)$ over the spectrum. Owing to energy conservation, the rate of change $\partial_t E$ is bounded $2\gamma_{\min} E \leq dE/dt \leq 2\gamma_{\max} E$ where the limiting values occur when the energy spectrum is concentrated on the $\mathbf{k}$ with $\gamma_{\min}$ or $\gamma_{\max}$. Suppose the system point tends to infinity, $\|Y(t)\| \to \infty$, then the nonlinear transfer dominates and the system evolves to equipartition of the modal energy. For equipartition the mean value E^* becomes

$$E^* = \frac{1}{N} E = \frac{1}{N} \sum_{l=1}^{N} E_{k_l} \quad \text{for } \|\mathbf{Y}\| \to \infty \quad (18)$$

Substituting Eq. (18) into Eq. (17) shows that for sufficiently large $\|\mathbf{Y}\|$ the total energy decays

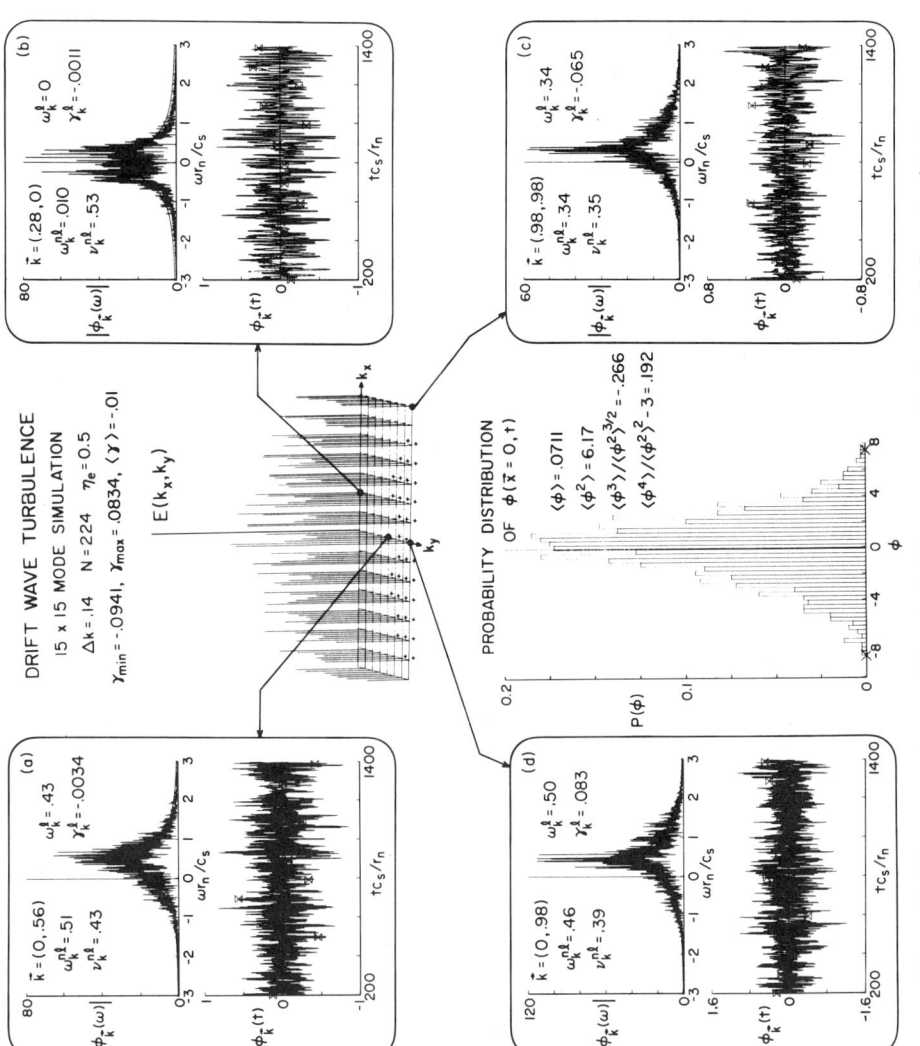

Figure 3. Energy spectrum and four typical frequency spectra in a simulation experiment.

$$\frac{dE}{dt} = \frac{2}{N}\sum_{\mathbf{k}} \gamma_{\mathbf{k}}^l E(t) = 2\langle\gamma\rangle E(t) \qquad (19)$$

provided $\langle\gamma\rangle < 0$.

A typical simulation experiment satisfying $\langle\gamma\rangle < 0$ is shown in Fig. 3. The parameters are $\delta = 1.0$, $\mu = 0.2$, and $\eta_e = 0.5$ with $\Delta k = 2\pi/L = 0.14$ on a 15×15 grid, and thus there are $2N = 224$ first-order equations for $\mathbf{Y}(t)$. For these parameters $\gamma_{\min} = -0.0941$, $\gamma_{\max} = 0.0834$, and $\langle\gamma\rangle = -0.012$.

The center of Fig. 3. shows the time-averaged energy spectrum $E(\mathbf{k})$. The guide lines to the four surrounding graphs show the time signals $\phi_{\mathbf{k}}(t)$ and the frequency spectra $|\phi_{\mathbf{k}}(\omega)|$ of four typical modes from the 112 in $\{E_{\mathbf{k}}\}$. The frequency spectrum $|\phi_{\mathbf{k}}(\omega)|$ is computed from

$$\phi_{\mathbf{k}}(\omega) = \frac{1}{T}\int_{t_0}^{t_0+T} dt\, \phi_{\mathbf{k}}(t) e^{i\omega t} = \frac{1}{M}\sum_{m=0}^{M-1} \phi_{\mathbf{k}}(t_0 + m\Delta t) \exp\left(\frac{i2\pi m\Delta t}{T}\right) \qquad (20)$$

where $T = 1200[r_n/c_s]$ and $M = 1024$. A typical correlation function $\langle\phi_{\mathbf{k}}(t)\phi_{\mathbf{k}}(t+\tau)\rangle$ for one of these modes is shown in Fig. 4 and has $\tau_{1/2} \simeq 5[r_n/c_s]$. The saturated amplitude of the turbulence follows from Fig. 3 with $y_{\mathrm{rms}} = (2E)^{1/2} = 2.6$ or $E = 3.38$.

As reported in earlier studies[3,4] the time series for each mode contains a wide range of frequencies owing to the chaotic attractor. For each mode we parametrize the frequency spectrum by the characteristic nonlinear frequency $\omega_{\mathbf{k}}$, line width $\nu_{\mathbf{k}}$, and the frequency-integrated spectral intensity $I_{\mathbf{k}}$. These are the quantities defined in renormalized turbulence theory, for example, Horton and Choi.[11,12] Two alternative parametrizations have been studied: (1) the Gaussian frequency distribution $G_{\nu_{\mathbf{k}}}(\omega - \omega_{\mathbf{k}}) = I_k(2\pi\nu_{\mathbf{k}}^2)^{-1/2} \exp[-(\omega - \omega_{\mathbf{k}})^2/2\nu_{\mathbf{k}}^2]$ and (2) the Lorentzian parametrization

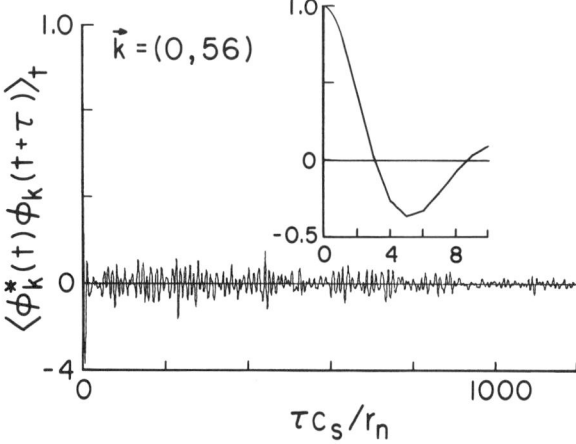

Figure 4. Correlation function for a typical $\phi_{\mathbf{k}}(t)$.

$$L_{\nu_k}(\omega - \omega_k) = I_k \frac{2\nu_k}{(\omega-\omega_k)^2 + \nu_k^2} \tag{21}$$

where $(2\pi)^{-1}\int d\omega G_{\nu_k}(\omega) = (2\pi)^{-1}\int d\omega L_{\nu_k}(\omega) = I_k$. Generally, we find that the Lorentzian, or perhaps the square of a Lorentzian, represents the frequency spectrum better than the Gaussian.

The four frequency spectra in Fig. 3 show both the observed values ω_k, ν_k derived from the nonlinear regression fit to the Lorentzian (21) and the linear frequency and growth-damping rate ω_k^l, γ_k^l from Eq. (9). Weak-turbulence theory[13] is based on ω_k^l, γ_k^l and renormalized turbulence theory[11,12,14,15] gives ω_k, γ_k from ω_k^l, γ_k^l and I_k.

Finally, to assess the applicability of the basic assumption in the statistical theories of plasma turbulence we introduce several measures of the correlation functions and the statistics of the interacting modes. The principal assumption common to different forms of statistical turbulence theory is that the statistics of the modes $\phi_k(t) = y_{2l-1} + iy_{2l}$ are near to those of a Gaussian or a normal probability distribution. It is this assumption, expressed in various forms, that allows the hierarchy of multifield correlations $\langle \phi_1 \phi_2 \cdots \phi_k \rangle$ to be closed in terms of a few low-order moments.

As a typical example we consider the closure of the hierarchy for the quadratic problem $\varepsilon_k \phi_k + \sum_{k_1} \varepsilon_{k_1,k_2}^{(2)} \phi_{k_1} \phi_{k_2} = 0$, where $k_2 = k - k_1$. Assuming that the skewness or three-field correlations are weak, one calculates perturbatively the correlations

$$\langle \phi_k^* \phi_{k_1} \phi_{k-k_1} \rangle = \langle \phi_k^{(1)*} \phi_{k_1}^{(0)} \phi_{k-k_1}^{(0)} \rangle + \langle \phi_k^{(0)*} \phi_{k_1}^{(1)} \phi_{k-k_1}^{(0)} \rangle + \langle \phi_k^{(0)*} \phi_{k_1}^{(0)} \phi_{k-k_1}^{(1)} \rangle \tag{22}$$

and closes the four-field correlation function

$$\langle \phi_k^* \phi_{k_1} \phi_{k_2} \phi_{k_3} \rangle = \langle \phi_k^* \phi_{k_1} \rangle \langle \phi_{k_2} \phi_{k_3} \rangle + \langle \phi_k^* \phi_{k_2} \rangle \langle \phi_{k_1} \phi_{k_3} \rangle + \langle \phi_k^* \phi_{k_3} \rangle \langle \phi_{k_1} \phi_{k_2} \rangle \tag{23}$$

assuming Gaussian statistics at this order. These assumptions are often stated as the "quasinormal approximation" or the "assumption of maximal randomness."

For an individual mode $\phi_k(t) = y_{2l-1}(t) + iy_{2l}(t)$, we compute the probability distribution from the cumulative distribution of $\{y_l(t + m\Delta t), m = 1, 2 \ldots M\}$. A typical result is shown in Fig. 5 for $M = 10^3$ sample points and 50 intervals. The distribution $P(y_l)$ has the first four moments given in the Fig. 5. From this and other samples we conclude that the experiments show that the mean and the variance are $\langle y_l \rangle \simeq \langle y_{2l-1} \rangle \simeq 10^{-3} - 10^{-2}$ and $\langle y_{2l-1}^2 \rangle \simeq 10^{-2} - 10^{-1}$ for typical modes and that the measures of non-gaussianity are typically

$$C_l^3 = \langle y_l^3 \rangle / \langle y_l^2 \rangle^{3/2} \sim \text{few tenths}$$

$$E_l = \langle y_l^4 \rangle / \langle y_l^2 \rangle^2 - 3 \sim \text{few tenths.} \tag{24}$$

In terms of the amplitude $a_l = (y_{2l-1}^2 + y_{2l}^2)^{1/2}$ and phase $\theta_l = \tan^{-1}(y_{2l}/y_{2l-1})$ the measured probability distributions are $P(\theta) = 1/2\pi$ and $P(a) =$

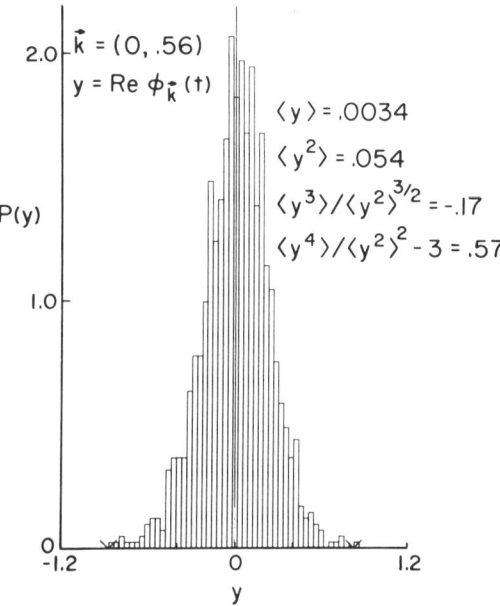

Figure 5. Probability distribution of a typical $\phi_{\mathbf{k}}(t)$.

$a \exp[-a^2/2\langle y^2\rangle]/\langle y^2\rangle$. Time sample points plotted in the $y_{2l-1} - y_{2l}$ phase plane of oscillator $\mathbf{k}_l$ give a Gaussian cloud.

We conclude that the non-Gaussianity is a subdominant, although non-negligible, feature of the fluctuations in a given mode—at least for the experiments with $\delta \sim \eta_e \sim 1$ and $\mu \ll 1$. We suspect that this implies that for the parameters of this experiment, statistical turbulence theories based on small C_l^3 and C_l^4 should yield qualitatively, but not quantitatively, correct predictions for quantities such as $E(\mathbf{k})$ and $\nu(\mathbf{k})$.

The statistics of the total signal $\phi(\mathbf{x}, t) = \sum_{\mathbf{k}} \phi_{\mathbf{k}}(t) \exp(i\mathbf{k} \cdot \mathbf{x})$ are also considered at a given space point $\mathbf{x}$. To the extent that the modes $\phi_{\mathbf{k}}(t)$ are statistically independent we would expect that the skewness of $\phi(\mathbf{x}, t)$ would decrease as $C_\phi^3 \sim NS_l/N^{3/2} \sim S_l/N^{1/2}$ and that $C_\phi^4 \sim C_l^4/N$ by the central-limit theorem. We observe from the experiments, however, that C_ϕ^3 and C_ϕ^4 are rather comparable to C_l^3 and C_l^4 indicating that there are strong, or at least significant, correlations among the $\{\phi_{\mathbf{k}}\}$ in the experiment. We have not been able to test the scaling with N at this time.

Finally, as another measure of the correlations we consider the spatial average $\langle \ \rangle_x \equiv (L_x L_y)^{-1} \int dxdy$ of $\phi^3(\mathbf{x}, t)$. The average of ϕ^3 is

$$\langle \phi^3(\mathbf{x}, t)\rangle_x = \sum_{\mathbf{k}j_1+\mathbf{k}_2+\mathbf{k}_3=0} \phi_{\mathbf{k}_1}(t)\phi_{\mathbf{k}_2}(t)\phi_{\mathbf{k}_3}(t)$$

$$= \sum_{\mathbf{k}} \phi_{\mathbf{k}}^*(t) \sum_{k_1} \phi_{\mathbf{k}_1}(t)\phi_{\mathbf{k}-\mathbf{k}_1}(t)$$

which relative to $\langle\phi^2\rangle_x^{3/2}$, is defined as

$$T(t) = \frac{\langle\phi^3\rangle_x}{\langle\phi^2\rangle_x^{3/2}} = \frac{\sum_k \phi_k^* \sum_{k_1} \phi_{k_1} \phi_{k-k_1}}{(\sum_k |\phi_k|^2)^{3/2}} \quad (25)$$

The mean value of $T(t)$ is approximately zero and its root-mean-square value is $\langle T^2 \rangle_t^{1/2} = 0.17$. Again the finite value of $T(t)$ shows that subdominant but significant correlations are present in the system.

4. RENORMALIZED WAVE–KINETIC EQUATION

Various statistical theories of plasma turbulence have been developed in works too numerous to reference in detail here. The works derive reduced turbulence equations from the primitive equations at various levels of rigor or with ad hoc phenomenological assumptions. In each case, however, the underlying assumption is that the turbulent fields are sufficiently random that only a few lower-order moments of the fluctuations are independent. In such theories the hierarchy of correlations that arise from the nonlinearity of the equations are truncated by neglecting higher-order intrinsic or irreducible correlations. In the usual procedure the fourth-order cumulant [approximation (23)] is neglected in the *renormalized* perturbation expansion and the three-field correlations are computed perturbatively [approximation (22)]. A recent review of the various theoretical formulations is given by Krommes.[14]

The numerical experiments presented in Section 3 suggest that the dissipative drift wave turbulence may be a problem for which the principal assumption of statistical turbulence theory is satisfied. In this section we apply turbulence theory to this problem with the objective of testing its prediction with the results of the simulation experiments. In $\mathbf{k}\omega$ space the mode-coupling equation for Eq. (8) is

$$(1 + k_\perp^2)[\omega - \omega_k^l - i\gamma_k^l]\phi_k - \sum_{k_1 + k_2 = k} \frac{i\mathbf{k}_1 \times \mathbf{k}_2 \cdot \hat{\mathbf{z}}}{2}(\chi_{k_2} - \chi_{k_1})\phi_{k_1}\phi_{k_2} = 0 \quad (26)$$

where the short-hand notation is $k = (\mathbf{k}, \omega)$ and

$$\sum_{k_1 + k_2 = k} = (2\pi)^{-2} \int d\omega_1 d\omega_2 \delta(\omega - \omega_1 - \omega_2) \sum_{k_1, k_2} \delta_{k_1 + k_2, k}$$

are introduced.

The mode coupling Eq. (26) is of the form

$$\varepsilon_k^l \phi_k + \sum_{k_1 + k_2 = k} \varepsilon_{k_1, k_2}^{(2)} \phi_{k_1} \phi_{k_2} = 0$$

which has been studied extensively in the plasma-physics literature. The weak-turbulence equation are given by Galeev and Sagdeev[13] in Eqs. (1.63) and (3.9).

The renormalized turbulence equations are given by Kadomtsev[15] in Eqs. (III.6) and (III.7); in Ref. 11 in Eqs. (3.40) and (3.41); in Ref. 12 in Eqs. (14) and (15); in Ref. 14 in Eqs. (118) and (123); in Ref. 16 in Eqs. (2.5) and (2.18) or (2.28); and in Ref. 17 in Eqs. (2.113) and (2.121). The basic result of these works is two reduced equations for the unknown nonlinear response function $\varepsilon_k(\omega)$ and the spectral distribution $I_k(\omega)$ of the potential ϕ_k fluctuation spectrum in terms of the linear-response function ε_k^l and the coupling function $\varepsilon_{k_1,k_2}^{(2)}$.

The two turbulence theory equations are

$$\varepsilon_k(\omega) = \varepsilon_k^l(\omega) - \sum_{k_1} \int_{-\infty}^{+\infty} \frac{d\omega_1}{2\pi} \frac{(k \times k_1 \cdot \hat{z})^2 (\chi_{k-k_1} - \chi_{k_1})(\chi_{-k_1} - \chi_k) I(k_1, \omega_1)}{\varepsilon_{k-k_1}(\omega - \omega_1)} \quad (27)$$

$$|\varepsilon_k(\omega)|^2 I(k, \omega) = \frac{1}{2} \sum_{k_1} \int_{-\infty}^{+\infty} \frac{d\omega_1}{2\pi} (k \times k_1 \cdot \hat{z})^2 |\chi_{k-k_1} - \chi_{k_1}|^2 I(k_1, \omega_1) I(k-k_1, \omega - \omega_1) \quad (28)$$

Equations (27) and (28) determine I_k and ε_k. The equations are non-Markovian depending on the time history of the fluctuations. Physically, Eqs. (27) and (28) follow from the selective summation to all orders of the most secular contributions in the small ϕ_k perturbation expansion of the mode-coupling equation. These equations are often called the DIA (direct interaction approximation) equation from the early fluid turbulence work of Kraichnan.[18]

4.1. Markovian Reduction

The simulation study in Section 3 shows that the turbulent fluctuations have a short correlation time $\tau_c = 1/\nu_k$ owing to the chaotic attractor. To describe the rapid fluctuations and the nonlinear evolution of the fields we introduce the relative time $\tau = t-t'$ and the centered time $T = \frac{1}{2}(t+t')$ variables in the two-time correlation function

$$\langle \phi_k^*(t) \phi_k(t') \rangle = \int_{-\infty}^{+\infty} \frac{d\omega}{2\pi} I(k, \omega, T) \exp(-i\omega\tau) \quad (29)$$

and observe that

$$\omega_k \sim \nu_k \sim \frac{1}{\tau_c(k)} \gg \frac{d}{dT} \quad (30)$$

from Figs. 3 and 4. Thus, we consider approximate solutions of Eqs. (27) and (28) of the form

$$\varepsilon_k(\omega) = (1 + k_\perp^2)(\omega - \omega_k + i\nu_k) \quad (31)$$

$$I_k(\omega) = I(k) \frac{2\nu_k}{(\omega - \omega_k)^2 + \nu_k^2} \quad (32)$$

with ω_k, ν_k, $I(k)$ having a local dependence on the slow time variable T. This ω parametrization is called the "pole approximation" by Dubois and Rose,[19] who study its validity for the Zakaharov fluid model of coupled Langmuir–ion acoustic turbulence.

Equations (31) and (32) are then a particularly simple frequency parametrization of the unknown functions $\varepsilon_k(\omega)$ and $I_k(\omega)$. In demanding that approximations (31) and (32) satisfy Eqs. (27) and (28) to the dominant order we obtain three reduced Markovian equations for the unknowns $\omega_k(T), \nu_k(T)$, and $I_k(T)$.

For the Lorentzian frequency distribution $L_{\nu_k}(\omega - \omega_k)$ defined in Eq. (21) the ω_1 intergrals in Eqs. (27) and (28) can be performed by contour integration to obtain

$$\int_{-\infty}^{+\infty} \frac{d\omega_1}{2\pi} \frac{L_{\nu_{k_1}}(\omega_1 - \omega_{k_1})}{\omega - \omega_1 - \omega_{k_2} + i\nu_{k_2}} = \frac{1}{\omega - \omega_{k_1} - \omega_{k_2} + i(\nu_{k_1} + \nu_{k_2})} \tag{33}$$

and

$$\int_{-\infty}^{+\infty} \frac{d\omega_1}{2\pi} L_{\nu_{k_1}}(\omega_1 - \omega_{k_1}) L_{\nu_{k_2}}(\omega - \omega_1 - \omega_{k_2}) = L_{\nu_{k_1} + \nu_{k_2}}(\omega - \omega_{k_1} - \omega_{k_2}) \tag{34}$$

We substitute Eqs. (31), (32), (33), and (34) into Eqs. (27) and (28) and evaluate the right-hand members at $\omega = \omega_k + i\nu_k$ the frequency about which the equation is peaked. The resulting equations for ω_k, ν_k and I_k are

$$\omega_k = \omega_k^l + \operatorname{Re} \sum_{k_1} \frac{(\mathbf{k} \times \mathbf{k}_1 \cdot \hat{z})^2 (\chi_{k-k_1} - \chi_{k_1})(\chi_k - \chi_{-k_1}) I_{k_1}}{(1 + k^2)(1 + k_2^2)[\omega_k - \omega_{k_1} - \omega_{k_2} + i(\nu_k + \nu_{k_1} + \nu_{k_2})]} \tag{35}$$

$$\nu_k = -\gamma_k^l - \operatorname{Im} \sum_{k_1} \frac{(\mathbf{k} \times \mathbf{k}_1 \cdot \hat{z})^2 (\chi_{k_2} - \chi_{k_1})(\chi_k - \chi_{-k_1}) I_{k_1}}{(1 + k^2)(1 + k_2^2)[\omega_k - \omega_{k_1} - \omega_{k_2} + i(\nu_k + \nu_{k_1} + \nu_{k_2})]} \tag{36}$$

and

$$2\nu_k(1 + k_\perp^2)^2 I_k = -\operatorname{Im} \sum_{k_1} \frac{(\mathbf{k} \times \mathbf{k}_1 \cdot \hat{z})^2 |\chi_{k_1} - \chi_{k_2}|^2 I_{k_1} I_{k_2}}{\omega_k - \omega_{k_1} - \omega_{k_2} + i(\nu_k + \nu_{k_1} + \nu_{k_2})} \tag{37}$$

where $\mathbf{k}_2$ is short-hand for $\mathbf{k} - \mathbf{k}_1 = \mathbf{k}_2$ under the $\mathbf{k}_1$ summation.

Equations (35), (36), and (37) describe the steady-state turbulence. The evolution and stability of the system on the slow time variable $T = \frac{1}{2}(t + t')$ defined in Eq. (29) is given by the evaluation of the frequency integral of Im $\varepsilon_k I_k$. We introduce the phase-averaged modal energy from Eq. (11)

$$E(\mathbf{k}, t = T) = (1 + k_\perp^2) I(\mathbf{k}, T) \tag{38}$$

and define the response function

$$R_{\mathbf{k},\mathbf{k}_1,\mathbf{k}_2} = \frac{1}{\omega_k - \omega_{k_1} - \omega_{k_2} + i(\nu_k + \nu_{k_1} + \nu_{k_2})}$$

$$= \frac{1}{-\omega_{k_1} - \omega_{k_2} - \omega_{k_3} + i(\nu_{k_1} + \nu_{k_2} + \nu_{k_3})} \tag{39}$$

which is symmetric in $\mathbf{k}_1$, $\mathbf{k}_2$, $\mathbf{k}_3$ where $\mathbf{k}_3 = -\mathbf{k}$.

From Eq. (37) the steady-state balance occurs when the source $S_\mathbf{k}$ function balances the decay rate $\nu_\mathbf{k}$

$$2\nu_\mathbf{k} E_\mathbf{k} = S_\mathbf{k}$$

where

$$S_\mathbf{k} = -\mathrm{Im}\sum_{\mathbf{k}_1} \frac{(\mathbf{k}\times\mathbf{k}_1\cdot\hat{z})^2|\chi_{\mathbf{k}_2}-\chi_{\mathbf{k}_1}|^2 R_{\mathbf{k},\mathbf{k}_1,\mathbf{k}_2} E_{\mathbf{k}_1} E_{\mathbf{k}_2}}{(1+k^2)(1+k_1^2)(1+k_2^2)} \quad (40)$$

where $\nu_\mathbf{k}$ is given by Eq. (36). Out of equilibrium, the slow time scale evolution of the system is given by

$$\frac{dE_\mathbf{k}}{dt} = -2\nu_\mathbf{k} E_\mathbf{k} + S_\mathbf{k}$$

$$= 2\gamma_\mathbf{k}^l E_\mathbf{k} + C_\mathbf{k}^{nl}(\{E_\mathbf{k}\}) \quad (41)$$

where the nonlinear modal interaction or "collision" operator $C_\mathbf{k}^{nl}(\{E_\mathbf{k}\})$ is given by

$$C_\mathbf{k}^{nl}(\{E_\mathbf{k}\}) = \mathrm{Im}\sum_{\mathbf{k}_1} \frac{(\mathbf{k}\times\mathbf{k}_1\cdot\hat{z})^2(\chi_{\mathbf{k}_2}-\chi_{\mathbf{k}_1})}{(1+k^2)(1+k_1^2)(1+k_2^2)} R_{\mathbf{k},\mathbf{k}_1,\mathbf{k}_2}$$

$$\times[-(\chi_{\mathbf{k}_2}^*-\chi_{\mathbf{k}_1}^*)E_{\mathbf{k}_1}E_{\mathbf{k}_2} + (\chi_\mathbf{k}-\chi_{\mathbf{k}_1}^*)E_{\mathbf{k}_1}E_\mathbf{k} - (\chi_\mathbf{k}-\chi_{\mathbf{k}_2}^*)E_{\mathbf{k}_2}E_\mathbf{k}] \quad (42)$$

where we have used Eqs. (36) and (40) in Eq. (41) to write out $C_\mathbf{k}^{nl}$.

Equation (41) is a Markovian description of the phase-averaged spectral dynamics where the evolution $C_\mathbf{k}^{nl}$ depends on the local nonlinear frequency $\omega_\mathbf{k}(T)$ and decay rate $\nu_\mathbf{k}(T)$. Equations (35), (36), and (41) are called the renormalized wave kinetic equation.

4.2. Properties of the Renormalized Wave–Kinetic Equation

The renormalized wave kinetic equation describes the nonlinear system in an N-dimensional state space with the following properties.

Equipartition of Modal Energy. The equilibrium solution of the undriven problem $\gamma_\mathbf{k}^l = 0$ is

$$E_\mathbf{k}(t) = \frac{1}{N}E_T = \mathrm{const} \quad (43)$$

The equipartition solution is a stable fixed point in the N-dimensional state space with $\dot{E}_\mathbf{k} = C^{nl}(\{E_\mathbf{k}\}) = 0$. The stability of the fixed point (43) can be shown by introducing the entropy production functional $\sigma(\{E_\mathbf{k}\}) \geq 0$ as the Lyapunov stability functional and showing that all perturbations from (43) result in $d\sigma/dt < 0$.

Conservation of Energy in the Nonlinear Transfer. The nonlinear interaction $C_\mathbf{k}^{nl}(\{E_\mathbf{k}\})$ acts on each triplet in such a way as to conserve wave energy. Consider the summation with $\mathbf{k}_3 = -\mathbf{k}$ of Eq. (42)

$$\sum_{\mathbf{k}} C_{\mathbf{k}}^{nl}(\{E_{\mathbf{k}}\}) = \text{Im} \sum_{\mathbf{k}_1+\mathbf{k}_2+\mathbf{k}_3=0} \frac{(\mathbf{k}_1 \times \mathbf{k}_2 \cdot \hat{\mathbf{z}})^2 R_{1,2,3}}{(1+k_1^2)(1+k_2^2)(1+k_3^2)}[-(\chi_2-\chi_1)(\chi_2^*-\chi_1^*)E_1E_2$$
$$+(\chi_2-\chi_1)(\chi_3^*-\chi_1^*)E_1E_3 - (\chi_2-\chi_1)(\chi_3^*-\chi_2^*)E_2E_3]$$

Interchanging 3 and 2 in the second term and 3 and 1 in the third term and using the symmetry of

$$R_{123} = \left(-\sum_i \omega_{\mathbf{k}_i} + i\sum_i \nu_{\mathbf{k}_i}\right)^{-1}$$

we obtain

$$\sum_{\mathbf{k}} C_{\mathbf{k}}^{nl}(\{E_{\mathbf{k}}\}) = \text{Im} \sum_{\mathbf{k}_1+\mathbf{k}_2+\mathbf{k}_3=0} \frac{(\mathbf{k}_1 \times \mathbf{k}_2 \cdot \hat{\mathbf{z}})^2 R_{1,2,3}}{(1+k_1^2)(1+k_2^2)(1+k_3^2)}$$
$$\times E_1E_2(\chi_2^*-\chi_1^*)[-\chi_2+\chi_1+\chi_3-\chi_1+\chi_2-\chi_3] = 0$$

If $\chi_{\mathbf{k}}$ is purely real or purely imaginary, then we can repeat this analysis for

$$\sum_{\mathbf{k}} \chi_{\mathbf{k}}' C_{\mathbf{k}}^{nl} \quad \text{or} \quad \sum_{\mathbf{k}} \chi_{\mathbf{k}}'' C_{\mathbf{k}}^{nl}$$

to show that the summation reduces to

$$E_1E_2(\chi_2-\chi_1)[-\chi_3(\chi_2-\chi_1)+\chi_2(\chi_3-\chi_1)+\chi_1(\chi_2-\chi_3)] \equiv 0$$

Thus, for example, for $\chi_{\mathbf{k}}'' = 0$ and $\chi_{\mathbf{k}}' = k_\perp^2$ the potential enstrophy

$$U = \sum_{\mathbf{k}} k_\perp^2 E_{\mathbf{k}}$$

is also a constant of the motion of the renormalized wave-kinetic equation.

Decoupling of k = 0. In the limit $\mathbf{k} \to 0$ the quantities $\omega_{\mathbf{k}} \to 0$, $\nu_{\mathbf{k}} \to 0$, and $S_{\mathbf{k}} \to 0$ so that the fluctuations completely decouple from the $\mathbf{k} = 0$ equations.

Small |k| Limit. For small $|\mathbf{k}|$ compared with the average wavenumber $\bar{k}$ in the spectrum, defined by

$$\bar{k} = \left(\sum_k k^2 I_k \Big/ \sum_k I_k\right)^{1/2}$$

the induced turbulent damping ν_k reduces to a nonlinear eddy viscosity varying as $\nu_n k^n$, where ν_n varies as E for $E < 1$ and $E^{1/2}$ for $E \geqslant 1$ as shown in Fig. 9 of Ref. 4. For small k the resonance function reduces to $R_{\mathbf{k},\mathbf{k}_1,\mathbf{k}-\mathbf{k}_1} \simeq (\omega_{\mathbf{k}} - \mathbf{k} \cdot \mathbf{v}_{\mathbf{k}_1} + 2i\nu_{\mathbf{k}_1})^{-1}$ where we define $\mathbf{v}_{\mathbf{k}_1} = \partial \omega_{\mathbf{k}_1}/\partial \mathbf{k}_1$.

For small k compared with $\langle k_\perp^2 \rangle^{1/2} = \bar{k}$ and an isotropic spectrum for $E_{\mathbf{k}_1}$ (for simplicity) we obtain two regimes for $\nu_{\mathbf{k}}$ and $S_{\mathbf{k}}$ depending on the strength of dissipation. For the nearly conservative system the small k limit becomes

$$\nu_k = 2\pi k^4 \int_{k^2}^{\infty} \frac{dk_1^2 k_1^4}{(1+k_1^2)^2} \frac{2\nu_{k_1}\frac{\partial}{\partial k_1^2}(k_1^2 E_{k_1})}{(\omega - \mathbf{k}\cdot\mathbf{v}_{k_1})^2 + 4\nu_{k_1}^2} \simeq \nu_c(E)k^4$$

The $\nu_k = \nu_c k^4$ limit applies when $|\omega - \mathbf{k}\cdot\mathbf{v}_{k_1}| < 2\nu_{\bar{k}}$ where $\nu_c \propto E^{1/2}$. The result of $\nu_c k^4$ is given in Ref. 20.

In contrast for dissipative systems the leading contribution is

$$\nu_k = 4\pi k^2 \int_{k^2}^{\infty} \frac{dk_1^2 k_1^2 (\chi''_{\mathbf{k}_1})^2 2\nu_{k_1} E_{\mathbf{k}_1}}{(1+k_1^2)^2[(\omega - \mathbf{k}\cdot\mathbf{v}_{k_1})^2 + 4\nu_{k_1}^2]} \cong \nu_d(E)k^2$$

where the $\nu_k = \nu_d k^2$ limit applies for $|\omega - \mathbf{k}\cdot\mathbf{v}_{k_1}| < 2\nu_{\bar{k}}$ and $\nu_d \propto E^{1/2}$. This analysis for the eddy viscosity shows that the dissipation changes the wavenumber scaling of the decay rate ν_k of the turbulent fluctuations.

5. CONCLUSIONS

The dissipative drift wave equations have stochastic solutions with broad frequency spectra for each $\mathbf{k}$ mode. The simplest explanation for this behavior is the presence of a chaotic attractor in the phase space of the system. The chaotic attractor exists in each (typical) triplet of the $\mathbf{k}$ spectrum as shown in the three-wave study of Terry and Horton.[3] The statistical properties of the signals on the attractor are independent of initial data owing to the exponential separation of neighboring orbits, a property also present in each triplet. The flow is mixing on the attractor.

The basin of attraction defined as the domain of data $\{\phi_\mathbf{k}(t_0)\}$, which are pulled into the attractor after a transient, may be very large as indicated from a few randomly selected initial data sets. No attempt has been made at determining the dimensionality or the boundary of the attractor.

On the attractor the time average appears equal to the phase average over the random-phase ensemble. Thus, the attractor gives rise to a well-defined turbulent steady state with a unique spectral distribution and a well-defined anomalous transport flux. The turbulent state has no obvious inconsistencies with the $\mathbf{k}\omega$ spectra and anomalous transport of the so-called microturbulence measured by electromagnetic wave scattering experiments in tokamaks such as in the Mazzucato experiment.[21]

The statistical properties of the solutions in the steady state are examined for deviations from Gaussian statistics. The cumulative distribution of the time series $\phi_\mathbf{k}(t)$ shows an approximately normal distribution. The deviations from Gaussianity are finite and characterized by the values given in Eqs. (24). The composite signal $\phi(\mathbf{x},t)$ has non-Gaussian features, Fig. 3, of the same order of magnitude as the individual $\phi_\mathbf{k}(t)$'s indicating significant correlations between the $\mathbf{k}$ modes. We conclude that the signals are sufficiently close to Gaussian to allow turbulence theories based on the quasinormal approximation (23) to be qualitatively, but perhaps not quantitatively, valid for these simulations.

Although we cannot rule out the possibility that these non-Gaussian features will scale out with increasing N to justify statistical turbulence theory, we find no evidence for this trend within present experiments with $N \leq 224$.

Statistical turbulence theory applied to these experiments is reduced to a Markovian description for $\omega_\mathbf{k}(T)$, $\nu_\mathbf{k}(T)$, and $I_\mathbf{k}(T)$ based on parametrization of the frequency spectra $\phi_\mathbf{k}(\omega)$ by a Lorentzian frequency distribution. The resulting system of Eqs. (35), (36), and (41) when solved numerically show stable steady states with properties in reasonable qualitative agreement with the time-averaged features of the phase-dependent equations. A similar demonstration and conclusion is given by Waltz[22] in his study of the dissipative drift wave equations.

The renormalized wave–kinetic equation predict broad frequency spectra with the nonlinear fluctuation decay rate $\nu_k \sim \omega_k$. The renormalized equations show that the scaling of ν_k with k at small k is approximately $k^2 E^{1/2}$, where E is the total turbulent energy density. The turbulence equation also predicts that the root-mean-square fluctuating amplitude $\tilde{\phi}$ increases with weak instability and saturates at the mixing length level $\langle (\mathbf{k} \cdot \mathbf{v}_E)^2 \rangle^{1/2} \sim |k_y v_{de}| \sim \nu_k \gg \gamma_k^l$ as also shown by Waltz.[22] The scaling of $\tilde{\phi}$ and ν_k predicted by the Terry–Horton model for dissipative drift wave turbulence is in marked contrast to the prediction of the dissipationless Hasegawa–Mima model where $\nu_k \propto k^4 \tilde{\phi}$ with $\tilde{\phi}$ an arbitrary constant of the motion.

ACKNOWLEDGMENTS

The author wishes to acknowledge the efficient work of Lee Leonard in performing the numerical simulations. This work was supported by the DOE Grant No. DE-FG05-80ET-53088.

REFERENCES

1. A. B. Mikhailovskii, *Theory of Plasma Instabilities* (Consultants Bureau, New York, 1974) Vol. III. Also, "Instabilities of an Ihomogeneous Plasma," in *Basic Plasma Physics Handbook I*, (North Holland, Amsterdam, to be published) Sec. VI.5
2. F. L. Hinton and C. W. Horton, *Phys. Fluids* **14**, 116 (1971).
3. P. W. Terry and W. Horton, *Phys. Fluids* **25**, 491 (1982).
4. P. W. Terry and W. Horton, *Phys. Fluids* **26**, 107 (1983).
5. D. H. E. Dubin, J. A. Krommes, C. Oberman, and W. W. Lee, Princeton Plasma Physics Laboratory Report (1983).
6. A. Hasegawa and K. Mima, *Phys. Rev. Lett.* **39**, 205 (1977).
7. V. D. Larichev and G. M. Reznik, *Oceanologia* **16**, 547 (1976).

8. G. R. Fied, V. D. Larichev, J. C. McWilliams, and G. M. Reznik, *Dynamics of Atmospheres and Oceans* **5**, 11 (1980).
9. D. Fyfe and D. Montgomery, *Phys. Fluids* **22**, 246 (1979).
10. R. H. Kraichnan and D. Montgomery, *Rep. Prog. Phys.* **43**, 547 (1980).
11. W. Horton and D. I. Choi, *Phys. Rep.* **49**, 273 (1979).
12. W. Horton, "Renormalized Plasma Turbulence Theory," in *Long-Time Prediction in Dynamics*, W. Horton, L. E. Reichl, and V. G. Szebehely, eds. (Wiley, New York, 1983), p. 301.
13. A. A. Galeev and R. Z. Sagdeev, "Nonlinear Plasma Theory," in *Reviews of Plasma Physics*, M. A. Leontovich, ed. (Consultants Bureau, New York 1973), Vol. 7. pp. 1–54.
14. J. A. Krommes, "Statistical Descriptions and Plasma Turbulence," in *Basic Plasma Physics Handbook I*, (North Holland, Amsterdam, to be published).
15. B. B. Kadomtsev, *Plasma Turbulence* (Academic Press, London, 1965).
16. D. F. Dubois and M. Espedal, *Plasma Physics* **20**, 1209 (1978).
17. V. N. Tsytovich, *Theory of Turbulent Plasma* (Consultants Bureau, New York, 1977), pp. 84–95.
18. R. N. Kraichnan, *J. Fluid Mech.* **5**, 497 (1959).
19. D. F. Dubois and H. A. Rose, *Phys. Rev. A* **24**, 1476 (1981).
20. W. Horton, *Physica* **2D**, 107 (1981).
21. E. Mazzucato, *Phys. Rev. Lett.* **48**, 1828 (1982).
22. R. E. Waltz, *Phys. Fluids* **26**, 169 (1983).

SOLITONS IN TURBULENT FLOW

J. D. Meiss
Institute for Fusion Studies, The University of Texas, Austin, Texas

COHERENT STRUCTURES IN TURBULENCE

In this chapter I will discuss the connection between two seemingly dissimilar concepts: turbulence and coherence. Historically, turbulent flow has been characterized by extreme incoherence or randomness, the most successful theoretical treatments assuming quasi-Gaussianity. At the opposite end of the spectrum of fluid motion are laminar flows such as isolated vortices, which we refer to as coherent structures. Perhaps the most coherent structure is the soliton, which maintains its integrity as a manifestation of the integrability of the field theory that describes it.

Over the last 10 years, there have been a large number of observations of organized flow in fluids commonly termed turbulent. An example is the experiments of Brown and Roshko[5] (and many others) on mixing layers (the interface between two fluids moving at different velocities). Visualizing the fluid flow with injected dyes or reflecting particles shows that even for very large Reynolds numbers, the turbulent shear layer is composed of vortexlike structures which roll up the interface between two fluids. The vorticies propagate downstream, interacting and combining. A similar case is the formation of "bursts" and turbulent spots in boundary layers.[6,7] These local patches begin to appear intermittently near the critical Reynolds number for transition to turbulence. There is some evidence that they persist even in fully turbulent situations. Finally, we mention the elegant experiments of Swinney, as reported at this Workshop, on Taylor–Couette flow. Here, remnants of Taylor vortices are observed even at extremely high Reynolds numbers.

Several examples of intermittent behavior in plasma turbulence may also be found. Satellite measurements of the electric fields in the auroral zone show localized pulses propagating parallel to the magnetic field, which may be interpreted as double layers and ion–acoustic solitary waves.[25] The most striking observation is the high degree of intermittency in the electric field. If we define the intermittency coefficient as

$$I = \frac{\text{Number of Pulses} \times \text{Pulse Width}}{\text{Time}} \qquad (1)$$

the measurements give $I \simeq 0.1$–0.2, as noted by Lotko.[13] Measurements made closer to home of turbulence in tokamaks have also begun to reveal a coherent component. Photographs of the edge region show localized filaments extended in the toroidal direction which are presumably regions of higher density.[29] Laser interferometry detects these density pulses and shows that they propagate poloidally and remain coherent for some distance, as reported by Surko and Slusher.[24] We have speculated that such coherent structures could be described as solitary drift waves.[14]

A major problem in determining the extent to which turbulent flow consists of coherent features is quantifying this idea. While most observations of coherence use flow visualization in distinguishing incoherent from coherent, it is not clear how mathematically to separate the two. One important, but vague, idea is the intermittency of the flow, as in Eq. (1). A better technique would be to construct the probability distribution of a field variable, say $P(\phi)$. A conventional turbulence theory would assume quasi-Guassianity, while the coherent structure probability distribution would have non-Gaussian features. As an example, solitary waves tend to form with a distinct sign, and thus would lead to a large third moment. Berman et al[4] have used this technique as a diagnostic for the existence of electron-holes in a plasma situation.

INTEGRABLE VERSUS NONINTEGRABLE FIELD THEORIES

In this chapter I will concentrate on the simplest models containing coherent structures: one-dimensional nonlinear fields. It is fitting that the premier coherent structure, the soliton, was discovered in the unsuccessful attempt to show numerically that nonlinear systems would be ergodic, and thus that statistical methods would apply.

Since the numerical discovery of the soliton by Zabusky and Kruskal,[27] analytical progress has shown that a class of nonlinear differential equations can be integrated using effectively linear techniques. These equations, the integrable theories, are distinguished by three properties (for reviews see Scott et al.[23] or Ablowitz and Segur[2]): the existence of an inverse-scattering transform—which allows formal solution of the initial value problem; a Backlund transformation—the nonlinear superposition principle; and an infinite set of independent conserved quantities. The soliton solutions of an integrable field theory are localized traveling waves that have the remarkable property that they are preserved upon collision. It can be shown that an arbitrary localized initial state evolves into a number of solitary waves and some small-amplitude (roughly) linear waves.

As illustrative examples of the contrast between integrable and nonintegrable systems, consider the Korteweg–de Vries (KdV) and regularized-long-wave (RLW) equations,[3]

$$\phi_t + \phi_x + \phi_{xxx} - \phi\phi_x = 0 \quad \text{(KdV)} \tag{2}$$

$$\phi_t + \phi_x - \phi_{xxt} - \phi\phi_x = 0 \quad \text{(RLW)} \tag{3}$$

Each of these equations has a solitary-wave solution of the form

$$\phi_s(x, t) = A \operatorname{sech}^2[k(x - ct)] \tag{4}$$

where the amplitude A, width k^{-1}, and speed c are related by

$$A = -3(c - 1)$$

$$k = \begin{pmatrix} \frac{1}{2}(c-1)^{1/2} & \text{(KdV)} \\ \frac{1}{2}(1 - 1/c)^{1/2} & \text{(RLW)} \end{pmatrix} \tag{5}$$

Note that for KdV, the speed must be greater than one; while for RLW, c may also be negative.

The two branches of the RLW solitary wave will be referred to as the KdV ($c > 1$) and plasma ($c < 0$) branches. The KdV branch solitary waves are quite similar to actual solutions of the KdV equation, since if we take the small-wavenumber, $k \ll 1$, limit of the RLW equation, it is equivalent to KdV. The plasma branch is so named because the RLW equation is applicable to drift waves in plasmas and the $c < 0$ solitary wave should be observable in this case.[14,22]

Each of these equations has three conserved quantities obtained by symmetry considerations: mass, momentum, and energy:

$$M = \int dx\, \phi$$

$$P = \begin{cases} \frac{1}{2} \int dx\, \phi^2 & \text{(KdV)} \\ \frac{1}{2} \int dx\, (\phi^2 + \phi_x^2) & \text{(RLW)} \end{cases}$$

$$H = \begin{cases} \int dx(-\frac{1}{2}\phi^2 + \frac{1}{2}\phi_x^2 + \frac{1}{6}\phi^3) & \text{(KdV)} \\ \int dx(-\frac{1}{2}\phi^2 + \frac{1}{6}\phi^3) & \text{(RLW)} \end{cases} \tag{6}$$

However, the KdV equation has an infinite sequence of such polynomial-conserved quantities,[17] each containing a term $1/n!\,\phi^n$. RLW, by contrast, has *only* three independent polynomial-conserved quantities.[20]

Both equations are Hamiltonian systems, and Eqs. (2) and (3) can be written

$$\phi_t = [\phi, H]$$

where the Poisson bracket is defined (for review see Morrison[18]) as

$$[A, B] = \int dx \frac{\delta A}{\delta \phi} \mathcal{O} \frac{\delta B}{\delta \phi}$$

$$\mathcal{O} = \begin{cases} \partial_x & \text{(KdV)} \\ \partial_x(1 - \partial_x^2)^{-1} & \text{(RLW)} \end{cases}$$

Use of the inverse-scattering transform allows one to show that the KdV equation is a completely integrable Hamiltonian system: it can be transformed to action-angle form as shown by Zakharov and Faddeev.[28] No such transformation has been found for the RLW equation, and we strongly suspect that none exists.

One of the most striking manifestations of the nonintegrability of the RLW equation appears when we allow two solitary waves to collide. These collisions are inelastic: besides the two incoming solitary waves, small-amplitude radiation, and sometimes new solitary waves are emitted.[19] Collisions between solitary waves with like-signed velocities are nearly elastic.[1] In fact, when the speeds are only moderately larger than one, the emitted radiation is difficult to detect. This is probably because, for these parameters, the RLW equation is near to the integrable KdV equation.

Collision between one positive- and one negative-velocity solitary wave can result in four outgoing solitary waves as well as radiation. Figure 1 shows a collision with $c_1 = 2.5$ and $c_2 = -1.5$. Note that during the collisions, sharp peaks are produced, and so care must be taken to ensure numerical convergence. After the collision, four peaks which obey the amplitude, width, and speed relations of Eq. (5) are seen, as well as a localized bundle of radiation. The resulting speeds are $c_1 = 2.31$, $c_2 = -1.20$, $c_3 = 1.32$, and $c_4 = -0.37$. The radiation packet spreads out roughly in accord with the linear group velocity.

We have found an entire region of the parameter space (c_1, c_2) above a certain threshold, for which two new solitary waves are produced.[19] Below that threshold only radiation is emitted.

These results can be stated in terms of the S-matrix, which relates the state at $t = -\infty$ to that at $t = +\infty$. For the KdV equation, the preservation of solitons implies the S-matrix is diagonal (at least in the soliton sector, treatment of the radiation is perhaps more difficult). For RLW there are nondiagonal terms in the S-matrix for solitary-wave production as well as radiation production. In addition, time-reversal symmetry implies that it is possible to construct new solitary waves from solitary waves and radiation, and possibly from radiation alone.

Similar considerations apply to other nonintegrable systems, such as the

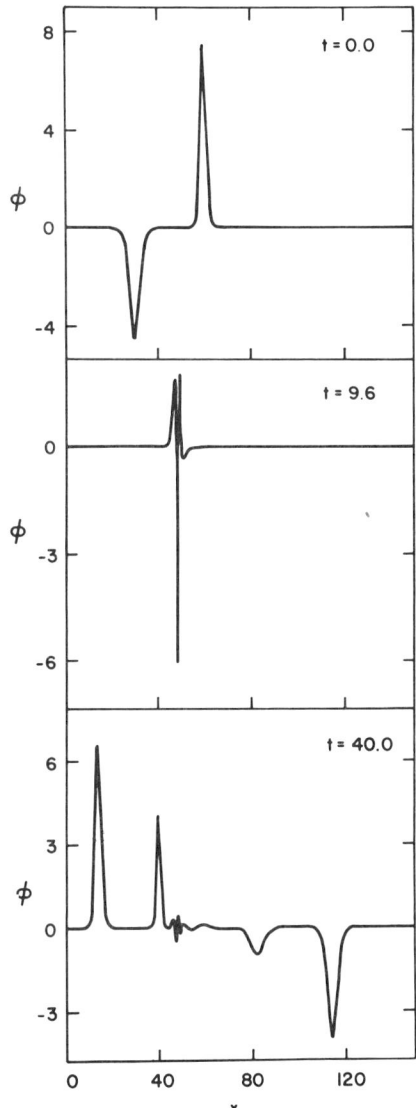

Figure 1. Collision of two RLW solitary waves, $c_1 = 2.5$, $c_2 = -1.5$, resulting in four outgoing solitary waves plus radiation.

Zakharov equations, where ion–acoustic radiation is formed during solitary-wave collisions,[26] and the double sine-Gordon equation for which bound states can be formed.[12] It is probable that inelasticity of this sort is a general property of field theories.

The primary question, for a statistical mechanic, is how can the long-time (equilibrium?) state be characterized. Consider the RLW equation on a periodic space $0 < x < L$ where $\phi(x, t) = \phi(x + L, t)$, and an arbitrary state with fixed values of M, P, and H: is the asymptotic state independent of the initial? We suspect that when the momentum is small (we use P because it is positive definite), the evolution is nearly integrable (KdV-like) and regular regions in phase space do not allow ergodicity. As the momentum increases, the regular regions will shrink, plasma branch solitary waves can be excited for $P \gtrsim 14.01$,[14] and the flow looks more ergodic. For small times, the number of solitary waves will fluctuate as collisions occur. The suspicion is that even for long times, the number of solitary waves will be nonzero.

Statistical mechanics of field theories has been discussed for the "ϕ^4" theory in the context of solid-state physics, and techniques for evaluating the partition function have been developed.[8] It can be seen that the free energy can be decomposed into a (dressed) solitary-wave part and a (dressed) phonon part. We have begun a similar calculation for RLW.[14]

TURBULENCE AND THE FORCED RLW EQUATION

So far we have confined our discussion to conservative systems, and were led to consider the validity of equilibrium statistical mechanics. A more common situation in plasma physics is the presence of an external energy source, which leads to linear instability. In this section, I will argue that the saturation of such an instability, in the context of the RLW equation, leads to the formation of coherent structures. The discussion is based on a collaboration with Kamimura and Horton.[16]

The model is Eq. (3) with an added source term

$$\phi_t + \phi_x - \phi_{xxt} - \phi\phi_x = \hat{S}\phi \tag{7}$$

where the linear operator $\hat{S}$ is given by its Fourier transform

$$\hat{S}\phi(x,t) = \frac{1}{2\pi} \int S\phi_{k_x} \exp(-ik_x x) dk_x$$

$$S = \delta \frac{k_x^4}{(1+k_x^2)^2} - \mu k_x^4 \tag{8}$$

The linearized version of Eq. (7) has the dispersion relation

$$\omega = \omega_r + i\gamma$$

$$\omega_r = \frac{k_x}{1+k_x^2}$$

$$\gamma = \frac{k_x^4}{1+k_x^2}\left[\frac{\delta}{(1+k_x^2)^2} - \mu\right] \tag{9}$$

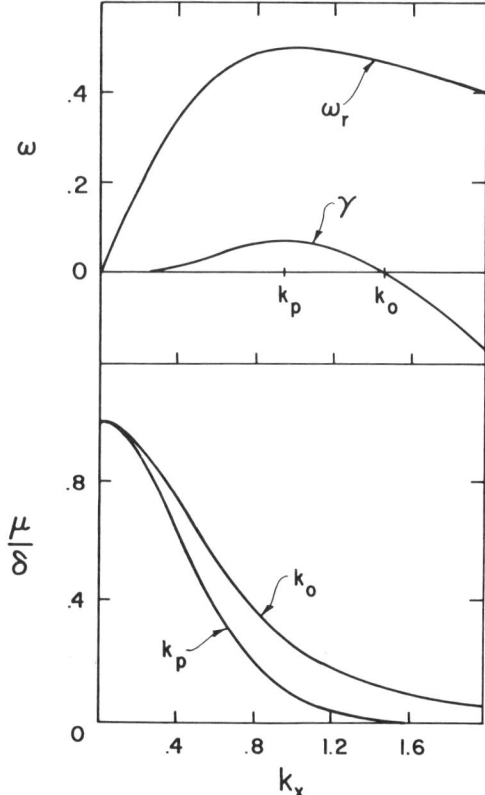

Figure 2. Model dispersion relation from Eq. (9) for $\delta = 1.0$, $\mu = 0.1$. Lower figure shows wavenumbers of peak growth rate, k_p, and zero growth rate as a function of μ/δ.

For $\delta > \mu$ the modes $k_x < k_0 = [(\delta/\mu)^{1/2} - 1]^{1/2}$ are unstable; large-wavenumber modes are damped at a rate proportional to μk_x^2. This growth rate, shown in Fig. 2, is a model for drift waves: the instability $\propto \delta$ arising from inverse electron dissipation and the damping $\propto \mu$ provided by ion viscosity. More generally, γ has features in common with most instabilities: forcing at low and damping at high wavenumbers. We expect the dynamics to be governed by the width of the unstable spectrum k_0, and the maximum growth rate $\gamma(k_p) = \gamma_p$.

Prior to the treatment of the full evolution implied by Eq. (7), consider the effect of $\hat{S}$ on a single solitary-wave solution. For $\mu < \delta \ll 1$, we can assume that the solitary wave maintains its form while its amplitude, width, and speed slowly change with time. The standard technique[21] for obtaining equations describing the adiabatic change of parameters derives from the modified conservation laws, Eq. (6), which now vary according to

$$\frac{d}{dt}M = 0$$

$$\frac{d}{dt}P = \int \phi \hat{S} \phi \, dx$$

$$\frac{d}{dt}H = \int [\int^x \phi_t(x') dx'] \hat{S} \phi \, dx \tag{10}$$

We assume that the field has the form

$$\phi = A(t) \operatorname{sech}^2\{k(t)[x - \theta(t)]\} + \delta\phi$$

where A, k, and $\dot{\theta} = c(t)$ are related by the unperturbed relations, Eq. (5), and $\delta\phi$ represents the nonadiabatic change. Substitution of this into Eq. (10) yields three ordinary differential equations for the parameters of ϕ. Mass conservation implies that any area change in the solitary wave is compensated for by $\delta\phi$:

$$\int \delta\phi \, dx = \frac{A(0)}{k(0)} - \frac{A(t)}{k(t)}$$

Typically, $\delta\phi$ takes the form of a "tail," or shelf, emitted behind the solitary wave. It takes a more-complete perturbation theory[10,11] to determine the structure of the tail. Unfortunately, these theories are based on the inverse-scattering transform, which is unavailable for the RLW equation.

Momentum conservation yields

$$\frac{dP}{dt} = \delta \frac{64}{21} A^2 k^3 \left(R(k) - \frac{\mu}{\delta} \right) \tag{11}$$

$$P = \frac{2}{3} \frac{A^2}{k} (1 + \tfrac{4}{5} k^2) \tag{12}$$

$$R(k) = 42\pi \int_0^\infty dz \frac{z^6 \operatorname{csch}(\pi z)}{[1 + (2kz)^2]^2} \tag{13}$$

The function $R(k)$ asymptotically obeys

$$R(k) = \begin{cases} 1 - \tfrac{56}{5} k^2 + \mathcal{O}(k^4) \\ \tfrac{7}{4} k^{-4} + \text{logarithmic corrections.} \end{cases}$$

Remarkably, the energy equation is identical to the momentum equation by virtue of the fact that, for a solitary wave, $dH/dP = -c$.[19]

Equation (11) implies that for a given μ/δ one particular solitary wave is in equilibrium with the forcing. This solitary wave has $k = k_s$ where $R(k_s) = \mu/\delta$. The saturation condition, shown in Fig. 3, is roughly given by $k_s \simeq k_0/2$. It is significant that saturation depends only on the ratio μ/δ (hence only on the width of the unstable spectrum k_0), and not on the magnitude of the growth rate. To attain saturation, the solitary wave picks a width, k_s^{-1}, so that

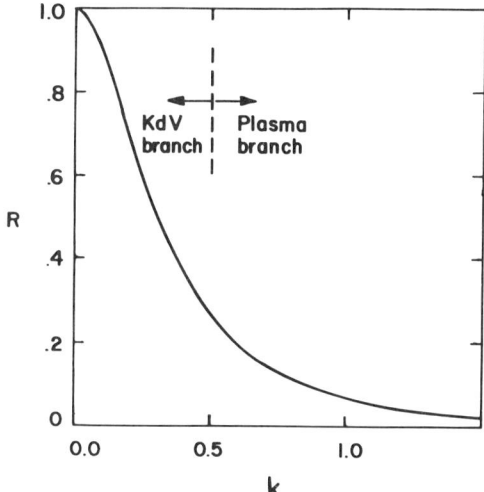

Figure 3. Source function, $R(k)$, from Eq. (13). The saturation width is determined by $R(k_s) = \mu/\delta$.

its spectrum has the property that the energy lost from high wavenumbers is exactly balanced by the input to low wavenumbers. The coupling between stable and unstable regions of the spectrum is provided by the coherent nonlinearity of the solitary wave. This picture is appropriate if the nonlinear time scale is shorter than the linear growth rate, so that the solitary wave can maintain its shape:

$$A \gg \frac{\gamma_p}{\omega_p}$$

The opposite condition, that the linear time scales are more rapid than nonlinear ones, would give incoherent turbulence.

Using the relationships between amplitude, width, and speed in Eq. (5), the momentum equation [Eq. (11)] can be expressed as an equation for c:

$$\dot{c} = \delta \frac{20}{7} \frac{(c-1)^3}{24c^2 - 8c - 1} \left(R[k(c)] - \frac{\mu}{\delta} \right) \tag{14}$$

It is easy to see that when $k_s < 1/2$ and $c > 1$, the point $k = k_s$ is a stable attractor of Eq. (14). Plasma branch ($c < 0$) solitary waves are attracted to the point

$$c = \frac{2 - \sqrt{10}}{12}$$

(where $\partial P/\partial c = 0$) but since $\dot{c} \to \infty$, Eq. (14) ceases to be valid. Probably such solitary waves are destroyed by the source.

When $k_s > 1/2$, or equivalently $\delta > 3.727\mu$, then even the KdV branch solitary waves are unstable according to Eq. (14). In fact, for this case, the velocity increases exponentially in time.

$$c(t) \propto \exp(\gamma t), \quad \gamma = 0.032(\delta - 3.727\mu)$$

for $c \gg 1$. Therefore, the value $\delta/\mu = 3.727$ is a critical value dividing the region of stable saturation from explosive instability.

From the chaotic perspective it is more relevant to study evolution from initial small-amplitude noise. To correspond with the simulations, we consider a periodic system of length L and assume $\phi(x, 0) = A_0 \sum_j \cos[(2\pi j x/L) + \theta_j]$, where $A_0 \ll 1$ and the θ_j are picked randomly. During the first few e-folding times (γ_p^{-1}) the stable modes will quickly damp away and the spectrum will become peaked about k_p with a width $\Delta k \sim (\gamma_p'' t)^{-1/2}$. As the peak mode reaches an amplitude of order one, mode coupling begins to spread energy into the stable region of wavenumber space.

At this point, providing $\delta/\mu < 3.727$ and $\gamma_p/\omega_p \ll 1$, it is plausible (and we observe numerically) that solitary waves will begin to form. Actually, since the system is periodic, it is more appropriate to assume that the cnoidal wave forms:

$$\phi = \phi_0 + A \operatorname{cn}^2[k(y - ct)|m]$$

$$A = \frac{12k^2}{4\lambda k^2 - 1}$$

$$c = \frac{1 - \phi_0}{1 + (1 - 2m)4k^2} \tag{15}$$

where $\lambda = m - 2 + 3[E(m)/K(m)]$, cn is a Jacobi elliptic function, and E and K are complete elliptic integrals. As $m \to 0$, this solution becomes a linear wave, obeying the correct dispersion relation, while as $m \to 1$ it becomes the solitary-wave solution. There are three independent parameters: m, ϕ_0, and k.

The conservation laws provide evolution equations for the parameters of the cnoidal wave in the same manner as before. Now ϕ_0 is determined by mass conservation and represents the "tail" formation of the solitary wave case. The equations predict a saturated state with an amplitude dependent only on μ/δ and L. The critical value of μ/δ for explosive instability is somewhat larger than before, due to periodicity, but if $L \gg 1$, this effect is relatively small.

Preliminary numerical experiments tend to confirm the cnoidal wave model for saturation. As nonlinear saturation occurs, the single mode spectrum $\phi \sim A_0 \exp(\gamma_p t) \cos(k_p x)$ begins to develop sharper valleys and flatter peaks, which is similar to the shape of a periodic array of KdV branch ($A < 0$) solitary waves. The wavenumber spectrum broadens, approaching that of the cnoidal wave

$$|\phi_j|^2(t) = \left\{ \frac{2\pi^2 j}{Lmk^2} \operatorname{csch}\left[\frac{2\pi j}{Lk} K'(m)\right] \right\}^2$$

Because the spectrum is not initially a single mode at k_p (or perhaps due to a modulational instability) the peaks of the cnoidal wave, once formed develop different amplitudes. Peaks with large amplitudes propagate more rapidly than those with small amplitudes and so collisions occur. Collisions are inelastic just as those between solitary waves, though now the inelasticity is enhanced by the source term. Smaller peaks lose energy at each collision and seem to have a lifetime of about eight collisions. The number of solitary peaks decreases linearly during this stage, and the field begins to look more like a rarefied gas of solitary waves. Further numerical work is in progress to determine the final steady state. In our preliminary runs the saturation amplitude seems to agree with the model, but more analysis is needed.[16]

Overall, the picture we propose can be formulated in terms of the two important parameters of the source; γ_p/ω_p and k_0. For a narrow unstable spectrum and large growth rates, linear terms in the RLW equation predominate and incoherent turbulence theory should apply. However, as the width of the unstable spectrum increases (increasing the saturation amplitude) or if γ_p/ω_p decreases, nonlinearity acts to create coherent structures outside the scope of ordinary turbulence theory. At this stage, the saturated state may be more appropriately described as a gas of localized coherent objects.

ACKNOWLEDGMENTS

Collaboration with P. J. Morrison, J. R. Cary, T. Kamimura, and W. Horton on various portions of this research is gratefully acknowledged. Support was provided by the United States Department of Energy, Grant No. DE-FG05-80ET-53088.

REFERENCES

1. K. Abdullov, J. Bogolubsky, and V. Makhankov, *Phys. Lett.* **56A**, 427 (1976).
2. M. J. Ablowitz, and H. Segur, *Solitons and the Inverse Scattering Transform* (SIAM, Philadelphia, 1981).
3. T. Benjamin, J. Bona, and J. Mahoney, *Phil. Trans. R. Soc. London*, **A272**, 47 (1972).
4. R. Berman, D. Tetrault, T. Dupree, and T. Boutros-Ghali, *Phys. Rev. Lett.* **48**, 1249 (1982).
5. G. Brown and A. Roshko, *J. Fluid Mech.* **64**, 775 (1974).
6. B. Cantwell, D. Coles, and P. Dimotakis, *J. Fluid Mech.* **87**, 641 (1978).
7. B. J. Cantwell, in *Annual Review of Fluid Mechanics 13*, M. Van Dyke, J. V. Wehausen, and J. L. Lumley, eds. (Annual Reviews, Palo Alto, 1981), p. 457.

8. J. F. Currie, J. A. Krumhansl, A. R. Bishop, and S. E. Trullinger, *Phys. Rev. B* **22**, 477 (1980).
9. T. H. Dupree, *Bull. Am. Phys. Soc.* **27**, 1038 (1982).
10. V. Karpman and E. Maslov, *Sov. Phys. JETP* **48**, 252 (1978).
11. J. P. Keener and D. W. McLaughlin, *Phys. Rev. A* **16**, 777 (1977).
12. P. Kumar and R. R. Holland, in *Nonlinear Problems: Present and Future*, A. R. Bishop, D. K. Campbell, and B. Nicolaenko, eds. (North Holland, Amsterdam, 1982), p. 229.
13. W. Lotko (1982, personal communication).
14. J D. Meiss and W. Horton, *Phys Fluids* **25**, 1838 (1982).
15. J. D. Meiss and W. Horton, *Phys. Fluids* **26**, 990 (1983).
16. J. D. Meiss, T. Kamimura, and W. Horton, "Turbulence in the Forced RLW Equation," (unpublished).
17. R. M. Miura, C. W. Gardner, and M. D. Kruskal, *J. Math. Phys.* **9**, 1204 (1968).
18. P. J. Morrison, in *Mathematical Methods in Hydrodynamics and Intergrability in Dynamical Systems*, M. Tabor and Y. Trieve, eds. (AIP, New York, 1982), p. 13.
19. P. J. Morrison, J. D. Meiss, and J. R. Cary, "Scattering of Regularized-Long-Wave Solitary Waves," (Institute for Fusion Studies, University of Texas at Austin, Austin, 1983), IFSR #80.
20. P. Olver, *Math. Proc. Camb. Phil. Soc.* **88**, 71 (1979).
21. E. Ott and R. M. Sudan, *Phys. Fluids* **13**, 1432 (1970).
22. V Petviashvili, *Sov. J. Plas. Phys.* **3**, 150 (1977)
23. A. C. Scott, F. Y. F. Chu, and D. W. McLaughlin, *Proc. IEEE* **61**, 1443 (1973).
24. C. M Surko and R. E. Slusher, *Bull. Am. Phys. Soc.* **27**, 937 (1982).
25. M. Temerin, K. Cerny, W. Lotko, and F. S. Mozer, *Phys. Rev. Lett.* **48**, 1175 (1982).
26. S. G. Thornhill and D. ter Haar, *Phys. Rep.* **43**, 45 (1978).
27. N. J. Zabusky and M. Kruskal, *Phys. Rev. Lett.* **15**, 240 (1965).
28. V. E. Zakharov and L. D. Faddeev, *Funct. Anal. Appl.* **5**, 280 (1971).
29. S. J. Zweben, P. C. Liewer, and R. W. Gould, *Bull. Am. Phys. Soc.* **27**, 973 (1982).

DYNAMIC ALIGNMENT AND SELECTIVE DECAY IN MHD

W. H. Matthaeus and D. Montgomery
Physics Department, College of William and Mary
Williamsburg, Virginia

Abstract. Under some circumstances, incompressible magnetohydrodynamic turbulence will evolve toward a state in which the velocity fields and magnetic fields are aligned or antialigned. We propose a mechanism for this effect and illustrate with numerical computations. Under some other circumstances, the energy appears to decay selectively toward a minimum energy state in which the kinetic energy has disappeared. It has not been possible so far to identify a boundary in the phase space which divides the two regimes.

INTRODUCTION

Turbulent, dissipative media obey a statistical mechanics whose central features are very far from clear. If the dissipative terms in the equations of motion are dropped, a truncated discretized representation of the remaining ideal equations seems to exhibit behavior that is well described by classical statistical mechanics: time averages are well represented by phase-space averages calculated by Gibbsian techniques[3]. But as soon as the dissipation (viscous or ohmic, for example) is turned on with even a very small coefficient, the picture changes drastically.

Typical dissipation amounts to a perfect or nearly perfect absorption in wavenumber space, if the discretization is accomplished by Fourier decomposition. The high-**k** Fourier coefficients are systematically drained, and phase points no longer spend times in phase volumes which are proportional to those volumes. Even though this fractional depletion of the time spent in volumes of phase space corresponding to the excitations at high **k** is not sharply characterized, it may not be the only thing happening in phase space which is nonstandard, according to the perspectives of classical conservative statistical mechanics. There may also be preferences for the phase points of systems to spend excessive fractions of their time on exotic sets in the phase space: strange attractors, fractals, and the like. Most discussion of these possibilities has occurred for model systems with very low numbers of degrees of freedom. Efforts to draw conclusions, from these low-order "dynamical systems," about realistic turbulent systems with many degrees of freedom have not always been very convincing. We have no candidates, for example, for a description of a set in the (Fourier) phase space that would lead to the Kolmogoroff spectrum.

An alternative to the "mappings" game, which characterizes increasingly remote-looking model systems, is to continue to deal with realistic turbulent systems, and to try for the present to give a *physical* characterization of the tendencies in their turbulent dynamics that appear to be general. This is also an inadequate characterization of their phase-space behavior, but at least it may be regarded as physically useful information.

An example of such a physical effect that we wish to remark upon here is the dynamic alignment of velocity fields and magnetic fields in magnetohydrodynamic (MHD) turbulence. It was observed some time ago by Belcher and Davis,[1] in comparing solar-wind magnetometer measurements with velocity fields inferred from particle-energy analyzers, that it often occurred that solar-wind turbulence near 1 A.U. found itself in a highly aligned ($\mathbf{v} = \mathbf{B}$) or antialigned ($\mathbf{v} = -\mathbf{B}$) state. (Here $\mathbf{v}$ is the velocity field in a zero-momentum frame, expressed in units of an rms Alfvén speed, and $\mathbf{B}$ is the turbulent magnetic field.) These alignment properties have been frequently observed since, though we attempt no survey of the literature here, and are often discussed rather inconclusively in terms of the properties of small-amplitude Alfvén waves. They have been seen also in a turbulence closure calculation by Grappin et al.[7] and have been computed dynamically by Pouquet et al.[13] and by Matthaeus et al.[10] An early important theoretical initiative on their role in the solar wind was due to Dobrowolny et al.[2] Our purpose here is to describe in print for the first time the computations of Matthaeus et al.[10] and to suggest a qualitative explanation for the observed alignment which differs from the one suggested by Pouquet et al.[13]

THE DYNAMICAL PROBLEM

We begin with the equations of incompressible MHD in a familiar dimensionless form:

$$\frac{\partial \mathbf{v}}{\partial t} + \mathbf{v} \cdot \nabla \mathbf{v} = -\nabla p + \mathbf{B} \cdot \nabla \mathbf{B} + \nu \nabla^2 \mathbf{v} \tag{1}$$

$$\frac{\partial \mathbf{B}}{\partial t} + \mathbf{v} \cdot \nabla \mathbf{B} = \mathbf{B} \cdot \nabla \mathbf{v} + \mu \nabla^2 \mathbf{B} \tag{2}$$

where p is the total pressure (mechanical plus magnetic) and where $\nabla \cdot \mathbf{v} = 0 = \nabla \cdot \mathbf{B}$. The constants μ^{-1}, ν^{-1} are the magnetic and mechanical Reynolds numbers, respectively. We work in rectangular, periodic, two-dimensional geometry, with all variables independent of the z coordinate: $\partial/\partial z \equiv 0$. We solve Eqs. (1) and (2) numerically, using a spectral-method code of the Orszag type[6,11,12] used previously[4,5,8,9,14]

We solve the initial-value problem starting from random initial conditions. These are specified by choosing a set of random initial values for the Fourier coefficients in the expressions

$$\mathbf{v} = \sum_{\mathbf{k}} \mathbf{v}(\mathbf{k}, t) \exp(i\mathbf{k} \cdot \mathbf{x})$$

$$\mathbf{B} = \sum_{\mathbf{k}} \mathbf{B}(\mathbf{k}, t) \exp(i\mathbf{k} \cdot \mathbf{x}) \quad (3)$$

and then following the evolution of the Fourier amplitudes by solving the Fourier-transformed version of Eqs. (1) and (2).

The degree of alignment may be measured by the ratio of two ideal invariants, the cross helicity, H_c, and the energy, E:

$$\frac{H_c}{E} = \frac{\frac{1}{2}\int \mathbf{v} \cdot \mathbf{B} \, d^2x}{\frac{1}{2}\int (\mathbf{v}^2 + \mathbf{B}^2) d^2x} = \frac{\sum_{\mathbf{k}} \mathbf{v}^*(\mathbf{k}, t) \cdot \mathbf{B}(\mathbf{k}, t)}{\sum_{\mathbf{k}} (|\mathbf{v}(\mathbf{k}, t)|^2 + |\mathbf{B}(\mathbf{k}, t)|^2)} \quad (4)$$

The ratio H_c/E must lie within the range $-1/2 \leq H_c/E \leq +1/2$, the limits indicating perfect antialignment and perfect alignment, respectively. H_c and E are both conserved by Eqs. (1) and (2) if we set $\mu = \nu = 0$, but decay in absolute value if $\mu \neq 0$, $\nu \neq 0$. The decay of $|H_c|$ need not be monotonic, but often is; the decay of E is monotonic.

THE COMPUTATION

The simulation results described here consist of four runs, designated A, B, C, and D. All four were obtained using a $(128)^2$ code; that is, wave vectors with integer components ranging from $k_{\min} = 1$ to $k_{\max} = 64$ and their negative counterparts are included. Initial nonzero $|\mathbf{B}(\mathbf{k}, 0)|^2$ and $|v(k, 0)|^2$ were confined to a ring in k space between $k^2 = 9$ and $k^2 = 25$. For runs A, B, and C there was twice as much magnetic energy as kinetic energy in each mode, but for run D, kinetic and magnetic energies were equal. In all runs the spectrum was independent of $\mathbf{k}$ within the ring, and phases of magnetic modes were randomly chosen. The amount of cross helicity initially present is determined by selecting phase angles for kinetic Fourier amplitudes to make a specified angle θ with respect to the magnetic amplitude at the same $\mathbf{k}$. This controls both the modal cross helicity, $H_c(\mathbf{k}) = \text{Re } \mathbf{v}^*(\mathbf{k}) \cdot \mathbf{B}(\mathbf{k})$ and the total cross helicity. Runs A and D are initially highly aligned, with $\cos^{-1}(0.95) = \theta$ while runs B and C are only moderately correlated with

$\cos^{-1}(0.3) = \theta$. In runs A, B, and D, the parameters $\mu = \nu = 1/400$, and in run C, $\mu = \nu = 1/1000$. Times are measured in units of Alfvén transit times of unit distance, with timesteps of 1/256 such units. Other runs, not described here, have shown results similar to these four.

PHYSICAL RESULTS

Typical behavior is exhibited in Fig. 1: a monotonic *increase* of $|H_c/E|$. If we start with H_c/E negative, it decreases with time.

A simple explanation is suggested by writing Eqs. (1) and (2) in the Elsässer variables $\mathbf{Z}^\pm = \mathbf{v} \pm \mathbf{B}$. We assume for the moment that interest lies in the low-**k** region in the Fourier space, where the dissipative terms may be neglected. Equations (1) and (2) may be combined to give

$$\frac{\partial}{\partial t}\mathbf{Z}^+ = -\mathbf{Z}^- \cdot \nabla \mathbf{Z}^+ - \nabla p \tag{5}$$

$$\frac{\partial}{\partial t}\mathbf{Z}^- = -\mathbf{Z}^+ \cdot \nabla \mathbf{Z}^- - \nabla p \tag{6}$$

The pressure p is, as usual, obtained by taking the divergence of either Eq. (5) or Eq. (6) and using $\nabla \cdot \mathbf{Z}^\pm = 0$ to yield a Poisson equation.

Now suppose that in some mean-square sense, $\mathbf{Z}^+$ is large compared to $\mathbf{Z}^-$: $|\mathbf{Z}^+| \gg |\mathbf{Z}^-|$. It follows that $\mathbf{Z}^+$ is nearly time independent, since its time derivative is of $O(\mathbf{Z}^-)$:

$$\frac{\partial \mathbf{Z}^+}{\partial t} \cong 0 \tag{5a}$$

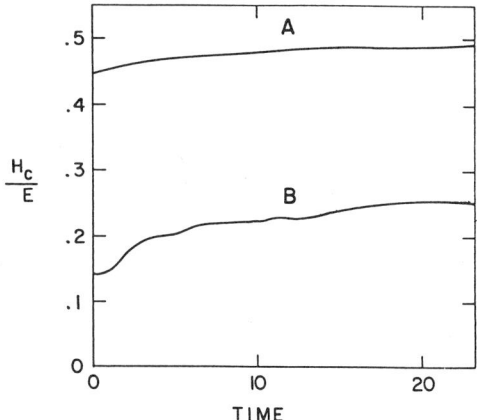

Figure 1. H_c/E for runs A and B vs time in Alfvén transit times of unit distance. The increase of H_c/E with time shows that dynamic alignment of **v** and **B** is proceeding.

(∇p is of the same order as $\mathbf{Z}^- \cdot \nabla \mathbf{Z}^+$). However, the evolution of $\mathbf{Z}^-$ is *not* slow, and is given to lowest significant order

$$\frac{\partial}{\partial t}\mathbf{Z}^- \cong -\mathbf{Z}^+(t=0) \cdot \nabla \mathbf{Z}^- - \nabla p \quad (6a)$$

where $\mathbf{Z}^+(0) \cdot \nabla \mathbf{Z}^-$ is not negligible. Thus the fractional rate of change $|\mathbf{Z}^+|^{-1}|\partial \mathbf{Z}^+/\partial t|$ is small, but $|\mathbf{Z}^-|^{-1}|\partial \mathbf{Z}^-/\partial t|$ is not.

No spectral transfer in the $\mathbf{Z}^+$ field is implied by Eq. (5a). Its spectrum is frozen at this order; if it is initially confined to wavenumbers where dissipation is negligible, it will remain there. However, at every time step, Eq. (6a) will transfer the $\mathbf{Z}^-$ spectrum to all additive combinations of wavenumbers that can be made up of those present in $\mathbf{Z}^+(0)$ and $\mathbf{Z}^-$ at the previous time step. Clearly, the *transfer* of the $\mathbf{Z}^-$ spectrum to higher wavenumbers will continue at a fractional rate that does not go to zero as $\mathbf{Z}^-$ becomes small. $\mathbf{Z}^-$ will be dissipated as it reaches the higher $\mathbf{k}$ values, thus enhancing the inequality $|\mathbf{Z}^+| \gg |\mathbf{Z}^-|$. (A somewhat different result was obtained by Dobrowolny et al.[2])

We believe this mechanism, by which the majority species cannibalizes the minority one by sending it to high wavenumbers to be dissipated, is the essential mechanism involved in dynamic alignment. (A pure $\mathbf{Z}^+$ field is aligned and a pure $\mathbf{Z}^-$ field is antialigned.) A more-complicated dynamical calculation is required to make the effect precise, particularly for $O(1)$ initial ratios of $|\mathbf{Z}^+|/|\mathbf{Z}^-|$.

Figure 2 shows the normalized cross-helicity spectrum $2H_c(\mathbf{k})/E(\mathbf{k})$ vs k at an early stage of the evolution of run A, $t \approx 0.19$. The initial value of this quantity was 0.895, near to its maximum possible value of 1.0, in the range of nonzero excitations, $3 < k < 5$. Figure 2 shows that the transfer of excitation to higher k's than those initially present consists almost entirely of *negative* cross-helicity contributions, corresponding to highly anticorrelated small-scale magnetic- and velocity-field fluctuations. This typical

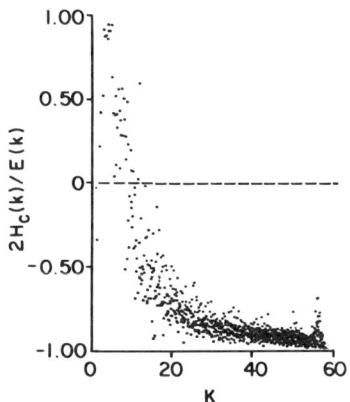

Figure 2. Normalized cross-helicity spectrum $2H_c(k)/E(k)$ for run A at $5 = 0.19$. $H_c(k)$ and $E(k)$ are modal spectra, that is, average values over all $\mathbf{k}$ with length k. Initially $H_c(k) = 0.895$ for $3 \le k \le 5$ and is zero elsewhere. Spectral transfer to higher k is shown to be almost entirely negative cross-helicity excitation (that is, the $\mathbf{Z}^-$ field).

behavior shows the transfer of Z^- to higher $\mathbf{k}$ where nothing was initially, and being dissipated there.

The picture seems, however, not to be as simple as the preceding argument would indicate. A competing process, the "selective decay" of energy relative to mean square vector potential $A \equiv \sum_{\mathbf{k}} |\mathbf{B}(\mathbf{k}, t)|^2/k^2$, also occurs under some circumstances, and at times competes with the increase of $|H_c/E|$ illustrated in Figs. 1 and 2. A is also an ideal invariant, and in several other computations,[8] the ratio E/A has been observed to decay monotonically, to close to its minimum value of $k_{min}^2 = 1$.

The alignment process and selective decay cannot simultaneously proceed to their limiting states, since the limits $|2H_c/E| \to 1$ and $E/A \to k_{min}^2$ are incompatible. Figure 3 shows that for initially highly aligned runs, A and D, the ratio $(1 - 2H_c/E)/(1 - A/E)$ decreases, indicating that the

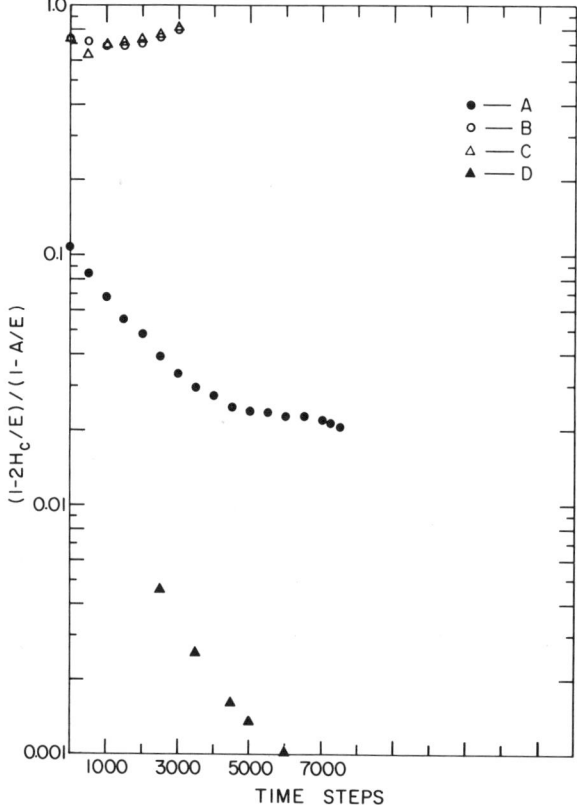

Figure 3. $(1 - 2H_c/E)/(1 - A/E)$ for runs A, B, C, D. Increase of this ratio indicates dominance of selective decay; decrease signals dynamic alignment. The suggestion is that initially highly aligned conditions leads to ultimate dominance of alignment, but no threshold has been established.

alignment process dominates (note that $k_{min} = 1$). However, the ratio increases for the less-aligned runs B and C. At this writing, we do not know of any general theoretical considerations that determine which process "wins" in the long-time high-Reynolds number limits, for arbitrary initial conditions.

We may conjecture a (complicated or simple) boundary in the phase space of initial conditions on opposite sides of which the tendency is toward alignment and toward a minimum energy state. Similar uncertainties exist with regard to the three-dimensional case, where magnetic helicity takes over the role of A, and the final selectively decayed state is the force-free "Taylor state." We have not succeeded, at this point, in identifying the boundary.

ACKNOWLEDGMENTS

We appreciate helpful conversations with Drs. Uriel Frisch and Annick Pouquet. This work was supported in part by the U.S. Department of Energy and in part by the National Aeronautics and Space Administration Grant NSG-7416.

REFERENCES

1. J. W. Belcher and L. Davis, *J. Geophys. Res.* **76,** 3534 (1971).
2. M. Dobrowolny, A. Mangeney, and P. L. Veltri, *Phys. Rev. Lett.* **45**, 144 (1980).
3. D. Fyfe and D. Montgomery, *J. Plasma Phys.* **16,** 181 (1976).
4. D. Fyfe, G. Joyce, and D. Montgomery. *J. Plasma Phys.* **17,** 317 (1977).
5. D. Fyfe, G. Joyce, and D. Montgomery, *J. Plasma Phys.* **17**, 369 (1977).
6. D. Gottlieb and S. A. Orszag, *Numerical Analysis of Spectral Methods* (Society for Industrial & Appl. Math., Philadelphia, 1977).
7. R. Grappin, U. Frisch, J. Léorat, and A. Pouquet, *Astron. & Astrophys.* **105**, 6 (1982).
8. W. H. Matthaeus and D. Montgomery, *Ann. N. Y. Acad. Sci.* **357**, 203 (1980).
9. W. H. Matthaeus and D. Montgomery, *J. Plasma Phys.* **25**, 11 (1981).
10. W. H. Matthaeus, M. L. Goldstein, and D. Montgomery, "Dynamic Alignment of Velocity and Magnetic Fields in Magnetohydrodynamic Turbulence," *EOS* (*Transactions of the AGU*) **63**, SS32-10 (1982).
11. S. A. Orszag, *Stud. Appl. Math.* **50,** 293 (1971).
12. G. S. Patterson and S. A. Orszag, *Phys. Fluids* **14**, 2358 (1971).
13. A. Pouquet, M. Meneguzzi, and U. Frisch, "The Growth of Correlations in MHD Turbulence," Observatorie de Nice, preprint (1983).
14. J. Shebalin, W. H. Matthaeus, and D. Montgomery, *J. Plasma Phys.* **29**, 525 (1983).

Part 5

CLUMPS IN PLASMA PHYSICS

A DIAGRAMMATIC APPROACH TO THE THEORY OF CLUMPS IN TURBULENT PLASMAS

R. Balescu

Faculté des Sciences CP231, Association Euratom-Etat Belge, Université Libre de Bruxelles, Brussels, Belgium

J. H. Misguich

Association Euratom-CEA sur la Fusion, BP6, 92260 Fontenay-aux-Roses, France

Abstract. *It is shown that much insight is gained by the use of a diagrammatic method in the analysis and classification of both the one-particle fluctuations and the two-particle correlations in a turbulent Vlasov plasma. The various types of renormalization are discussed. The kinetic equation for the average background distribution function is discussed in some detail.*

1. INTRODUCTION

In a *quiescent* plasma, the microscopic behavior is completely described by two distribution functions $f^\alpha(\mathbf{x}, \mathbf{v}; t)$ ($\alpha = e, i$), obeying *deterministic* kinetic equations of evolution. In contrast, owing to external or internal factors (instabilities), the distribution functions in a *turbulent* plasma are considered as *stochastic* quantities. We introduce the usual decomposition of the distribution functions into an average part and a fluctuation:

$$f^\alpha(\mathbf{x}, \mathbf{v}; t) = \langle f^\alpha(\mathbf{x}, \mathbf{v}; t)\rangle + \delta f^\alpha(\mathbf{x}, \mathbf{v}; t) \tag{1}$$

We also assume spatial homogeneity and electroneutrality on the average, which implies that $\langle f^\alpha \rangle$ only depends on the velocity and on time:

$$\langle f^\alpha(\mathbf{x}, \mathbf{v}; t)\rangle = F^\alpha(\mathbf{v}; t) \tag{2}$$

Moreover, the time dependence of $F^\alpha(\mathbf{v}; t)$ is assumed to be slow as compared to the one of $\delta f^\alpha(\mathbf{x}, \mathbf{v}; t)$.

The problem of turbulence is the systematic study of the fluctuations, their origin, their evolution in time, and their consequences on observable quantities.

2. ONE-PARTICLE FLUCTUATIONS

We consider here the simpler problem of an electron–ion plasma in the absence of a magnetic field. Let $\delta f^\alpha_{\mathbf{k}z}(\mathbf{v})$ be the Fourier transform (in space) and Laplace transform (in the fast time) of $\delta f^\alpha(\mathbf{x}, \mathbf{v}; t)$: it is a solution of the Vlasov equation [combined with Eq. (1)]

$$-i(z - \mathbf{k} \cdot \mathbf{v}) \delta f^\alpha_{\mathbf{k}}(\mathbf{v}) = \frac{e_\alpha}{m_\alpha} \mathop{\mathbf{S}}_{\mathbf{k}'} \phi_{\mathbf{k}-\mathbf{k}'} i(\mathbf{k}-\mathbf{k}') \cdot \partial \delta f^\alpha_{\mathbf{k}'}(\mathbf{v})$$
$$+ \frac{e_\alpha}{m_\alpha} \phi_\mathbf{k} i\mathbf{k} \cdot \partial F^\alpha(\mathbf{v}) + \gamma^\alpha_\mathbf{k}(\mathbf{v}) \quad (3)$$

which is coupled to the Poisson equation for the electrostatic potential ϕ_k:

$$k^2 \phi_k = \sum_\beta 4\pi e_\beta \int d\mathbf{v}\, \delta f^\beta_\mathbf{k}(\mathbf{v}) \quad (4)$$

e_α, m_α are, respectively, the charge and mass of a particle of species α, $\partial \equiv (\partial/\partial \mathbf{v})$, a subscript k denotes the set $(\mathbf{k}, z)$, and

$$\mathop{\mathbf{S}}_{\mathbf{k}} \equiv \sum_\mathbf{k} \frac{1}{2\pi} \int dz$$

The first term in the right-hand side of Eq. (3) contains the *nonlinearity*, the second is a *source* term depending on the average function F^α, and the third term contains the *initial condition*

$$\gamma^\alpha_\mathbf{k}(\mathbf{v}) \equiv \delta f^\alpha_\mathbf{k}(\mathbf{v}; t=0) \quad (5)$$

Among several methods for solving this nonlinear equation, we choose to use here a formal perturbation expansion, in powers of ϕ_k, carried out to all orders. This expansion is conveniently supported by a *diagram technique*, the idea of which was first put forward by Altshul' and Karpman,[1] and subsequently developed by Horton and Choi.[2]

A typical contribution to $\delta f^\alpha_\mathbf{k}(\mathbf{v})$, resulting from a multiple iteration of Eq. (3) (say, of the fourth order) is represented by the diagram:

The translation of this diagram into a mathematical expression is obtained by the following rules of correspondence:

Horizontal segment:

$$\xrightarrow[k]{\quad\quad} \rightarrow g_k^{0\alpha}(v) \equiv [-i(z - \mathbf{k} \cdot \mathbf{v})]^{-1}. \tag{6}$$

Vertical line (intermediate):

$$\rightarrow \frac{e_\alpha}{m_\alpha} \underset{k}{S} \phi_k i\mathbf{k} \cdot \partial$$

Terminal vertical lines are of two kinds:

$$\rightarrow \frac{e_\alpha}{m_\alpha} \phi_k i\mathbf{k} \cdot \partial F^\alpha(\mathbf{v}) \quad \text{(source)}$$

$$\rightarrow \gamma_\mathbf{k}^\alpha(\mathbf{v}) \quad \text{(initial condition)}.$$

In order to proceed, we need a *statistical assumption*. We adopt here the simplest of these, although this would be generalized if necessary. We assume that ϕ_k is a Gaussian process (hence, any average of n factors ϕ_k is expressed as a combination of second moments); moreover, we assume the process to be stationary and homogeneous, which implies:

$$\langle \phi_{\mathbf{k}\omega} \phi_{\mathbf{k}'\omega'} \rangle = \delta(\mathbf{k} + \mathbf{k}') \delta(\omega + \omega') I_{\mathbf{k}\omega} \tag{7}$$

where $I_{\mathbf{k}\omega}$ is the spectral density of the potential fluctuations, or simply the "spectrum."

This discussion introduces a natural operation of *statistical coupling*. Among the diagrams, we select those in which a pair of vertical lines has opposite wave vectors and opposite (real parts) of the frequency: the corresponding pair of potential factors is replaced by its average according to Eq. (7):

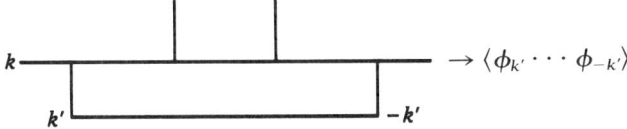

The importance of this operation comes from the fact that diagrams such as

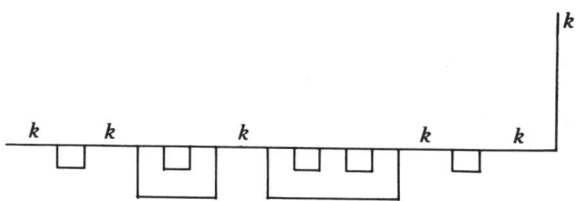

correspond to contributions growing secularly in time, because of the presence of a multiple pole in $z = \mathbf{k} \cdot \mathbf{v}$. Therefore, in order to obtain finite results for long times, the most divergent terms in each order must be summed. This operation leads to the well-known *renormalization procedure* (see, for example, Horton and Choi[2]).

The next step in the analysis of fluctuations involves a separation into *coherent* and *incoherent* fluctuations (Dupree[3]; Boutros-Ghali and Dupree[4]):

$$\delta f^\alpha_{\mathbf{k}z}(\mathbf{v}) = \delta f^{\alpha(c)}_{\mathbf{k}z}(\mathbf{v}) + \delta \tilde{f}^\alpha_{\mathbf{k}z}(\mathbf{v}) \tag{8}$$

The coherent part $\delta f^{\alpha(c)}_{\mathbf{k}z}(\mathbf{v})$ is, by definition, proportional to $\phi_{\mathbf{k}z}$ (with the same $\mathbf{k}$ and z), with a coefficient involving an arbitrary number of statistically coupled pairs of potential factors. This definition is easily translated into diagrams. The coherent fluctuations are represented by the set of diagrams having one single uncoupled vertical line. There are two classes of such diagrams, illustrated by diagrams A and B below:

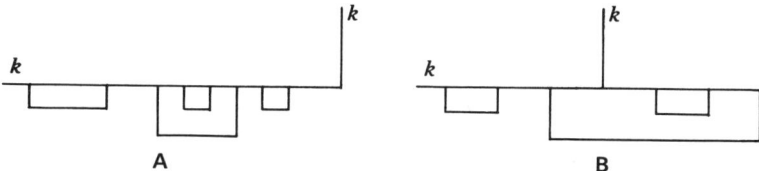

Class A: The Uncoupled Line is Terminal. The formal summation of these diagrams is well known (Horton and Choi[2]): it amounts to a *renormalization of the propagator*. The result of this operation can be expressed as follows:

In the expression of the first-order diagram:

$$\delta f^{\alpha(1)}_{\mathbf{k}z}(\mathbf{v}) = g^{0\alpha}_{\mathbf{k}z}(\mathbf{v}) \frac{e_\alpha}{m_\alpha} \phi_{\mathbf{k}z} i\mathbf{k} \cdot \partial F^\alpha(\mathbf{v}) \tag{9}$$

replace the unperturbed propagator $g_{kz}^{0\alpha}(v)$, Eq. (6), by a renormalized propagator $g_{kz}^{\alpha}(v)$:

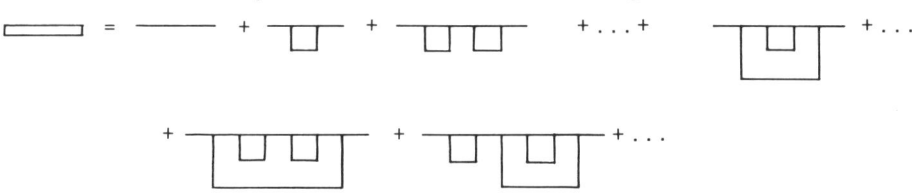

$$\delta f_{kz}^{\alpha}(v) = g_{kz}^{\alpha}(v)\frac{e_\alpha}{m_\alpha}\phi_k z i\mathbf{k}\cdot\partial F^\alpha(v) \tag{10}$$

The double line symbolizes the infinite sum of diagrams:

From this series, an integral equation for the renormalized propagator is easily derived (Horton and Choi[2]):

$$g_{kz}^{\alpha}(v) = g_{kz}^{0\alpha}(v) + g_{kz}^{0\alpha}(v)C_{kz}^{\alpha}(v)g_{kz}^{\alpha}(v) \tag{11}$$

where $C_{kz}^{\alpha}(v)$ is a diffusionlike operator, involving the unknown function:

$$C_{kz}^{\alpha}(v) = \sum_{k'}\frac{1}{2\pi}\int dz' \frac{e_\alpha^2}{m_\alpha^2} I_{k'z'}\mathbf{k}'\cdot\partial g_{k-k',z-z'}^{\alpha}(v)\mathbf{k}'\cdot\partial \tag{12}$$

Class B: The Uncoupled Line is Not Terminal. The occurrence of this class of contributions was apparently overlooked in earlier theories of turbulence. Its importance in the drift wave problem was first pointed out by Dupree and Tetreault[5] (see also Boutros-Ghali and Dupree[4]). In order to understand the structure of these terms, consider the third-order contribution:

$$\delta f_k^{\alpha(3)} = g_k^{0\alpha}\mathbf{S}_{k'}\frac{e_\alpha}{m_\alpha}\phi_{k'} i\mathbf{k}'\cdot\partial g_{k-k'}^{0\alpha}\left(\frac{e_\alpha}{m_\alpha}\phi_k i\mathbf{k}\right)\cdot\partial$$

$$\times g_{-k'}^{0\alpha}\left(\frac{e_\alpha}{m_\alpha}\right)\phi_{-k'}(-i\mathbf{k}'\cdot\partial)F^\alpha \tag{13}$$

Commuting the bracketed factors to the left of the summation sign, this expression can be written as

$$\delta f_{\mathbf{k}z}^{\alpha(3)} = g_k^{0\alpha} \frac{e_\alpha}{m_\alpha} \phi_{\mathbf{k}z} i\mathbf{k} \cdot \xi_{\mathbf{k}z}^{\alpha(2)} F^\alpha(\mathbf{v}) \tag{14}$$

This expression is of the same form as Eq. (9), provided we replace the velocity gradient operator ∂ in the source term by the operator $\xi_{\mathbf{k}z}^{\alpha(2)}(\mathbf{v})$:

$$\xi_k^{\alpha(2)}(\mathbf{v}) = \frac{e_\alpha^2}{m_\alpha^2} \mathop{\mathbf{S}}_{k'} I_{k'}(\mathbf{k}' \cdot \partial) g_{k-k'}^{0\alpha}(\mathbf{v}) \partial g_{-k'}^{0\alpha}(\mathbf{v})(\mathbf{k}' \cdot \partial) \tag{15}$$

This replacement of the operator ∂ by ξ_k^α will be quite naturally called a *vertex renormalization*. It should be noted that the present interpretation is different from the one of Boutros-Ghali and Dupree.[4] These authors (and others) interpret this class of coherent contributions in terms of a modified background distribution. In order to follow their interpretation, Eq. (13) should be rearranged as follows:

$$\delta f_k^{\alpha(3)} = g_k^{0\alpha} \frac{e_\alpha}{m_\alpha} \phi_k i k_r \partial_s \bar{F}_{rs}^\alpha$$

with

$$\bar{F}_{rs}^\alpha = \mathop{\mathbf{S}}_{k'} \frac{e_\alpha}{m_\alpha} I_{k'} k'_s g_{k-k'}^{0\alpha} \partial_r g_{-k'}^{0\alpha} \mathbf{k}' \cdot \partial F^\alpha$$

This shows that the "modified distribution function" $\bar{F}_{rs}^\alpha$ is a *tensor*, and hence makes no physical sense. Only in one-dimensional problems is such an interpretation possible.

We now proceed with the complete renormalization of both the propagator and the vertex, by summing all the diagrams of the form:

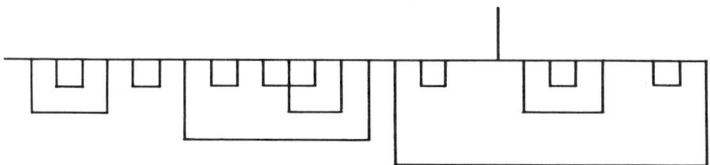

The sum of all the diagrams of this type will be represented by the single diagram

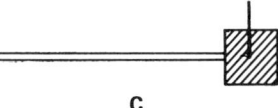

c

By straightforward algebra it is shown that the corresponding mathematical expression for the *complete coherent fluctuation* is

$$\delta f_{\mathbf{k}z}^{\alpha(c)}(\mathbf{v}) = g_{\mathbf{k}z}^\alpha(\mathbf{v}) \frac{e_\alpha}{m_\alpha} \phi_{\mathbf{k}z} i\mathbf{k} \cdot \xi_{\mathbf{k}z}^\alpha(\mathbf{v}) F^\alpha(\mathbf{v}) \tag{16}$$

where the renormalized propagator is defined by Eq. (11) and the renormalized vertex operator is defined by

$$\xi^\alpha_{\mathbf{k}z}(\mathbf{v}) = \partial + \sum_{\mathbf{k}'} \frac{1}{2\pi} \int dz' \frac{e^2_\alpha}{m^2_\alpha} I_{\mathbf{k}'z'}(\mathbf{k}' \cdot \partial) g^\alpha_{\mathbf{k}-\mathbf{k}',z-z'}(\mathbf{v}) \partial g^\alpha_{-\mathbf{k}',-z'}(\mathbf{v})(\mathbf{k}' \cdot \partial)$$
(17)

The physical significance of the coherent fluctuations is seen upon substituting Eq. (16) instead of $\delta f^\alpha_k(\mathbf{v})$ into the Poisson equation [Eq. (4)]. Since $\delta f^{\alpha(c)}_k$ is proportional to ϕ_k, the latter equation takes the form:

$$\varepsilon_{\mathbf{k}z} \phi_{\mathbf{k}z} = 0$$
(18)

where the *renormalized dielectric function* is defined as

$$\varepsilon_{\mathbf{k}z} = 1 - \sum_\beta \frac{4\pi e^2_\beta}{m_\beta k^2} \int d\mathbf{v}\, g^\beta_{\mathbf{k}z}(\mathbf{v}) i\mathbf{k} \cdot \xi^\beta_{\mathbf{k}z}(\mathbf{v}) F^\beta(\mathbf{v})$$
(19)

Equation (18) immediately yields a *homogeneous*, nonlinear equation for the spectrum $I_{\mathbf{k}\omega}$: there is no source term for the fluctuations in this coherent approximation.

We now turn to the *incoherent part* of the fluctuations, $\delta \tilde{f}^\alpha_{\mathbf{k}z}(\mathbf{v})$. In the spirit of Dupree's[3] theory, these are not evaluated in detail at this stage. However, it is useful for the subsequent treatment to identify and classify their diagrams. One easily convinces oneself that the contributions to δf^α_k that have not been summed under $\delta f^{\alpha(c)}_k$ are of the following types:

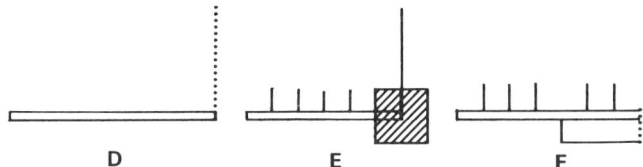

Diagrams of class D represent a *ballistic propagation* of the initial condition $\gamma^\alpha_k(\mathbf{v})$: their contribution will be denoted by $\delta f^{\alpha(b)}_{\mathbf{k}z}(\mathbf{v})$.

Diagrams of group E are denoted pictorially as *uncombed diagrams*: they contain uncoupled factors $\phi_{\mathbf{k}'}$.

The contributions of type F, involving the correlation function $\langle \phi_{\mathbf{k}'} \gamma^\alpha_{\mathbf{k}} \rangle$ will be neglected here.

If we substitute the complete fluctuations in Eq. (8) into Poisson's equation [Eq. (4)], the latter takes the form

$$\varepsilon_{\mathbf{k}z} \phi_{\mathbf{k}z} = \sum_\beta \frac{4\pi e_\beta}{k^2} \int d\mathbf{v}\, \delta \tilde{f}^\beta_{\mathbf{k}z}(\mathbf{v}) \equiv \tilde{\phi}_{\mathbf{k}z}$$
(20)

or

$$\phi_{\mathbf{k}z} = \frac{\tilde{\phi}_{\mathbf{k}z}}{\varepsilon_{\mathbf{k}z}}$$
(21)

This equation shows that the incoherent fluctuations act as a *source term* in Poisson's equation, leading to a broadening of the spectrum.

3. KINETIC EQUATION FOR THE AVERAGE DISTRIBUTION FUNCTION

An equation for the (slow) time evolution for $F^\alpha(\mathbf{v}; t)$ is easily obtained by averaging the Vlasov equation:

$$\partial_t F^\alpha(\mathbf{v}; t) = \frac{e_\alpha}{m_\alpha} \sum_\mathbf{k} \frac{1}{2\pi} \int dz\, i\mathbf{k} \cdot \boldsymbol{\partial} \langle \phi_{-\mathbf{k}-z} \delta f^\alpha_{\mathbf{k}z}(\mathbf{v}) \rangle \qquad (22)$$

Using Eqs. (8), (16), (19), (20), and (21) one obtains

$$\partial_t F^\alpha(\mathbf{v}; t) = \sum_\beta \frac{4\pi e_\alpha e_\beta}{m_\alpha} \frac{1}{2\pi} \int dz \int d\mathbf{k} \int d\mathbf{v}' \frac{1}{k^2 |\varepsilon_{\mathbf{k}z}|^2} \mathbf{k} \cdot \boldsymbol{\partial}$$

$$\times \left[\langle \tilde{\phi}_{-\mathbf{k}-z} \delta \tilde{f}^\beta_{\mathbf{k}z}(\mathbf{v}') \rangle g^\alpha_{\mathbf{k}z}(\mathbf{v}) \frac{e_\alpha}{m_\alpha} \mathbf{k} \cdot \boldsymbol{\xi}^\alpha_{\mathbf{k}z}(\mathbf{v}) F^\alpha(\mathbf{v}) \right.$$

$$\left. - \langle \tilde{\phi}_{-\mathbf{k}-z} \delta \tilde{f}^\alpha_{\mathbf{k}z}(\mathbf{v}) \rangle g^\beta_{\mathbf{k}z}(\mathbf{v}') \frac{e_\beta}{m_\beta} \mathbf{k} \cdot \boldsymbol{\xi}^\beta_{\mathbf{k}z}(\mathbf{v}') F^\beta(\mathbf{v}') \right] \qquad (23)$$

This equation, obtained in a simpler form by Dupree,[3] and later by Boutros-Ghali and Dupree[4] (including vertex renormalization), was called by these authors a "Balescu–Lenard" equation. In other words, they say that it is equivalent to an equation of the form

$$\partial_t F^\alpha = \boldsymbol{\partial} \cdot \mathbf{D}^a(\mathbf{v}) \cdot \boldsymbol{\partial} F^\alpha + \boldsymbol{\partial} \cdot \boldsymbol{\mathcal{F}}^\alpha(\mathbf{v}) F^\alpha(\mathbf{v}) \qquad (24)$$

involving a diffusion term and a friction term. They then interpret this equation as describing collisions of particles with "clumps."

My claim is that such an interpretation cannot be supported without further analysis. It is true that the first bracketed term in Eq. (23) [originating from $\delta f^{\alpha(c)}$] is indeed of diffusion type (forgetting the vertex renormalization); but the second one (originating from $\delta \tilde{f}^\alpha$) cannot be interpreted as a friction term, unless

$$\langle \tilde{\phi}_{-\mathbf{k},-z} \delta \tilde{f}^\alpha_{\mathbf{k}z}(\mathbf{v}) \rangle \sim F^\alpha(\mathbf{v}) \qquad (25)$$

More generally speaking, Eq. (23) cannot be called a kinetic equation unless it is shown that the correlations involved in its right-hand side are functionals of the average distribution functions (kinetic regime). We must, therefore, start a program that can be defined by the following steps:

1. Identify $\langle \delta \tilde{f}^\alpha_{\mathbf{k}z} \delta \tilde{f}^\beta_{-\mathbf{k}-z} \rangle$ as part of the complete correlation function $\langle \delta f^\alpha_{\mathbf{k}z} \delta f^\beta_{-\mathbf{k}-z} \rangle$.
2. $\langle \delta f^\alpha_{\mathbf{k}z} \delta f^\beta_{-\mathbf{k}-z} \rangle$ is the Laplace transform of the *two-time correlation function* $\langle \delta f^\alpha_\mathbf{k}(t+\tau) \delta f^\beta_{-\mathbf{k}}(t) \rangle$ with respect to the (fast) time τ.
3. The determination of the latter function requires the knowledge of the

one-time correlation function $\langle \delta f_{\mathbf{k}}^{\alpha}(t) \delta f_{-\mathbf{k}}^{\beta}(t) \rangle \equiv c_{\mathbf{k}}^{\alpha\beta}(t)$ as an initial condition. An equation must be derived for this function, and solved: at this stage, the *clump effect* will manifest itself.

4. From $c_{\mathbf{k}}^{\alpha\beta}(t)$ one derives the incoherent part $\tilde{c}_{\mathbf{k}}^{\alpha\beta}(t)$, which determines $\langle \delta \tilde{f}_{\mathbf{k}}^{\alpha} \delta \tilde{f}_{-\mathbf{k}}^{\beta} \rangle$; the latter then provides an explicit form for the kinetic equation.

4. ANALYSIS OF $\langle \delta f_{\mathbf{k}z}^{\alpha}(\mathbf{v}_1) \delta f_{-\mathbf{k}-z}^{\beta}(\mathbf{v}_2) \rangle$

The insight into the nature of these correlations is greatly facilitated by using the diagram technique. The contributions to the correlation function are represented by two superposed lines, one for each one-particle fluctuation: these lines are of any of the types C, D, or E. The averaging involves a statistical coupling of all the vertical lines of these diagrams. We now introduce the following classification of the contributions:

Purely Coherent Correlations—$\langle \delta f_{\mathbf{k}z}^{\alpha(c)}(1) \delta f_{-\mathbf{k}-z}^{\beta(c)}(2) \rangle$. These diagrams are constructed by superposing two lines of type C and introducing a statistical coupling of the two vertical lines:

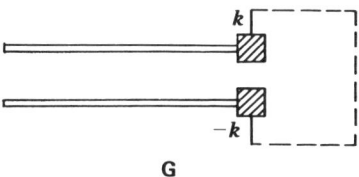

G

Their expression is immediately obtained from Eq. (16):

$$c_{\mathbf{k}z}^{\alpha\beta(cc)}(12) = g_{\mathbf{k}z}^{\alpha}(1) g_{-\mathbf{k}-z}^{\beta}(2) I_{\mathbf{k}z} s_{\mathbf{k}z}^{\alpha}(1) s_{-\mathbf{k}-z}^{\beta}(2) \tag{26}$$

where the one-particle source terms are defined by

$$s_{\mathbf{k}z}^{\alpha}(4) = \frac{e_{\alpha}}{m_{\alpha}} i\mathbf{k} \cdot \boldsymbol{\xi}_{\mathbf{k}z}^{\alpha}(1) F^{\alpha}(1) \tag{27}$$

One notes that these diagrams involve factorized propagators and factorized source terms.

Mixed Correlations—$\langle \delta f_{\mathbf{k}z}^{\alpha(c)}(1) \delta \tilde{f}_{-\mathbf{k}-z}^{\beta}(2) \rangle$. The only possible diagrams of this type originate from a superposition of a line of type C and a line of type D:

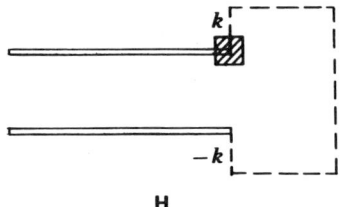

H

Its contribution is

$$c_{kz}^{\alpha\beta(\tilde{c})}(12) = g_{kz}^{\alpha}(1) g_{-k-z}^{\beta}(2) \frac{e_\alpha}{m_\alpha} i\mathbf{k} \cdot \xi_{kz}^{\alpha}(1) F^{\alpha}(1) \langle \phi_{kz} \gamma_{-k}^{\beta}(2) \rangle \tag{28}$$

It will be noted that a diagram of type E cannot be coupled to C, because the "uncombed" lines will give zero upon averaging.

Purely Incoherent Correlations—$\langle \delta \tilde{f}_{kz}^{\alpha}(1) \delta \tilde{f}_{-k-z}^{\beta}(2) \rangle$. In this case we have several types of contributions. The simplest are those obtained from a coupling of two lines of type D:

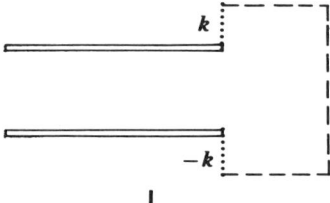

I

giving

$$\tilde{c}_{kz}^{\alpha\beta(1)}(12) = g_{kz}^{\alpha}(1) g_{-k-z}^{\beta}(2) \langle \gamma_{k}^{\alpha}(1) \gamma_{-k}^{\beta}(2) \rangle \tag{29}$$

This contribution represents independent ballistic propagation of the initial condition.

We may now also construct diagrams involving two uncombed lines of type E: they may give nonvanishing contributions through *statistical cross couplings*. Two simple examples of such contributions are shown here:

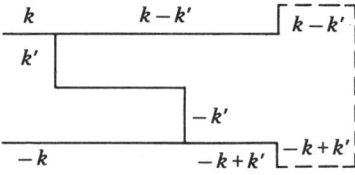

This diagram is interpreted in terms of a cross-renormalized propagator. This trajectory correlation effect is very important for the clump concept (Misguich and Balescu[6,7]). A different type of cross coupling results in a cross-renormalized vertex:

A Diagrammatic Approach to the Theory of Clumps in Turbulent Plasmas 305

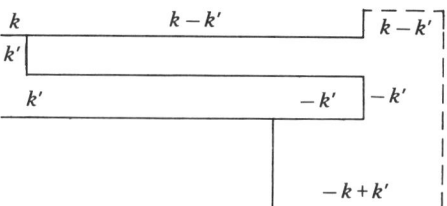

The systematic summation of diagrams of this type leads to the following three classes of cross-renormalized diagrams:

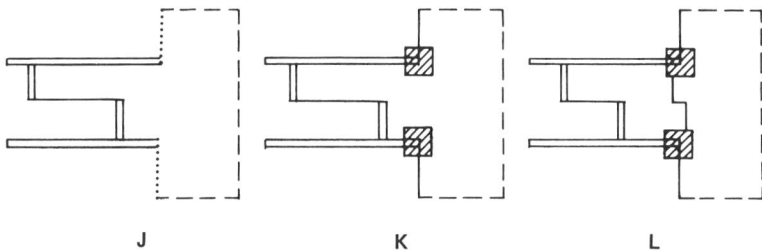

Diagrams of class J represent *correlated ballistic propagation* of the initial condition:

$$\tilde{c}_{\mathbf{k}z}^{\alpha\beta(b)}(12) = \frac{1}{2\pi} \int dz' \int d\mathbf{k}' g_{k/k-k'}^{\alpha\beta}(12) \langle \gamma_{\mathbf{k}-\mathbf{k}'}^{\alpha}(1) \gamma_{-\mathbf{k}+\mathbf{k}'}^{\beta}(2) \rangle \qquad (30)$$

where the *correlated two-particle propagator* is defined by

$$g_{k/k-k'}^{\alpha\beta}(12) = g_k^{\alpha}(1) g_{-k}^{\beta}(2) \delta(k') + \frac{e_\alpha e_\beta}{m_\alpha m_\beta} \underset{k''}{\mathbf{S}} I_{k''} g_k^{\alpha}(1) g_{-k}^{\beta}(2)$$
$$\times (\mathbf{k}'' \cdot \boldsymbol{\partial}_2) g_{k-k''/k-k''-k'}^{\alpha\beta}(12) \qquad (31)$$

(Note that the diagrams in class J include those of class I.)

Diagrams of class K are similar to class G, but involve the correlated propagator

$$\tilde{c}_k^{\alpha\beta(c1)} = \underset{k'}{\mathbf{S}} [g_{k/k-k'}^{\alpha\beta}(12) - \delta(k') g_k^{\alpha}(1) g_{-k}^{\beta}(2)] I_{k-k'} s_{k-k'}^{\alpha}(1) s_{-k+k'}^{\beta}(2) \qquad (32)$$

Finally, the diagrams of class L are the most completely renormalized ones as they involve both a correlated propagator and a correlated source:

$$\tilde{c}_k^{\alpha\beta(L)}(12) = \underset{k'}{\mathbf{S}} g_{k/k-k'}^{\alpha\beta}(12) s_{k-k'}^{\alpha\beta}(12) \qquad (33)$$

where the cross-renormalized source obeys the following complicated equation (this expression includes the diagrams G and K):

$$s_k^{\alpha\beta}(12) = I_k s_k^{\alpha}(1) s_{-k}^{\beta}(2) + \frac{e_\alpha e_\beta}{m_\alpha m_\beta} \mathbf{S} \, I_{k''}$$
$$\times \{[-i\mathbf{k}'' \cdot \partial_1 g_{k-k''}^{\alpha}(1) s_{k-k''}^{\alpha}(1)][i(\mathbf{k} - \mathbf{k}'') \cdot \partial_2 g_{-k''}^{\beta}(2) s_{-k''}^{\beta}(2)]$$
$$+ [-i(\mathbf{k} - \mathbf{k}'') \cdot \partial_1 g_{k''}^{\alpha}(1) s_{k''}^{\alpha}(1)][i\mathbf{k}'' \cdot \partial_2 g_{-k+k''}^{\beta}(2) s_{-k+k''}^{\beta}(2)]\} I_{k-k''}$$
(34)

This class of diagrams will not be retained in this chapter.

5. EXPLICIT FORM OF THE KINETIC EQUATION

We now briefly sketch the derivation of the kinetic equation. In order to complete point (3) of the program of Section 3, one starts from the Vlasov equation [Eq. (3)], which yields

$$[\partial_t + i\mathbf{k} \cdot (\mathbf{v}_1 - \mathbf{v}_2)] \langle \delta f_{\mathbf{k}}^{\alpha}(1; t) \delta f_{-\mathbf{k}}^{\beta}(2; t) \rangle$$
$$= \frac{e_\alpha}{m_\alpha} \sum_{\mathbf{k}'} i\mathbf{k}' \cdot \partial_1 \langle \phi_{\mathbf{k}'}(t) \delta f_{\mathbf{k}-\mathbf{k}'}^{\alpha}(1; t) \delta f_{-\mathbf{k}}^{\beta}(2; t) \rangle$$
$$+ \frac{e_\alpha}{m_\alpha} \langle \phi_{\mathbf{k}}(t) \delta f_{-\mathbf{k}}^{\beta}(2; t) \rangle i\mathbf{k} \cdot \partial_1 F^{\alpha}(1; t) + \{\alpha \to \beta, \mathbf{k} \leftrightarrow -\mathbf{k}, 1 \leftrightarrow 2\} \quad (35)$$

The triple correlations are analyzed by using a diagram technique similar to the one developed in the previous paragraph. The result of this operation is to transform the right-hand side into a functional of the *two-time* correlation function $c_{\mathbf{k}}^{\alpha\beta}(12; t, \tau)$: this equation is therefore not closed.

In order to close it, a particular form of a *Markovian approximation* is applied: this method follows closely a procedure described in Boutros-Ghali and Dupree.[4] The resulting equation is Fourier transformed and expressed in terms of relative variables $\mathbf{x}_- = \mathbf{x}_1 - \mathbf{x}_2$ and center of mass variables $\mathbf{x}_+$ (with similar definitions for $\mathbf{v}_-, \mathbf{v}_+$). After a few more approximations described in Dupree[3] and Misguich and Balescu,[7] one obtains

$$\{\partial_t + \mathbf{v}_- \cdot \nabla_- + \partial_- \cdot \mathbf{D}_- \cdot \partial_-\} c^{\alpha\beta}(12; t) = \Sigma^{\alpha\beta}(12; t) \quad (36)$$

an equation slightly more general than Dupree's.[3] Here,

$$\mathbf{D}_- = \mathbf{D}^{\alpha\alpha}(\mathbf{v}_1) + \mathbf{D}^{\beta\beta}(\mathbf{v}_2) - \mathbf{D}^{\alpha\beta}(\mathbf{x}_-, \mathbf{v}_2) - \mathbf{D}^{\beta\alpha}(\mathbf{x}_-, \mathbf{v}_1) \quad (37)$$

where $\mathbf{D}^{\alpha\alpha}(\mathbf{v}_1)$ is an "ordinary" diffusion coefficient, well known from the theory of turbulent plasmas [see Eq. (12)], whereas $\mathbf{D}^{\alpha\beta}(\mathbf{x}_-, \mathbf{v}_2)$ is a *relative* diffusion coefficient, originating from the trajectory correlations, that is, from the diagrams of class K.

The source term $\Sigma^{\alpha\beta}$, including one-particle vertex renormalization (but not vertex cross renormalization) is

$$\Sigma^{\alpha\beta}(12; t) = [\xi^\alpha F^\alpha(1)] \cdot \mathbf{D}^{\alpha\beta}(\mathbf{x}_-, \mathbf{v}_2) \cdot [\xi^\beta F^\beta(2)]$$
$$+ \frac{e_\alpha}{m_\alpha} \frac{m_\beta}{e_\beta} [\xi^\alpha F^\alpha(1)] \cdot \mathscr{F}^\beta(\mathbf{x}_-, \mathbf{v}_2) F^\beta(2) + [\alpha \leftrightarrow \beta, 1 \leftrightarrow 2] \quad (38)$$

where $\mathscr{F}^\beta(\mathbf{x}_-, \mathbf{v}_2)$ bears the same relation to $\mathscr{F}^\beta(\mathbf{v}_2)$ in Eq. (24) as $\mathbf{D}^{\alpha\beta}(\mathbf{x}_-, \mathbf{v}_2)$ to $\mathbf{D}^\beta(\mathbf{v}_2)$. The remarks after Eq. (24) apply here as well: the notation $\mathscr{F}^\beta F^\beta$ is artificial, and does not necessarily imply that $\mathscr{F}^\beta$ can be interpreted as a friction force.

At this stage, the *clump effect* is quite manifest. In a steady situation ($\partial_t c = 0$), and for $\alpha = \beta$, when $\mathbf{x}_- \to 0$, $\mathbf{v}_- \to 0$, the left-hand side of Eq. (37) goes to zero (because $\mathbf{D}^{\alpha\beta} \to \mathbf{D}^{\beta\beta}$), but the source term does not. Hence, there is a very strong *enhancement of the correlation function at small relative velocity and distance*. This is understood from the fact that such particles (of the same species) feel the same fluctuating fields, hence travel together for a long time. Their *relative diffusion is anomalously low*.

Equation (37) can be solved (approximately), and after subtracting the coherent and mixed contributions (Sections 4A and 4B), we are left with an expression for the *incoherent correlation function*:

$$\tilde{c}^{\alpha\alpha}(12; t) = \tilde{c}^{\alpha(b)}(12; t) + \tilde{c}^{\alpha(cl)}(12; t) \quad (39)$$

The first term represents the *ballistic part* of the correlations (diagram J) and is shown to be simply expressed as follows (Pelletier and Pomot[8]):

$$\tilde{c}^{\alpha(b)}(12; t) = \delta(\mathbf{x}_1 - \mathbf{x}_2)\delta(\mathbf{v}_1 - \mathbf{v}_2) F^\alpha(\mathbf{v}_1; t) \quad (40)$$

The second term represents the *clump correlations*: it is obtained [by neglecting the second term in Eq. (38)] in the following form (Misguich and Balescu[6,7,9]), corresponding to the diagram K:

$$\tilde{c}^{\alpha(cl)}(12; t) = \int d\mathbf{k} \, \exp(i\mathbf{k} \cdot \mathbf{x}_-) \tau_\mathbf{k}^\alpha(\mathbf{x}_-, \mathbf{v}_-) \mathbf{d}_\mathbf{k}^\alpha : [\xi_1^\alpha F^\alpha(1)][\xi_2^\alpha F^\alpha(2)] \quad (41)$$

where $\mathbf{d}_\mathbf{k}^\alpha$ is the Fourier transform of the *relative diffusion coefficient* $[\mathbf{D}^{\alpha\alpha}(\mathbf{x}_-, \mathbf{v}_1) + \mathbf{D}^{\alpha\alpha}(\mathbf{x}_-, \mathbf{v}_2)]$:

$$\mathbf{d}_\mathbf{k}^\alpha(\mathbf{v}_1, \mathbf{v}_2) = \frac{e_\alpha^2}{m_\alpha^2} \frac{1}{2\pi} \int dz \, I_{\mathbf{k}z} \mathbf{k}\mathbf{k}[g_{\mathbf{k}z}^\alpha(1) + g_{-\mathbf{k}-z}^\alpha(2)] \quad (42)$$

We introduced the "*duration of a clump*":

$$\tau_\mathbf{k}^\alpha(\mathbf{x}_-, \mathbf{v}_-) = \int_0^\infty d\tau \, \exp(-i\mathbf{k} \cdot \mathbf{v}_- \tau)\{\exp[-\tfrac{1}{2}\mathbf{k}\mathbf{k} : \langle \delta\mathbf{x}_-(-\tau)\delta\mathbf{x}_-(-\tau)\rangle]$$
$$- \exp[-\tfrac{1}{2}\mathbf{k}\mathbf{k} : \langle \delta\mathbf{x}_-(-\tau)\delta\mathbf{x}_-(-\tau)\rangle_{\text{unc}}]\} \quad (43)$$

This function contains the essence of the clump effect. It is only nonzero if there is a difference between true relative diffusion $\langle \delta\mathbf{x}_-(-\tau)\delta\mathbf{x}_-(-\tau)\rangle$ and uncorrelated diffusion [where $\delta\mathbf{x}_-(\tau)$ is the relative distance advanced by a time τ from its initial value $\mathbf{x}_-$]. In agreement with our qualitative discussion, it goes to infinity as $\mathbf{x}_- \to 0$, $\mathbf{v}_- \to 0$. It was first estimated rather roughly by Dupree,[3] and more thoroughly discussed in Misguich and Balescu.[7]

Equation (41) is the starting point for a rather straightforward derivation of a kinetic equation for $F^\alpha(1; t)$. We cannot give the details of this calculation in the restricted space available here. The result is (neglecting vertex renormalization):

$$\partial_t F^\alpha(1) = \sum_\beta \frac{1}{2\pi} \int d\omega \int d\mathbf{k} \int d\mathbf{v}_2 \frac{8\pi^3 e_\alpha^2 e_\beta^2}{k^4 m_\alpha} \frac{1}{|\varepsilon_{\mathbf{k}\omega}|^2} \mathbf{k} \cdot \partial_1$$

$$\times \Big(g_{\mathbf{k}\omega}^\alpha(1)[g_{\mathbf{k}\omega}^\beta(2) + g_{-\mathbf{k}-\omega}^\beta(2)]$$

$$\times [F^\beta(2) m_\alpha^{-1} \mathbf{k} \cdot \partial_1 F^\alpha(1) - m_\beta^{-1} \mathbf{k} \cdot \partial_2 F^\beta(2) F^\alpha(1)]$$

$$+ \int d\mathbf{v}_3 \int d\mathbf{l} \{ g_{\mathbf{k}\omega}^\alpha(1)[g_{\mathbf{k}\omega}^\beta(2) + g_{-\mathbf{k}-\omega}^\beta(3)] \hat{\tau}_{\mathbf{k}-\mathbf{l}}^\beta(\mathbf{l}, \mathbf{v}_{23})$$

$$\times \mathbf{d}_{\mathbf{k}-\mathbf{l}}^\beta(\mathbf{v}_2\mathbf{v}_3) : [\partial_2 F^\beta(2)][\partial_3 F^\beta(3)] m_\alpha^{-1}[\mathbf{k} \cdot \partial_1 F^\alpha(1)]$$

$$- g_{\mathbf{k}\omega}^\beta(3)[g_{\mathbf{k}\omega}^\alpha(1) + g_{-\mathbf{k}-\omega}^\alpha(2)] \hat{\tau}_{\mathbf{k}-\mathbf{l}}^\alpha(\mathbf{l}, \mathbf{v}_{12}) m_\beta^{-1} \mathbf{k} \cdot \partial_3 F^\beta(3)$$

$$\times \mathbf{d}_{\mathbf{k}-\mathbf{l}}^\alpha(\mathbf{v}_1\mathbf{v}_2) : [\partial_2 F^\alpha(2)][\partial_2 F^\alpha(1)] \} \Big) \tag{44}$$

where $\hat{\tau}_{\mathbf{k}}^\alpha(\mathbf{l}, \mathbf{v}_{12})$ is the Fourier transform (in $\mathbf{x}_-$) of $\tau_{\mathbf{k}}^\alpha(\mathbf{x}_-, \mathbf{v}_1 - \mathbf{v}_2)$.

The first group of terms in this collision integral (enclosed in the second pair of square brackets) is the contribution of the ballistic terms $\tilde{c}^{\alpha(b)}$: it was obtained by Pelletier and Pomot.[8] If the clump contribution were neglected, it would be precisely of the Balescu–Lenard form [Eq. (24)]: the only difference is the occurrence of renormalized propagators and dielectric function. This rather simple equation would deserve much more attention than it has had in the past.

The contribution of the clumps is much more complicated. One fact is, however, immediately striking: both terms in the third pair of brackets are of *diffusive type*, there is no term involving $F^\alpha(1)$ (rather than its derivatives). Hence, Dupree's claim that the incoherent fluctuations produce friction is not founded.

In conclusion, the clump effect is certainly an important feature of the particle correlations in a turbulent plasma, which has many implications in such problems as the determination of turbulent spectra and anomalous transport. The calculation of these effects on real observable quantities is, however, still in its beginning stage.

REFERENCES

1. L. M. Altshul' and V. I. Karpman, *Sov. Phys. JETP* **22**, 361 (1966).
2. W. Horton and D-I. Choi, *Phys. Rep.* **49**, 273 (1979).

3. T. H. Dupree, *Phys. Fluids* **15**, 334 (1972).
4. T. Boutros-Ghali and T. H. Dupree, *Phys. Fluids* **24**, 1839 (1981).
5. T. H. Dupree and D. J. Tetreault, *Phys. Fluids* **21**, 425 (1978).
6. J. H. Misguich and R. Balescu, *J. Plasma Physics* **19**, 147 (1978).
7. J. H. Misguich and R. Balescu, *Plasma Phys.* **20**, 781 (1978).
8. G. Pelletier and C. Pomot, *J. Plasma Phys.* **14**, 153 (1975).
9. J. H. Misguich and R. Balescu, *Plasma Phys.* **24**, 289 (1982).

GLOBAL ANALYSIS OF NONLINEAR TRANSIENT PHENOMENA NEAR THE INSTABILITY POINT

Masuo Suzuki

Department of Physics, Faculty of Science, University of Tokyo, Tokyo, Japan

Abstract. The purpose of this chapter is to review briefly the main idea of the scaling theory of unstable transient phenomena proposed by the present author[7-9] and to explain the recent extension of the scaling theory to the whole time region, namely, the global analysis of nonlinear unstable transient phenomena.[10] As an interesting example of applications of the above scaling theory, we discuss the relative diffusion in turbulent plasma, namely, we present the scaling theory[11] of clumps[2,3] in plasma. Relative diffusions in turbulent magnetic fields and turbulent velocity fields will be also discussed.

1. SCALING THEORY OF UNSTABLE TRANSIENT PHENOMENA

A general strategy to treat the relaxation and fluctuation from or near the instability point is given by the scaling theory of transient phenomena proposed by the present author. In the present section we explain the main idea of the scaling theory. Schematically we consider the potential[7,8] shown in Fig. 1. If an external random force is applied to a particle in the potential shown in Fig. 1, the particle can move stochastically as a Brownian particle. At the very beginning of time, it moves extremely slowly because of the unstable equilibrium property of the system. After the particle once moves appreciably far from the instability point a, then it moves rather quickly, and finally it approaches the stable equilibrium point again very slowly. Thus, the whole time region can be divided qualitatively into the three regimes; namely, (*a*) the initial regime in which the linear approximation (or Gaussian treatment) is valid, (*b*) the nonlinear scaling regime in which the

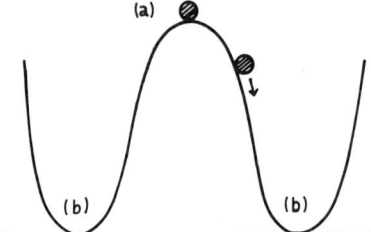

Figure 1. Schematic potential showing an instability: (a) unstable point, (b) stable points.

scaling law holds asymptotically in the limit of the small random force, and (c) the final regime in which the Gaussian treatment may be valid again. This situation is shown in Fig. 2. It should be remarked that this division of time region is not sharp but just qualitative as is shown in Fig. 3.

A given original stochastic equation can be simplified in each time region by extracting the most dominant contribution in each region asymptotically for a small random force. Then, we unify three solutions of the simplified equations thus obtained. This process can be substituted equivalently by a linearization approximation of an equation obtained by an appropriate nonlinear transformation[7,8] of a variable. One of the most remarkable points in this formulation is that a scaling property appears in the second nonlinear regime in the sense that

$$\langle Q(x(t)) \rangle \simeq \langle Q(x) \rangle_{st} f^{(sc)}(\tau) \tag{1}$$

for an arbitrary quantity $Q(x)$, where τ is the scaling variable defined by

$$\tau = \varepsilon e^{2\gamma t} \tag{2}$$

Here, $\langle Q(x) \rangle_{st}$ is the stationary value of $\langle Q(x(t)) \rangle$, ε denotes the strength of the random force, and γ is the growing rate of the system.

The reason why we call the above formulation scaling theory is that the normalized quantity $\langle Q(x(t)) \rangle / \langle Q(x) \rangle_{st}$ is invariant under the transformation that

$$\varepsilon' = b\varepsilon \quad \text{and} \quad t' = t + \frac{1}{2\gamma} \log\left(\frac{1}{b}\right) \tag{3}$$

That is, the relaxation and fluctuation of the system are universal in the scaling limit that $\varepsilon \to 0$, $t \to \infty$, $\tau =$ fixed.

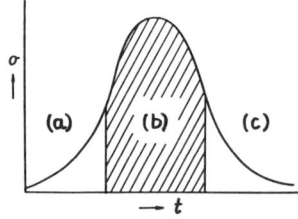

Figure 2. The division of the whole time region into three regimes; (a) initial, (b) scaling, and (c) final regimes.[7-9]

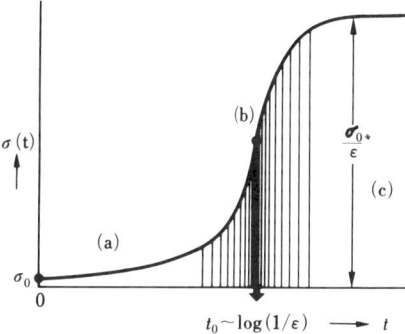

Figure 3. Each time region changes gradually.

This scaling property of unstable transient phenomena has been found in many interesting physical systems.[9] In particular, this scaling property is very useful to define the characteristic time or onset time t_0 around which the fluctuation of the system changes into a macroscopic order from a microscopic one of the order of ε, namely,

$$t_0 \simeq \frac{1}{2\gamma}\log\left(\frac{1}{\varepsilon}\right) \qquad (4)$$

which is determined by setting $\tau \simeq 1$ in Eq. (2).

Here, we give a simple instructive example described by the following nonlinear Langevin equation:[7-9]

$$\frac{d}{dt}x = \gamma x - gx^3 + \eta(t) \qquad (5)$$

where $\eta(t)$ denotes the Gaussian white noise satisfying

$$\langle \eta(t)\eta(t')\rangle = 2\varepsilon\delta(t-t') \qquad (6)$$

The scaling solution of the fluctuation $\langle x^2(t)\rangle$ for Eq. (5) is given by[7]

$$\frac{\langle x^2(t)\rangle}{\langle x^2\rangle_{st}} \simeq f^{(sc)}(\tau) = \frac{1}{\sqrt{2\pi}}\int_{-\infty}^{\infty} e^{-\xi^2/2}\frac{\xi^2\tau}{1+\xi^2\tau}d\xi \qquad (7)$$

where $\langle x^2\rangle_{st} = \gamma/g$ and τ is now given by

$$\tau = \frac{g}{\gamma}\left(\langle x^2(0)\rangle + \frac{\varepsilon}{\gamma}\right)e^{2\gamma t} \qquad (8)$$

This definition of τ is essentially equivalent to Eq. (2) for small initial fluctuation $\langle x^2(0)\rangle$. The probability distribution function $P(x, t, \varepsilon)$ also takes the following scaling form:

$$P(x, t, \varepsilon) \simeq P^{(sc)}(x, \tau) = \frac{1}{\sqrt{2\pi\tau}}\left(1 - \frac{g}{\gamma}x^2\right)^{-3/2}\exp\left(-\frac{x^2}{2\tau[1-(g/\gamma)x^2]}\right) \qquad (9)$$

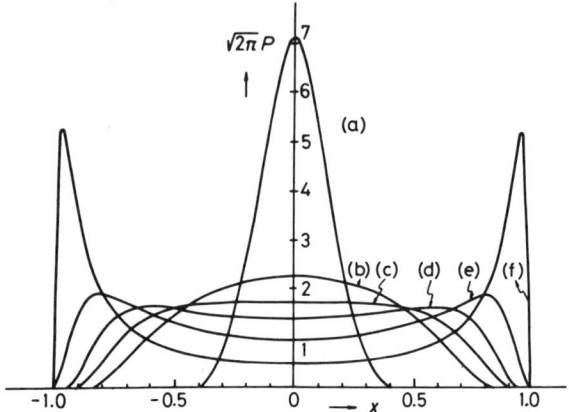

Figure 4. Scaling distribution function $P^{(sc)}(x, \tau)$ for the nonlinear Langevin equation (1.5): (a) $\tau = 0.02$, (b) $\tau = 0.2$, (c) $\tau = \tau_0 = \frac{1}{3}$, (d) $\tau = 0.5$, (e) $\tau = 4$, where τ is defined by Eq. (1.8) for $g = \gamma$.[7]

except near $x = x_{st} = \pm \gamma/g$. This is shown in Fig. 4. It is easily understood from Fig. 4 that the initial single peak changes into double peaks around the onset time t_0. This explains well the mechanism of formation of macroscopic order in nonequilibrium systems.

2. UNIFIED FORMULATION OF UNSTABLE TRANSIENT PHENOMENA

In the present section, we explain the unified formulation[10] of unstable transient phenomena, which includes a new derivation of the scaling theory in the scaling limit.

For simplicity, we consider here the following Fokker–Planck equation:

$$\frac{\partial}{\partial t} P(x, t) = \mathscr{L} P(x, t) \equiv \left(-\frac{\partial}{\partial x} \alpha(x) + \varepsilon \frac{\partial^2}{\partial x^2} \right) P(x, t) \qquad (10)$$

where $\alpha(x)$ denotes the deterministic part of the nonlinear Langevin equation corresponding to Eq. (10) to describe unstable transient phenomena. The formal solution of Eq. (10) is given by

$$P(x, t) = e^{t\mathscr{L}} P(x, 0) = e^{t\mathscr{L}_{\text{drift}} + t\mathscr{L}_{\text{dif}}} P(x, 0) \qquad (11)$$

where $\mathscr{L}_{\text{drift}}$ and $\mathscr{L}_{\text{dif}}$ denotes the drift and diffusion terms of the Fokker–Planck operator $\mathscr{L}$, respectively, as follows:

$$\mathscr{L}_{\text{drift}} = -\frac{\partial}{\partial x} \alpha(x) \quad \text{and} \quad \mathscr{L}_{\text{dif}} = \varepsilon \frac{\partial^2}{\partial x^2} \qquad (12)$$

Our global approximation method (so-called GAM) is to replace Eq. (11) by the following product operators:

$$P^{(GAM)}(x, t) = e^{g(t)\mathscr{L}_{dif}} e^{t\mathscr{L}_{drift}} e^{h(t)\mathscr{L}_{dif}} P(x, 0) \qquad (13)$$

Here $g(t)$ and $h(t)$ will be determined later by optimizing the solution under the conditions that (a) $g(t) + h(t) \simeq t$ for small t in the initial regime, (b) $P^{(GAM)}(x, t)$ includes the scaling solution in the nonlinear regime, and (c) $P^{(GAM)}(x, t)$ approaches the correct stationary state asymptotically for $t \to \infty$.

The above decoupling [Eq. (13)] corresponds to the separation of the deterministic drift effect and fluctuation effect in the time coordinate, namely, the diffusion effect is first taken into account in the initial regime through the operator $\exp[h(t)\mathscr{L}_{dif}]$, the nonlinear drift effect is important in the second regime and the random noise is again relevant in the final regime.

The usefulness of the decoupling (13) is quite clear from the observation that

$$e^{h\mathscr{L}_{dif}} f(x) = \frac{1}{\sqrt{4\pi\varepsilon h}} \int_{-\infty}^{\infty} \exp\left(-\frac{(\xi - x)^2}{4\varepsilon h}\right) f(\xi) d\xi \qquad (14)$$

and

$$e^{t\mathscr{L}_{drift}} f(x) = \frac{\alpha(\xi(x, t))}{\alpha(x)} f(\xi(x, t), t) \qquad (15)$$

where $\xi(x, t)$ is the solution of the equation $d\xi/dt = -\alpha(\xi)$ with the initial condition $\xi = x$ at $t = 0$ [i.e., $\xi(x, t)$ is the inverse function of the solution $x(\xi, t)$ of the equation $dx/dt = \alpha(x)$ with $x = \xi$ at $t = 0$].

By the help of the concept of the Lie algebra, we arrive[10] at the following conclusion that the global approximation $P^{(GAM)}(x, t)$ is given by

$$P^{(GAM)}(x, t) = \exp\left[\beta(t)(e^{2\gamma t} - 1)\left(\frac{\varepsilon}{2\gamma}\right)\frac{\partial^2}{\partial x^2}\right] \exp(t\mathscr{L}_{drift})$$
$$\times \exp\left[\{1 - \beta(t)\}(1 - e^{-2\gamma t})\left(\frac{\varepsilon}{2\gamma}\right)\frac{\partial^2}{\partial x^2}\right] P(x, 0) \qquad (16)$$

where $\beta(t) = (\sigma_e \gamma) e^{-2\gamma t}$ and σ_e denotes the variance of x at the stationary state x_{st}, that is, $\langle(x - x_{st})^2\rangle_{st} = \varepsilon \sigma_e$.

If we make use of the formulas (14) and (15), then we find easily an integral representation of $P^{(GAM)}(x, t)$ for an arbitrary drift term.

The above result is reduced to the scaling theory in the second nonlinear regime for the scaling limite as follows:

$$P^{(GAM)}(x, t) = P^{(sc)}(x, t) = e^{t\mathscr{L}_{drift}} e^{(\varepsilon/2\gamma)\partial^2/\partial x^2} P(x, 0) \qquad (17)$$

It is also easily confirmed that the expression (16) describes correctly the fluctuations both at the initial and final regimes. Thus, we obtain the required global approximate solution of the nonlinear Fokker–Planck equation (10).

The above formulation is easily extended to more-general time-dependent potentials [i.e., systems with a time-dependent $\alpha(x, t)$]. This will be reported in detail elsewhere.

3. SCALING THEROY OF CLUMPS

As an interesting application of the above scaling theory, we discuss here the scaling treatment[11] of clumps in turbulent plasma. We are interested in the relative spatial diffusion $y(t) = \langle r^2(t) \rangle$ of particles having an initial distance $r(0) = x_1(0) - x_2(0)$ and a relative velocity $g(0) = v_1(0) - v_2(0)$.

We study how two charged particles separate in time in a fluctuating electric field $E(x, t)$ when $r(0)$ and $g(0)$ are very small. Our starting equation of motion is given by

$$\dot{x}(t) = v(t) \quad \text{and} \quad \dot{v}(t) = \frac{q}{m} E(x(t), t) \tag{18}$$

where q and m denote the charge and mass, respectively. By introducing the following self-consistent Gaussian–white treatment[11] of the turbulent electric field

$$\langle E(x_1(t_1), t_1) E(x_2(t_2), t_2) \rangle = \tau_c \delta(t_1 - t_2) E^2 \exp\{-\kappa^2 \langle [x_1(t_1) - x_2(t_2)]^2 \rangle\} \tag{19}$$

we obtain our fundamental self-consistent equation of the relative spatial diffusion $y(t) \equiv \langle r^2(t) \rangle = \langle [x_1(t) - x_2(t)]^2 \rangle$ in the form

$$y(t) = y_0(t) + 2\left(\frac{qE}{m}\right)^2 \tau_c \int_0^t dt' \int_0^{t'} ds (t-s)(1 - e^{-\kappa^2 y(s)}) \tag{20}$$

with $y_0(t) = [r(0) + g(0)t]^2$. Here, τ_c denotes the effective correlation time of electric-field temporal fluctuations and κ^{-1} ($\equiv \xi$) denotes the correlation length of electric-field spatial fluctuations. The above integral equation (20) is easily reduced to the following nonlinear differential equation:

$$\kappa^2 \tau_0^3 \frac{d^3}{dt^3} y(t) = 1 - \exp[-\kappa^2 y(t)] \tag{21}$$

where τ_0 is the diffusion characteristic time[3] defined by

$$\tau_0^{-3} = 4\tau_c (\kappa qE/m)^2 \tag{22}$$

Here we introduce dimensionless quatities as

$$f(t) \equiv \kappa^2 y(t) \quad \text{and} \quad t/\tau_0 \to t \tag{23}$$

Then, we arrive at the following nonlinear differential equation:[11]

$$\frac{d^3}{dt^3} f(t) = 1 - e^{-f(t)} \tag{24}$$

under the initial condition that

$$f(0) = a^2 = [\kappa r(0)]^2$$
$$f'(0) = 2ab = 2\kappa^2 r(0) g(0) \tau_0 \qquad (25)$$
$$f''(0) = 2b^2 = 2[\kappa g(0) \tau_0]^2$$

namely

$$f(t)(a + bt)^2 + O(t^3) \qquad (26)$$

As mentioned above, we are interested in the situation in which the initial separations in phase space, $r(0)$ and $g(0)$, are very small, that is, $a \ll 1$ and $b \ll 1$. The intrinsic spatial diffusion $z(t)$ defined by

$$z(t) = f(t) - (a + bt)^2 \qquad (27)$$

satisfies the following equation:

$$\frac{d^3}{dt^3} z(t) = 1 - e^{-(a+bt)^2} e^{-z(t)} \qquad (28)$$

This equation is similar to that obtiained by Misguich and Balescu[3] whose equation takes the form

$$\frac{d^3}{dt^3} z(t) = 1 - \cos[\sqrt{2}(a+bt)] e^{-z(t)} \qquad (29)$$

in our notation. That is, their equation contains an oscillatory factor. However, both of Eqs. (28) and (29) are reduced to the following equation:

$$\frac{d^3}{dt^3} z(t) = 1 - e^{-z(t)} + (a+bt)^2 e^{-z(t)} \qquad (30)$$

for small a and b. It is easily confirmed that the above simplified equation is sufficient to find a scaling solution in the scaling limit that $a \to 0$, $b \to 0$, $t \to \infty$ and the scaling time τ fixed. Thus, our scaling solution of Eq. (24) is also the scaling one of Misguich and Balescu's equation (24).

According to the general scaling theory[7,9] of unstable transient phenomena brieflly reviewed in Section 2, we divide the whole time region into three regimes as follows.

Linear Regime. If we linearize Eq. (30) for small $z(t)$, we obtain the following Dupree equation:

$$\frac{d^3}{dt^3} z_l(t) = z_l(t) + (a+bt)^2 \qquad (31)$$

with the initial condition (25). The solution of Eq. (31) takes the following asymptotic form:

$$z_l(t) \simeq \tfrac{1}{3}(a^2 + 2ab + 2b^2) e^t \qquad (32)$$

for small a and b and for large t. Here, we introduce the following smallness parameter:

$$\delta = \tfrac{1}{3}(a^2 + 2ab + 2b^2) \tag{33}$$

that is, we have

$$z_l(t) \simeq \delta e^t \tag{34}$$

This suggests that the scaling variable τ in the nonlinear regime should be $\tau = \delta e^t$.

Nonlinear Scaling Regime. As was suggested in the linear approximation, we put $\tau = \delta e^t$ as our scaling variable, and we transform our original equation (30) in terms of the scaling variable τ in the scaling limit that $\delta \to 0$, $t \to \infty$, and $\tau =$ fixed and, consequently, we obtain[11]

$$\left(\tau \frac{d}{d\tau}\right)^3 f^{(sc)}(\tau) = 1 - \exp[-f^{(sc)}(\tau)] \tag{35}$$

for the scaling function $f^{(sc)}(\tau)$, under the initial condition that

$$f^{(sc)}(\tau) = \tau + O(\tau^2) \tag{36}$$

The solution of Eq. (35) is given in a power series of τ as

$$f^{(sc)}(\tau) = \sum_{n=1}^{\infty} a_n \tau^n; \quad a_1 = 1 \tag{37}$$

where a_n is given by

$$a_n = -\frac{1}{n!(n^3-1)}\left\{\frac{d^n}{d\tau^n} \exp\left(-\sum_{k=1}^{n-1} a_k \tau^k\right)\right\}_{\tau=0} \tag{38}$$

Thus, we obtain explicitly

$$f^{(sc)}(\tau) = \tau - \frac{\tau^2}{2(2^3-1)} + \frac{5\tau^3}{3(2^3-1)(3^3-1)} - \frac{447\tau^4}{4(2^3-1)^2(3^3-1)(4^3-1)}$$
$$+ \frac{59817\tau^5}{4 \times 5(2^3-1)^2(3^3-1)(4^3-1)(5^3-1)} \cdots \tag{39}$$

It will be an interesting open problem to find a compact solution of Eq. (35) or Eq. (24).

Final Regime. In the final regime that $t \to \infty$ ($\tau \to \infty$), Eq. (30) is reduced to

$$\frac{d^3}{dt^3} z(t) = 1 \tag{40}$$

Then, the solution of Eq. (40) takes the following form:

$$z(t) \simeq \tfrac{1}{6} t^3 \tag{41}$$

Global Analysis of Nonlinear Transient Phenomena near the Instability Point 319

This asymptotic behavior is also obtained from the scaling solution $f^{(sc)}(\tau)$ as

$$f^{(sc)}_{\text{asym}}(\tau) = \tfrac{1}{6}(\log \tau)^3 \simeq \tfrac{1}{6}t^3 \qquad (42)$$

Here, $f^{(sc)}_{\text{asym}}(\tau)$ is the asymptotic form of the scaling solution $f^{(sc)}(\tau)$, which should be obtained, in priciple, from Eq. (37) or Eq. (39), and more simply which satisfies

$$\left(\tau \frac{d}{d\tau}\right)^3 f^{(sc)}_{\text{asym}}(\tau) = 1 \qquad (43)$$

Thus, we can describe the characteristic feature of clumps in turbulent plasma for the whole region of time.

4. RELATIVE DIFFUSION IN RANDOM MAGNETIC FIELDS

We consider here how two charged particles separate in a random magnetic field $\mathbf{B}(\mathbf{r}, t)$. Our starting equation of motion is given by

$$m\frac{d^2\mathbf{r}}{dt^2} = q\frac{d\mathbf{r}}{dt} \times \mathbf{B}(\mathbf{r}, t) \qquad (44)$$

where m and q denote the mass and electric charge of particles. As before, we introduce here the following self-consistent Gaussian–white treatment:

$$\langle \mathbf{B}(\mathbf{r}_1(t), t)\mathbf{B}(\mathbf{r}_2(t'), t')\rangle = B^2 \tau_c \delta(t - t') \exp[-\kappa^2 \langle (\mathbf{r}_1(t) - \mathbf{r}_2(t'))^2 \rangle] \qquad (45)$$

We are now interested in the relative diffusion of the velocity $y(t) \equiv \langle (\mathbf{v}_1(t) - \mathbf{v}_2(t))^2 \rangle$ and its integrated diffusion

$$f(t) = \int_0^t y(s)\,ds \qquad (46)$$

It is easily shown as before that $f(t)$ satisfies the following nonlinear equation:

$$\frac{d^2 f(t)}{dt^2} = \frac{q^2 B^2 \tau_c}{2m^2}\left(2v^2 - \frac{df}{dt}\right)(1 - \exp\{-\kappa^2[\sigma_0 + \tau'_c f(t)]\}) \qquad (47)$$

by the help of the approximation (45) where $\sigma_0 = [\mathbf{r}_1(0) - \mathbf{r}_2(0)]^2$, v is the magnitude of the velocity, and τ'_c denotes the effective correlation time in the velocity space as

$$\langle [\mathbf{v}_1(t) - \mathbf{v}_2(t)][\mathbf{v}_1(s) - \mathbf{v}_2(s)]\rangle \simeq \tau'_c \delta(t - s)\langle [\mathbf{v}_1(t) - \mathbf{v}_2(t)]^2 \rangle \qquad (48)$$

According to the general scaling idea, we divide again the whole time region into three regimes as follows.

Linear Regime. By solving the linearized equation of (47), we obtain

$$f(t) \simeq \delta e^{\gamma t} \qquad (49)$$

for large t, where

$$\gamma = \frac{qB v \kappa}{m} (\tau_c \tau_c')^{1/2} \exp(-\tfrac{1}{2}\kappa^2 \sigma_0) \qquad (50)$$

and

$$\delta = y(0)/2\gamma \qquad (51)$$

This suggests such a choice of the scaling time τ as $\tau = \delta e^{\gamma t}$.

Scaling Regime. In the nonlinear regime, we obtain again

$$f(t) \equiv f(t, \delta) \simeq f^{(sc)}(\tau); \quad \tau = \delta e^{\gamma t} \qquad (52)$$

where $f^{(sc)}(\tau)$ is rather complicated and is omitted here.

Final Regime. In the final regime, $f(t)$ takes the following asymptotic form:

$$f(t) \simeq 2v^2 t \qquad (53)$$

because $\mathbf{v}_1(t)$ and $\mathbf{v}_2(t)$ move independently for $t \to \infty$.

5. SCALING THEORY OF RELATIVE DIFFUSION IN TURBULENT VELOCITY FIELDS

5.1 Scaling Theory of Relative Diffusion in Gaussian-White Turbulent Fields

Before studying the relative diffusion in realistic situations, we discuss a simple model of Gaussian–white turbulent fields. This model is valid in the long-time limit outside of the inertia range. However, we discuss here this model as an instructive example of the scaling theory.

Our equation of motion is given by

$$\frac{d}{dt} x_j(t) = v(x_j(t), t) \qquad (54)$$

where $v(x, t)$ denotes the turbulent field at a position x and time t. This is integrated formally as

$$x_j(t) = \int_0^t v(x_j(t'), t') dt' + x_j(0) \qquad (55)$$

The relative diffusion defined by

$$y(t) = \langle [x_1(t) - x_2(t)]^2 \rangle \qquad (56)$$

is expressed by the following integral:

$$y(t) = y_0 + \int_0^t dt' \int_0^{t'} dt'' \langle [v(x_1(t'), t') - v(x_2(t'), t')] \\ \times [v(x_1(t''), t'') - v(x_2(t''), t'')] \rangle \quad (57)$$

As a simple approximation, we introduce the following self-consistent Gaussian–white treatment:

$$\langle v(x_1(t), t) v(x_2(t'), t') \rangle \simeq v^2 \tau_c \delta(t - t') e^{-\kappa^2 \langle (x_1(t) - x_2(t))^2 \rangle} \quad (58)$$

Here τ_c denotes the correlation time of the velocity field and v is the maximum amplitude of the velocity field. With the help of Eq. (58), we obtain

$$\frac{d}{dt} y(t) = 4\tau_c v^2 (1 - e^{-\kappa^2 y(t)}) \quad (59)$$

By putting

$$\tau_0^{-1} = 4\tau_c v^2 \kappa^2, \quad \kappa^2 y(t) = f(t), \quad t/\tau_0 \to t \quad (60)$$

we arrive at the following simple dimensionless nonlinear equation:

$$\frac{d}{dt} f(t) = 1 - e^{-f(t)}; \quad f(0) = \kappa^2 y_0 \quad (61)$$

This is easily solved exactly to give a solution of the form

$$f(t) = \log[1 + (e^{f(0)} - 1)e^t] \quad (62)$$

That is,

$$y(t) = \kappa^{-2} \log[1 + (e^{\kappa^2 y_0} - 1)e^{t/\tau_0}] \quad (63)$$

in our original notation.

This takes the following scaling form:

$$f(t) \to f^{(sc)}(\tau) = \log(1 + \tau) \quad (64)$$

in the scaling limit

$$y_0 \to 0, \quad t \to \infty, \quad \text{and} \quad \tau \equiv \delta e^t = \kappa^2 y_0 e^{t/\tau_0} = \text{fixed} \quad (65)$$

(a) For $\tau \ll 1$ (linear regime), we have

$$f^{(sc)}(\tau) \simeq \delta e^{t/\tau_0} \quad (66)$$

(b) In the second scaling regime, the scaling solution is given by Eq. (64).

(c) For $\tau \gg 1$ (final regime), we obatin

$$y(t) \simeq \kappa^{-2} \log \tau \simeq t/\kappa^2 \tau_0 \quad (67)$$

as it should be.

5.2. Scaling Theory of Relative Diffusion in Realistic Turbulent Fields

In general, we introduce here the following two-time relative diffusion

$$y(t_1, t_2, \delta) = \langle (x_1(t_1) - x_2(t_1))(x_1(t_2) - x_2(t_2)) \rangle \tag{68}$$

for the initial condition $y(0, 0, \delta) = y_0 = \delta^2$. This is expressed by the following integral:

$$y(t_1, t_2, \delta) = y_0 + \int_0^{t_1} ds_1 \int_0^{t_2} ds_2 F(s_1, s_2) \tag{69}$$

where

$$F(t, t') = \langle [v(x_1(t), t) - v(x_2(t), t)] \times [v(x_1(t'), t') - v(x_2(t'), t')] \rangle \tag{70}$$

Here, we introduce again the following self-consistent treatment, namely the velocity-correlation function is approximated by some function of $y(t, t', \delta)$ as

$$2\langle v(x_1(t), t) v(x_2(t'), t') \rangle = F(y(t, t', \delta)) \tag{71}$$

Thus, the integrand in Eq. (69) is expressed by

$$F(t, t') = F(y(t, t', 0)) - F(y(t, t', \delta)) \tag{72}$$

By differentiating Eq. (69) with respect to t_1 and t_2, we obtain

$$\frac{\partial}{\partial t_1}\frac{\partial}{\partial t_2} y(t_1, t_2, \delta) = F(y(t_1, t_2, 0)) - F(y(t_1, t_2, \delta)) \tag{73}$$

under the initial condition that

$$y(t_1, t_2, \delta) = \delta^2 + O(t_1 t_2) \tag{74}$$

According to the general strategy of the scaling theory,[7,9] we discuss the solution of Eq. (73) in three different regimes.

Linear regime. By linearizing Eq. (73), we obtain

$$\frac{\partial}{\partial t_1}\frac{\partial}{\partial t_2} y_l(t_1, t_2, \delta) = \gamma^2 y_l(t_1, t_2, \delta) + F(y(t_1, t_2, 0)) - F(0); \quad \gamma^2 \equiv -F'(0) \tag{75}$$

This has the following solution

$$y_l(t_1, t_2, \delta) = \delta^2 e^{\gamma(t_1+t_2)} = (\delta e^{\gamma t_1})(\delta e^{\gamma t_2}) \tag{76}$$

for the initial condition (74)

Scaling Regime. In the nonlinear second regime, the two-time relative diffusion $y(t_1, t_2, \delta)$ is shown perturbationally to take the following asymptotic form:

$$y(t_1, t_2, \delta) \simeq f^{(sc)}(\delta e^{\gamma t_1} \times \delta e^{\gamma t_2}) \tag{77}$$

Therefore, the equal-time relative diffusion $y(t) \equiv y(t, t, \delta)$ takes the following scaling form:

$$y(t) = f^{(sc)}(\delta^2 e^{2\gamma t}) = f^{(sc)}(\tau) \qquad (78)$$

where $\tau = \delta^2 e^{2\gamma t}$. Furthermore, if we put $F(0) - F(y) \equiv \gamma^2 G(y)$, then the scaling function $f^{(sc)}(\tau)$ satisfies the following equation:

$$\left(\tau \frac{d}{d\tau}\right)^2 f^{(sc)}(\tau) = G(f^{(sc)}(\tau)) \qquad (79)$$

under the initial condition that $f^{(sc)}(\tau) = \tau + O(\tau^2)$.

Final Regime. This region is defined by the long-time region inside of the Kolmogorov inertia range in fully developed turbulence. Outside of the Kolmogorov inertia range, $y(t)$ takes the asymptotic form $y(t) \simeq t/\kappa^2 \tau_0$ as was previously shown.

Here, we assume that the energy spectrum $E(k)$ in the Kolmogorov inertia range takes the following form:

$$E(k) \sim k^{-n} \qquad (80)$$

asymptotically as usual. In the Kolmogorov theory,[4,6] we have $n = 5/3$. By the use of Eq. (80), we can estimate the asymptotic form of $F(y)$ in Eq. (71) as

$$F(y) \sim y^\phi; \quad \phi = \tfrac{1}{2}(n-1) \qquad (81)$$

Thus, $y(t_1, t_2, \delta)$ satisfies

$$\frac{\partial}{\partial t_1} \frac{\partial}{\partial t_2} y(t_1, t_2, \delta) \sim y(t_1, t_2, \delta)^\phi \qquad (82)$$

asymptotically. This has the following solution:

$$y(t_1, t_2, \delta) \sim (t_1 t_2)^{1/(1-\phi)} \qquad (83)$$

In particular, the equal-time relative diffusion $y(t)$ takes the following asymptotic form:

$$y(t) \sim t^{2/(1-\phi)} \sim t^{4/(3-n)} \qquad (84)$$

for large t but for $t \ll t_c$, where t_c denotes the correlation time of the velocity field. For more details, see Ref. 12. Therefore, for the Kolmogorov spectrum $n = 5/3$, we obtain

$$y(t) \sim t^3 \qquad (85)$$

namely, the Richardson t^3-law.[4]

In general, the value of n is larger than $n = 5/3$ owing to the intermittency effect,[5] the exponent $2/(1-\phi)$ in Eq. (84) is larger than 3.

Now we compare the above result with other theories.[6,13] According to Takayoshi–Mori theory, we have $y(t) \sim t^{4/(3-n-\mu)}$. Here, μ denotes the intermittency exponent and it is approximately equal to 1/3. This μ is related[5] to n through

$$n = \tfrac{1}{3}(5 + \mu) \tag{86}$$

6. SCALING THEORY OF CHAOS

As the essence of chaos is the instability of orbits, the scaling theory can be also applied to the relative diffusion

$$y(n, \delta) = \langle (x_1(n) - x_2(n))^2 \rangle \tag{87}$$

for the solution $x(n)$ of the following general difference equation:

$$x(n+1) = T(x(n)) \tag{88}$$

We define the scaling variable τ by

$$\tau = \delta e^{\lambda n} \tag{89}$$

where λ denotes the Lyapunov exponent[1] and δ is the initial difference defined by

$$\delta^2 = \langle [x_1(0) - x_2(0)]^2 \rangle \tag{90}$$

Under some appropriate conditions, $y(n, \delta)$ takes the scaling form

$$y(n, \delta) = f^{(sc)}(\tau) \tag{91}$$

and the scaling function itself becomes chaotic for large τ with respect to n.

For example, the solution $x(n)$ of the following quadratic map

$$x(n+1) = 4x(n)[1 - x(n)] \tag{92}$$

is given by[1,14]

$$x(n) = \sin^2[2^n \sin^{-1}\sqrt{x(0)}] \tag{93}$$

Then, we obtain

$$f^{(sc)}(\tau) = 1 - \cos \tau \tag{94}$$

where

$$\tau = \delta \times 2^n; \quad \delta^2 = \langle |\sin\sqrt{x_1(0)} - \sin\sqrt{x_2(0)}|^2 \rangle \tag{95}$$

More detailed arguments will be discussed elsewhere.[12]

7. SUMMARY

In this chapter we have reviewed the essential idea of the scaling theory of unstable transient phonomema as well as the unified method in the whole region of time. We have applied the general strategy of the above scaling theory to clumps in turbulent plasma and relative diffusions in turbulent magnetic fields and in turbulent velocity fields and chaos. We have found that the concept of scaling in unstable transient phenomena is very powerful in understanding the above various physical phenomena.

ACKNOWLEDGMENTS

The author would like to thank Professor R. Kubo, Professor R. Balescu, Professor H. Mori, and Professor J, Krommes for useful discussions. The present work is partially financed by the Scientific Research Fund of the Ministry of Education.

REFERENCES

1. P. Collet and J. P. Eckmann, *Iterated Maps on the Interval as Dynamical Systems* Birkhäuser, Boston (1980).
2. T. H. Dupree, "Theory of Phase Space Density Granulation in Plasma," *Physics Fluid* **15**, 334–344 (1972).
3. J. H. Misguich and R. Balescu, "On Relative Spatial Diffusion in Plasma and Fluid Turbulences: Clumps, Richardson's Law and Intrinsic Stochasticity," *Plasma Physics* **24**, 289–318 (1982).
4. A. S. Monin and A. M. Yaglom, *Statistical Fluid Mechanics* (MIT Press, Cambridge, 1971), Vol. 1.
5. M. Nelkin, "Turbulence, Critical Fluctuations, and Intermittency," *Phys. Rev. A* **9**, 388–395 (1974).
6. P. H. Roberts, "Analytical Theory of Turbulent Diffusion," *J. Fluid Mech.* **11**, 257–283 (1961).
7. M. Suzuki, "Scaling Theory of Non-Equilibrium Systems near the Instability Point. I–III," *Prog. Theor. Phys.* **56**, 77–94, 477–493 (1976); **57**, 380–392 (1976).
8. M. Suzuki, "Scaling Theory of Transient Phenomena near the Instabiltiy Point," *J. Stat. Phys.* **16**, 11–32 (1977).
9. M. Suzuki, "Passage from an Initial Unstable State to a Final Stable State," *Advance Chem. Phys.* **46**, 195–278 (1981).
10. M. Suzuki, "New Unified Formulation of Transient Phenomena Near the Instability Point on the Basis of the Fokker–Planck Equation," *Physica* **117A**, 103–108 (1983).
11. M. Suzuki, "Scaling Property of the Relative Diffusion of Charged Particles in Turbulent Electric Fields I," *Prog. Theor. Phys.* **71**, No. 2 (1984).
12. M. Suzuki, "Scaling Theory of Relative Diffusion in Chaos and Turbulence," KSI Proceedings, Y. Kuramoto, ed. (Springer).
13. F. Takayoshi and H. Mori, "Diffusion of Particles in Fully-Developed Turbulence," *Prog. Theor. Phys.* **68**, 439–447 (1982).
14. S. M. Ulam and J. V. Neumann, "On Combinations of Stochastic and Deterministic Processes," *Bull. Amer. Math. Soc.* **53**, 1120 (1947).

CLUMP KINETICS IN TURBULENT PLASMAS

M. Kono
Research Institute for Applied Mechanics, Kyushu University, Fukuoka, Japan

1. INTRODUCTION

One of the difficulties in constructing plasma turbulence theory is the appearance of a time scale proportional to the fractional power of wave amplitudes, that is, a trapping time. This implies the absence of analyticity, which requires, instead of higher-order perturbation theory, some new concepts and approaches that are more closely related to strong turbulence. Among a number of theoretical constructions, the notion of localized structures in phase space called clumps has long been discussed in expectation of going further beyond the weak-turbulence theory. The clumps as macroparticles or granulations in phase space were introduced by Dupree[1] to explain anomalous resistivity and at the same time by Kadomtsev and Pogutse[3] in order to study certain anomalies in velocity relaxation.

Dupree[2] showed that stochastic fields introduce strong correlations in the trajectories of initially nearby particles in such a way that the relative diffusion is seriously reduced owing to the cancellation of the cross diffusion with the self-diffusion. In other words, initially nearby particles have approximately the same fields, which in turn create the trajectory correlation. They behave like a single large discrete particle. It is then reasonable to expect that many plasma processes depending on particle discreteness such as radiation emission and absorption, resistivity, relaxation, diffusion, and so on can be greatly enhanced by the tendency of the plasma to form clumps. In fact, the enhancement factor of the transport coefficients will be shown to be $(N/n)z^2$, where z is the number of particles in a clump and N and n are the clump and particle number density, respectively. Since the clump size is of the order of the correlation length of the turbulent waves, which is supposed to be larger than the Debye length, clumps are naturally expected to be larger than the Debye spheres, indicating z is very large. The enhancement factor, therefore, could be quite large even though N/n is small.

In Section 2 we study the correlation of neighboring particles in phase space, showing that clump lifetime is determined by the trajectory instability of the relative motion. In Section 3 decomposing the resonant region of phase space into cells so that each of them represents a clump, and eliminating the intracell degrees of freedom, we derive equations describing cell dynamics. The construction of clump kinetics is now straightforward. In Section 4 we derive a kinetic equation for the average clump distribution function, which has the Balescu–Lenard collision term with an effective charge of macroparticles in addition to the conventional quasilinear diffusion term. We give a brief estimation of the enhancement factor of the transport coefficients.

2. INTRACLUMP CORRELATION AND CLUMP LIFETIME

Since the clump is formed due to the balance between the conversion of average velocity gradients into phase-space granulations by turbulent mixing and the destruction of local structures by trajectory instability, we first study the evolution of a short-range two-point correlation function, which corresponds to an intraclump correlation.

Starting with the Vlasov–Poisson equations

$$\left(\frac{\partial}{\partial t} + v \cdot \nabla + \frac{e_a}{m_a} E \cdot \frac{\partial}{\partial v}\right) F_a = 0 \tag{1}$$

$$\nabla \cdot E = \sum 4\pi e_a n_a \int dv F_a \tag{2}$$

we derive the equation for the two-point distribution function

$$\left(\frac{\partial}{\partial t} + v_1 \cdot \nabla_1 + v_2 \cdot \nabla_2 + \frac{e_a}{m_a} E(1) \cdot \frac{\partial}{\partial v_1} + \frac{e_b}{m_b} E(2) \cdot \frac{\partial}{\partial v_2}\right) f_{ab}(1,2)$$

$$= -\frac{e_a}{m_a} E(1) \cdot \frac{\partial F_a^{(0)}}{\partial v_1} f_b(2) - \frac{e_b}{m_b} E(2) \cdot \frac{\partial F_b^{(0)}}{\partial v_2} f_a(1) \tag{3}$$

where

$$f_{ab}(1,2) = f_a(1)f_b(2), \qquad f_a = F_a - F_a^{(0)}$$

and $F^{(0)}$ is an equilibrium distribution function. In the following we restrict our attention to the correlation of like particles. Introducing barycentric and relative coordinates

$$\begin{aligned} R = (x_1 + x_2)/2, & \quad U = (v_1 + v_2)/2 \\ r = x_1 - x_2, & \quad u = v_1 - v_2 \end{aligned} \tag{4}$$

Eq. (3) is converted to the following form:

$$\left(\frac{\partial}{\partial t} + U \cdot \frac{\partial}{\partial R} + u \cdot \frac{\partial}{\partial r} + \frac{e_a}{m_a} E^{(+)} \cdot \frac{\partial}{\partial U} + \frac{e_a}{m_a} E^{(-)} \cdot \frac{\partial}{\partial u}\right) f_{aa}(R, U; r, u)$$

$$= 2\left(\frac{e_a}{m_a}\right)^2 \sum_{KK'} \int dt' g_{K'}(t; t') E(K, t) E(K', t') e^{i(K+K')R}$$

$$\times \cos\left[\frac{(K-K') \cdot r}{2} - \frac{K' \cdot u(t-t')}{2}\right] \frac{\partial F_a^{(0)}}{\partial U} \frac{\partial F_a^{(0)}}{\partial U} \quad (5)$$

where

$$E^{(\pm)} = E\left(R + \frac{r}{2}, t\right) \pm E\left(R - \frac{r}{2}, t\right),$$

$$g_K(t; t') = \exp\left[iKU(t-t') - \frac{K^2 D(t-t')^3}{3}\right] \quad (6)$$

and

$$D = \left(\frac{e}{m}\right)^2 \sum_K |E(K)|^2 \mathrm{Re} \int dt' g_K(t; t') e^{i\omega(t-t')}$$

Since we are concerned with the initially nearby particles, a dipole approximation (that is, $|R| \gg |r|$ and $|U| \gg |u|$) is a good approximation, which reduces Eq. (5) to the simple form:

$$\left(\frac{\partial}{\partial t} + U \cdot \frac{\partial}{\partial R} + \frac{e_a}{m_a} E(R, t) \cdot \frac{\partial}{\partial U} + u \cdot \frac{\partial}{\partial r} + \frac{e_a}{m_a} r \cdot \frac{\partial}{\partial R} E(R, t) \cdot \frac{\partial}{\partial u}\right)$$

$$\times f_{aa}(R, U; r, u) = 2\left(\frac{e_a}{m_a}\right)^2 \sum_{KK'} \int dt' g_{K'}(t; t') E(k, t) E(k', t') e^{i(K+K')R}$$

$$\times \cos\left[\frac{(K-K') \cdot r}{2} - \frac{K' \cdot u(t-t')}{2}\right] \left(\frac{\partial F_a^{(0)}}{\partial U}\right)^2 \quad (7)$$

The structure of Eq. (7) suggests that the barycentric motion is independent of the relative motion and is described by the Vlasov equation within the validity of the dipole approximation. It follows, therefore, that the characteristic time for the barycentric motion to become diffusive is

$$t_G = \left(\frac{K^2 D}{3}\right)^{-1/3} \quad (8)$$

Thus we may put

$$f_{aa}(R, U; r, u) = F_a(R, U) C_a(r, u; R, U) \quad (9)$$

obtaining the equation for the intraclump correlation C_a as

$$\left(\frac{\partial}{\partial t} + u \cdot \frac{\partial}{\partial r} + \frac{e_a}{m_a} r \cdot \frac{\partial}{\partial R} E(R, t) \frac{\partial}{\partial u}\right) C_a(r, u; R, U)$$

$$= 2\left(\frac{e_a}{m_a}\right)^2 \sum_{KK'} \int dt' g_{K'}(t, t') E(K, t) E(K', t') e^{i(K+K')R}$$

$$\times \cos\left[\frac{(K-K') \cdot r}{2} - \frac{K' \cdot u(t-t')}{2}\right] \frac{(\partial F_a^{(0)}/\partial U)^2}{F_a^{(0)}} \quad (10)$$

which is supplemented by Eqs. (1) and (2). The right-hand side of Eq. (10) is a source term of producing clumps in such a way that the average velocity gradients in the resonant region cause the mixing of the phase-space density constrained by the constancy of the distribution function along a particle orbit, being converted into the phase-space granulations. It should be noted that the source term is not zero only for a group of the resonant particles because of the resonant function (6) and is zero for particles of large relative distance in phase space.

The clump lifetime is estimated through the trajectory instability described by the characteristics of Eq. (10):

$$\frac{dr}{dt} = u, \quad \frac{du}{dt} = \frac{e_a}{m_a} r \cdot \left(\frac{\partial E}{\partial R}\right) \tag{11}$$

Here, for simplicity, the barycentric motion is assumed to be a free streaming $R = R_0 + U_0 t$, where U_0 is in the resonant region $(\omega/K)_{min} < |U_0| < (\omega/K)_{max}$. Laplace transforming Eq. (11) and collecting the most secular terms in an iterated series expansion, we obtain

$$r(\omega) = -\omega^{-2} Z(\omega)^{-1} \Bigg(u_0 + i\omega r_0 - \frac{e_a}{m_a} \sum_K \int d\omega' K E(K, \omega') e^{iKR_0}$$

$$\times Z^{-1}(\omega - \omega' + K \cdot U)(u_0 + i(\omega - \omega' + K \cdot U_0) r_0) + \cdots \Bigg) \tag{12}$$

where r_0 and u_0 are initial values and

$$Z(\omega) = 1 - \left(\frac{e_a}{m_a}\right)^2 \sum_K \int d\omega' \frac{KK' |E(K, \omega')|^2}{\omega^2 (\omega - \omega' + K \cdot U_o)^2} \tag{13}$$

The time evolution of the trajectory is determined by zeros of $Z(\omega)$. Since the instability is due to wanderings from one resonance island to another, we may use Mathiue approximation in which neighboring waves to the main resonance $[\mu(K_0) = \omega(K_0) - K_0 \cdot U_0 = 0]$ are retained. The growth rate is then expressed as

$$\gamma_{cl} = t_{cl}^{-1} = 6^{-1/2} \left|\left(\frac{d\mu}{dK}\right)_0 \delta k\right|^{-1} (\varepsilon/2)^{1/6} s^{-2} \left[\left(\frac{s}{s_c}\right)^4 - 1\right]^{1/2} \quad \text{at } s \cong s_c$$

$$= \left|\left(\frac{d\mu}{dK}\right)_0 \delta k\right|^{-1} \varepsilon^{1/2} \quad \text{for } s > s_c \tag{14}$$

where δk is a spacing in wavenumber, ε is the ratio of the neighboring wave amplitudes, $|E(K_0 + \delta k)/E(K_0)|^2$, s is a stability parameter defined by the ratio of the width of the resonance island to the phase-velocity spacing, $(2e_a K_0 |E(K_0)|/m_a)^{1/2}/|(d\mu/dK)_0 \delta k|$, and $s_c = [(1 + 6(\varepsilon/2)^{1/3}]^{-1/4}$. The growth rate of this instability gives a measure of the time when initially nearby particles begin to separate appreciably after they move, sticking together under the action of approximately the same forces. Although Eq.

(14) is obtained for the free-streaming approximation of the barycentric orbit, numerical analyses show that the result is not seriously affected by the inclusion of the full orbit of the barycentric motion. From Eqs. (8) and (14) we have

$$\frac{t_{cl}}{t_G} \cong \frac{s}{\varepsilon^{1/2}} \gg 1 \quad \text{for } s \gg s_c \tag{15}$$

implying that we may neglect dynamical effects of the clump disintegration on the clump kinetics.

3. CELL DECOMPOSITION AND CLUMP KINETICS

It turned out that the resonant particles close each other in phase space behave in a group. In order to describe this discrete nature of the clumps, we decompose the resonant region of phase space into cells so that each of them represents a clump. Obviously, the cell decomposition itself is not quite equivalent to a discrete macroparticle description. For the macroscopic properties of plasmas, however, intracell degrees of freedom can be renormalized to physical quantities, giving an effective charge and mass. The elimination of the intracell motion is naturally followed by an appearance of the discrete macroparticles.

We introduce cell variables as

$$\begin{pmatrix} R_j(t) \\ U_j(t) \end{pmatrix} = \frac{n_a}{z_j} \int_{D_j} dx\, dv \begin{pmatrix} x \\ v \end{pmatrix} T(j; x, v) F_a(x, v, t) \tag{16}$$

where $T(j; x, v)$ is a weight function in the conversion of the fluid–particle problem into the cell–particle problem, and F_a is a distribution function that obeys the Vlasov equation. The integration is performed inside the jth clump, indicated by D_j. The number of particles in the jth clump is represented by z_j, defined through

$$z_j = n_a \int_{D_j} dx\, dv\, T(j; x, v) F_a(x, v, t). \tag{17}$$

The z_j is also given by the intracell correlation

$$z_j = n_a \int_{D_j} dr\, du\, C_a(r, u, t_{cl}; j) \tag{18}$$

which defines the operator T. Thus, the intracell correlation plays an essential role both in the grouping of neighboring fluid particles into cells, whereby the length scale of the transformation is defined and the choice of the weight function. From Eqs. (1), (2), and (16), we obtain equations for cell dynamics:

$$\frac{dR_j}{dt} = U_j \tag{19}$$

$$\frac{dU_j}{dt} = -\sum_b \frac{e_a e_b}{m_a} \sum_i \frac{\partial}{\partial U_j} \frac{z_i^{(b)}}{|R_j - R_i|} - \frac{nae_a^2}{m_a} \int_{D_j} drdu \left(\frac{\partial}{\partial r} \frac{1}{|r|} C_a(r, u; j)\right) \tag{20}$$

The second term on the right-hand side of Eq. (20) is identically zero when clumps are point particles. Here it does not vanish because of the temporal deformation of the clumps, though it is very small. In the following we neglect this self-field. Now we introduce the number density of clumps in (R, U)-space,

$$H(R, U, t) = \sum_j \delta[R - R_j(t)]\delta[U - U_j(t)] \tag{21}$$

The equation for $H(R, U, t)$ is readily obtained by Eqs. (20) and (21),

$$\left(\frac{\partial}{\partial t} + U \cdot \frac{\partial}{\partial R}\right)H_a - \sum_b z_b \frac{e_a e_b}{m_a} \frac{\partial}{\partial U} \int dR' dU' \left(\frac{\partial}{\partial R} \frac{1}{|R - R'|}\right) H_b(R', U') H_a = 0 \tag{22}$$

which is nothing but the Klimontovich equation for clumps. The average number of clumps is roughly defined as the ratio of the average exclusive volume of the clump to the volume of the resonant region in phase space:

$$N = V_{\text{res}}/(\langle r^2 \rangle \langle u^2 \rangle)^{3/2} \tag{23}$$

where

$$\binom{\langle r^2 \rangle}{\langle u^2 \rangle} = \frac{\int drdu \binom{r^2}{u^2} C(r, u; t_{cl})}{\int drdu\, C(r, u; t_{cl})} \tag{24}$$

and V_{res} is the phase-space spread

$$V_{\text{res}} \cong \left(\frac{\omega}{K}\right)_{\min}^2 \left[\left(\frac{\omega}{K}\right)_{\max} - \left(\frac{\omega}{K}\right)_{\min}\right] \tag{25}$$

The standard procedure leads to the clump kinetic equation, in which in order to see the lowest effect of the clump formation we neglect the mode–mode coupling and the cumulative effect of the wave–particle scattering on the particle propagator:

$$\frac{\partial h_a}{\partial t} = \frac{\partial}{\partial U} \sum \pi \left(\frac{e_a}{m_a}\right)^2 KKS(K)\delta[\omega(K) - K \cdot U]\frac{\partial}{\partial U} h_a$$

$$+ \pi \frac{e_a^2}{m_a} \sum_b N_b z_b e_b^2 \frac{\partial}{\partial U} \sum_K \int d\omega \frac{KK}{K^2 |\varepsilon(K, \omega)|^2}$$

$$\times \int dU' \left(\frac{z_b}{m_a}\frac{\partial}{\partial U} - \frac{z_a}{m_b}\frac{\partial}{\partial U'}\right) \delta(\omega - K \cdot U') h_a(U) h_b(U') \tag{26}$$

where h_a is the average distribution function and $S(K)$ is a structure factor. Thus the kinetic equation is characterized by the Balescu–Lenard collision term proportional to discrete parameters in such a way as $N(ze)^2(em)^2$, which does not vanish in the Vlasov limit because N and ze are preserved constants.

Now the enhancement factor of the collision term due to the clump formation is given as

$$\Gamma = \frac{N}{n} z^2 \tag{27}$$

The intraclump correlation and the number of particles in a clump are estimated as

$$C \cong (\omega_{pe} t_G)^2 \frac{t_{cl}}{t_G} \frac{|E|^2}{4\pi n T} \left(\frac{\omega}{K v_{Te}}\right)^2 \frac{1}{v_{Te}^3} e^{-(\omega/K v_{Te})^2/2}$$

and

$$z \cong \frac{n_{res}}{N} \frac{t_{cl}}{t_G} \left(\frac{|E|^2}{4\pi n T}\right)^{1/2} (K\lambda_D)^{-3},$$

where n_{res} is the number of resonant particles defined by

$$n_{res} = n \frac{V_{res}}{v_{Te}^3} \exp[-0.5(\omega/K v_{Te})^2]$$

Then Γ is roughly estimated as

$$\Gamma \cong n\lambda_D^3 \left(\frac{n_{res}}{n}\right)^2 \left(\frac{t_{cl}}{t_G}\right)^2 \left(\frac{|E|^2}{4\pi n T}\right)^{7/4} (K\lambda_D)^{-15/2} \tag{28}$$

which is possibly quite large.

4. DISCUSSION

In this article we discussed that the intracorrelation is a key to describe the formation and destruction of clumps. In order to represent the discrete nature of clumps, we decompose the phase space into cells and eliminate the intracell degrees of freedom. The resultant kinetic equation is characterized by not only the conventional quasilinear diffusion term, but also the Balescu–Lenard collision term with a possibly large enhancement factor.

One of the points necessary for refinement of the present approach is related to the estimation of the trajectory instability on which the neglection of the clump disintegration is based. In spite of the dramatic progress in studies of transition to chaos in dynamical systems, it is still complicated to analyze the trajectory evolution even in prescribed random fields characterized by a broad spectrum.

Another point is to reformulate the present approach by taking into account self-consistency, which is supposed to cause a kind of self-organization of turbulence.

Although there are several experiments revealing certain anomalies in transport phenomena, an experimental identification of clumps has not been reported. Experimental verifications of the foundations and conclusions of the clump theory are eagerly desired.

REFERENCES

1. T. H. Dupree, "Theory of Resistivity in Collisionless Plasma," *Phys. Rev. Letters* **25**, 789–792 (1970).
2. T. H. Dupree, "Theory of Phase Space Density Granulation in Plasma," *Phys. Fluids* **15**, 334–344 (1972).
3. B. B. Kadomtsev and O. P. Pogutse, "Collisionless Relaxation in Systems with Coulomb Interaction," *Phys. Rev. Letters* **25**, 1155–1157 (1970).

IMPACT OF CLUMPS ON PLASMA STABILITY AND THE NATURE OF TURBULENCE IN A SATURATED STATE

P. W. Terry and P. H. Diamond

Institute for Fusion Studies, The University of Texas at Austin, Austin, Texas

Abstract. *Phase-space density granulations (clumps) are studied using the theory of two-point phase-space density correlation. A novel mechanism of extraction of expansion-free energy is described. This mechanism affects questions pertaining to nonlinear stability. Theories for two-point correlation for the universal mode in a slab geometry with shear and trapped electrons in toroidal geometry are discussed. Results are presented that show destabilization of the universal mode and enhancement of the trapped electron growth rate. An analytic formula for the width of the frequency spectrum is obtained. By specifying the collective resonance damping mechanism, the wavenumber spectrum is also calculated. A formula for energy flux illustrates the impact of clumps on transport and energy confinement.*

1. INTRODUCTION

In recent years, it has become increasingly apparent that a description of plasma behavior solely in terms of collective normal modes is incomplete, particularly in the turbulent state or near an instability in the presence of fluctuations.[1-4] The inadequacy of the normal-mode description of plasma behavior, basically a legacy of linear theory, is underscored by experimental measurements of the fluctuation spectrum of low-frequency turbulence in tokamak plasma.[5] These measurements of broad frequency spectra at fixed wavenumbers are inconsistent with the normal-mode picture in which at saturation the dielectric is zero, the modes are marginally stable (nonlinearly), and, hence, the collective-mode resonance width is very narrow. Absent from this picture are fluctuations that are not phase coherent with the potential and, hence, are not modelike in nature. These incoherent fluctuations are produced by the mode coupling associated with the nonlinearity. Represented mathematically in Fourier-transform space as a convolution with the potential, the nonlinearity thus provides a component of the distribution at wavenumber k, which is proportional to the potential at some other wavenumber k'. When fluctuations resonate with particles, the incoherent fluctuations become particle-like in nature: they are a granulation in phase space resulting from a mixing process in which the conservation of phase-space density along the particle orbit inhibits the intermixing of different densities. This can be understood by considering the correlation of neighboring particles in the presence of the turbulent mixing. Neighboring trajectories diverge due to the mixing process. However, particles closely separated in phase space remain correlated for a longer time since the turbulent potential each particle sees is nearly the same. The particles in a sufficiently small volume in phase space stay correlated with each other for a time exceeding the typical correlation time of the turbulence and are thus scattered turbulently as a macroparticle. The mixing process continues, however, and the granulation or clump eventually decays in time. The decay process is offset by the continuous creation of microscale structure caused by the turbulent rearrangement of the average distribution, which has a gradient. The granulations are thus said to have a source. Systems having such a scale-dependent mixing process and a gradient-driven source are common in plasma physics. They include systems in which, for example, ion sound waves and drift instabilities might occur, as well as turbulent plasmas existing in conditions below the threshold of these instabilities, but which, nevertheless, have a free-energy source. Although historically the description of clumps and associated phenomena has been couched in a kinetic formalism, these considerations and processes are also germane to fluid descriptions, including both one-fluid (MHD)[6] and two-fluid models.[7]

The decay of the microscale correlation, formulated in terms of a relative diffusion process, has been extensively studied; a number of investigations have dealt exclusively with relative diffusion.[8] Here, we seek to emphasize the crucial role played by the source. The source is proportional to the rate of relaxation of the average distribution and, hence, to the rate of extraction of expansion-free energy. Through the source, the incoherent fluctuations have access to the expansion-free energy. The incoherent fluctuations act as noise, exciting the collective modes. In the saturated state, the modes are necessarily damped in order to balance the noise excitation. This damping, then, constitutes the width of the collective resonance centered at the frequency $\omega(k)$ of the collective resonance (mode). The standard view of turbulence in the saturated state as consisting of a spectrum of waves is thus replaced by a description in terms of collective resonances broadened by incoherent noise emission.

The relationship of the source of microscale correlation to the extraction of expansion-free energy provides a new accessibility mechanism and resultant clump-induced instability. The width of the frequency spectrum, as the collective mode's damping response to the instability, is thus strongly related and may be considered a signature of the clump-induced instability. Considerations of free-energy extraction and relaxation of the average distribution also point to a strong effect on transport arising from clumps. Thus, a consideration of the source introduces the issues of plasma stability and transport into the study of clump-related phenomena as well as alters the classical view of steady-state turbulence.

2. EVOLUTION OF THE TWO-POINT CORRELATION

We have already described the action of the turbulent mixing on neighboring phase-space trajectories and the scale-dependent correlation function that results. This correlation is a two-point phase-space density correlation. We seek an equation that describes the evolution of two-point phase-space density correlation under the influence of the turbulent mixing and the source. An equation that correctly describes this evolution cannot be obtained from a standard one-point renormalization theory. A two-point equation constructed from a renormalized one-point equation incorrectly predicts that two points will diffuse independently even when their separation is very small. Following Dupree,[1,2] we start with the evolution equation for two-point phase-space density correlation obtained from the Vlasov hierarchy and renormalize the triplet nonlinearity. It is then readily ascertainable that the relative diffusion indeed reflects correlation at small separation.

We have already alluded to measurements of the fluctuation spectrum of low-frequency turbulence in tokamaks[5] as motivation for considering incoherent fluctuations. We formulate the theory of two-point correlation for drift-wave fluctuations, widely considered responsible for anomalous transport processes in tokamaks. In contrast to Dupree, we treat clumps in the presence of collective modes — the dielectric is zero and the collective resonance shields the clumps, as governed by Poisson's equation. Drift-wave fluctuations occur at frequencies approximately given by the electron diamagnetic drift frequency, ω_{*e}. Incoherent fluctuations, which propagate ballistically, that is, are resonant, are therefore generated in the electron species. Collisionless electron dynamics are described by the gyrokinetic equation,

$$\left(\frac{\partial}{\partial t}+\mathbf{V}_d\cdot\nabla_\perp+v_\|\hat{n}\cdot\nabla\right)g-\frac{c}{B_0}\nabla\Phi\times\hat{n}\cdot\nabla_\perp g$$
$$=-\frac{|e|}{T_e}\langle f\rangle\frac{\partial\Phi}{\partial t}-\frac{c}{B_0}\nabla_\perp\Phi\times\hat{n}\cdot\nabla_\perp\langle f\rangle \quad (1)$$

where g is the nonadiabatic part of the fluctuating electron distribution, Φ is the potential, $\langle f\rangle$ is the average distribution, and $\hat{n}$ is the unit vector in the direction of the magnetic field. The velocity $\mathbf{V}_d$ represents the ∇B and curvature drifts. The last term on the left-hand side is the $\mathbf{E}\times\mathbf{B}$ drift, which is the dominant nonlinearity in the problem. Simplifications of Eq. (1) are appropriate for the problems we shall consider herein. When g is taken to describe a distribution of trapped electrons, a bounce average is performed. When the bounce frequency represents the fastest timescale, the parallel gradient term averages to zero and only the ∇B and curvature-drift resonance remains. When g describes the electron distribution for the universal mode, the ∇B and curvature drifts are negligible and the parallel resonance remains. For concreteness, our description of the theory will focus on the later application. Details for the former case are presented in Ref. 4.

From Eq. (1) we construct the Vlasov hierarchy equation for two-point phase-space correlation:

$$\left(\frac{\partial}{\partial t}+\frac{v_{\|_1}}{Rq(r_1)}\frac{\partial}{\partial\eta_1}+\frac{v_{\|_2}}{Rq(r_2)}\frac{\partial}{\partial\eta_2}\right)\langle\hat{g}(\eta_1,\phi_1)\hat{g}(\eta_2,\phi_2)\rangle$$
$$+\sum_n\sum_{n'}\sum_{n''}\left\langle\exp(in\phi_1)\exp(in'\phi_1)\exp(in''\phi_2)\frac{c}{B_0}\sum_{m'}k_\theta k'_\theta\hat{s}(2\pi m')\right.$$
$$\left.\times\exp(-2\pi in'qm')\hat{\Phi}_{n'}(\eta_1+2\pi m')\hat{g}_n(\eta_1)\hat{g}_{n''}(\eta_2)\right\rangle$$
$$=\sum_{\substack{n'\\ \omega'}}\frac{i|e|}{T_e}\exp[in'(\phi_1-\phi_2)]\langle f(1)\rangle(\omega'-\omega'_{*e})\langle\hat{\Phi}_{-n'}(\eta_1)\hat{g}_{n'}(\eta_2)\rangle \quad (2)$$

where $k_\theta=rq/r$. We have employed the ballooning representation,[9] given by the transformation

$$\begin{Bmatrix}\Phi\\ g\end{Bmatrix}=\sum_n\exp(in\phi)\sum_m\exp(-im\theta)\int d\eta\,\exp[i(m-nq)\eta]\begin{Bmatrix}\hat{\Phi}_n(\eta)\\ \hat{g}_n(\eta)\end{Bmatrix} \quad (3)$$

where ϕ is the toroidal angle and η is the variation in the direction of the magnetic field. This eikonal representation extracts the radial variation in $(m-nq)$ as rapid variation; the remaining explicit radial dependence is slow. Thus, for lowest-order theory, the radial variation is contained in η: $k_r=-k_\theta\hat{s}\eta$, and r enters as a parameter. A compact form for the $\mathbf{E}\times\mathbf{B}$ nonlinearity is so obtained,[10] in which the interaction is between n and n' at r along η.

It is necessary to close the two-point equation, since its nonlinearity contains a three-point correlation. This is accomplished by renormalizing the nonlinearity using the direct interaction approximation. We anticipate the separation of variables into average motion $\phi_+ = \phi_1 + \phi_2$ and relative motion $\phi_- = \phi_1 - \phi_2$. The averaging is performed by integrating over ϕ_+ to yield $\delta(n + n' + n'')$. The nonlinear term T_{12} thus becomes

$$T_{12} = \sum_n \sum_{n'} \left[\exp(in\phi_-) \sum_{m'} k_\theta k'_\theta \hat{s}(2\pi m') \exp(2\pi i n'qm') \right.$$

$$\times \langle \hat{\Phi}_{-n'}(\eta_1 + 2\pi m') \hat{g}_{n+n'}(1) \hat{g}_{-n}(2) \rangle - \exp(in\phi_- + in'\phi_-) \sum_m k_\theta k'_\theta$$

$$\left. \times \hat{s}(2\pi m') \exp(2\pi in'qm') \langle \hat{\Phi}_{-n'}(\eta_1 + 2\pi m') \hat{g}_{-n'}(\eta_1) \hat{g}_{n+n'}(\eta_2) \rangle \right] + (1 \to 2) \tag{4}$$

where shielding effects of a third term, $\langle \hat{\Phi}_{n+n'} \hat{g}_{-n'} \hat{g}_n \rangle$ are neglected. Wave-particle resonances dominate, and the clumps themselves are resonant so that

$$\hat{g}_{\substack{n+n' \\ \omega+\omega'}} \cong \hat{g}^{(2)}_{\substack{n+n' \\ \omega+\omega'}}$$

where

$$\hat{g}^{(2)}_{\substack{n+n' \\ \omega+\omega'}} = L_{\substack{n+n' \\ \omega+\omega'}} \frac{c}{B_0} \sum_{n''} \sum_{m'} (-k''_\theta)(k_\theta + k'_\theta + k''_\theta) \hat{s}$$

$$\times \exp(2\pi in''qm')[\hat{\Phi}_{-n''}(\eta_1 + 2\pi m') \hat{g}_{n+n'+n''}] \tag{5}$$

and

$$L_{\substack{n+n' \\ \omega+\omega'}} = [-i(\omega + \omega') + i(k_\| v_\| + k'_\| v_\|)]$$

Proximity to resonance permits the Markovian approximation, $L_{n+n'} \to L_{n'}$. Selecting from the sum over n'', the directly interacting triplet ($n'' = -n'$), we arrive at the renormalized kinetic equation for the two-point phase-space density correlation,

$$\left[\frac{\partial}{\partial t} + \left(\frac{v_{\|-}}{Rq(r)} + \frac{v_{\|+}r_-}{Rqr(r)} \right) \frac{\partial}{\partial \eta_-} - D_- \frac{\partial^2}{\partial y_-^2} \right] \langle \hat{g}(1) \hat{g}(2) \rangle$$

$$= \sum_{\substack{k' \\ \omega'}} \frac{i|e|}{T_e} \exp(ik'y_-)(\omega' - \omega^*_{*e}) \langle f(1) \rangle \langle \hat{g}(2) \hat{\Phi}(1) \rangle_{\substack{n' \\ \omega'}} \tag{6}$$

where $y = r\phi/q$, $D = 2D - D^{(2,1)} - D^{(1,2)}$,

$$D = \frac{c^2}{B_0^2} \sum_{\substack{k' \\ \omega'}} k'^2_\theta \hat{s}^2 (\mathrm{Re}\, L_{\substack{k' \\ \omega'}}) \sum_m (2\pi m)^2 \langle \hat{\Phi}(\eta + 2\pi m)^2 \rangle_{\substack{k' \\ \omega'}} \tag{7}$$

and

$$D^{(1,2)} = \frac{c^2}{B_0^2} \sum_{\substack{k' \\ \omega'}} \exp(ik'y_-) k_\theta'^2 \hat{s}^2 (\mathrm{Re}\, L_{k'}) \sum_m (2\pi m)^2 \exp(2\pi i m r_- k' \hat{s})$$
$$\times \langle \hat{\Phi}(\eta_1 + 2\pi m) \hat{\Phi}(\eta_2 + 2\pi m) \rangle \tag{8}$$

The diffusion coefficient D represents independent diffusion which is derived from a one-point renormalization, and hence characterizes coherent mode coupling. The coefficients $D^{(1,2)}$ and $D^{(2,1)}$ represent the correlated diffusion caused by incoherent mode coupling. Correlated diffusion vanishes for large relative separation and approaches $2D$ as the separation goes to zero. The relative diffusion D_- thus tends to zero as the separation of neighboring orbits approaches zero. The scales over which $D^{1,2} \neq 0$ as determined from Eq. (8) define the scale of the clump. For phase-space separations small compared to the clump scale, the relative diffusion is quadratic in the relative phase-space variables,

$$D_- = k_0^2 \left(y_-^2 + \hat{s}^2 r_-^2 + \frac{\eta_-^2}{k_0^2 \Delta \eta^2} \right) \tag{9}$$

where k_0^{-1}, $k_0^{-1} \hat{s}^{-1}$, and $\Delta \eta$ are the clump scales for the toroidal, radial, and parallel variations, respectively.

The evolution of the separation of neighboring trajectories permits a determination of the lifetime of miscroscale correlations. The time required for the separation to reach clump scales from an initially smaller separation defines the clump lifetime, τ_{cl};

$$\tau_{cl} = \frac{\tau_c}{2} \ln \left\{ k_0^{-2} \left[y_-^2 + \hat{s}^2 r_-^2 + \frac{\eta_-^2}{k_0^2 \Delta \eta^2} + \frac{\tau_c}{k_0^2 (\Delta \eta)^2} \eta_- \left(v_\| + \frac{v_{\|+} r_-}{r} \right) \frac{1}{Rq(r)} \right. \right.$$
$$\left. \left. + \frac{\tau_c^2}{2 k_0^2 \Delta \eta^2} \frac{1}{[Rq(r)]^2} \left(v_\| + \frac{v_{\|+} r_-}{r} \right)^2 \right]^{-1} \right\} \tag{10}$$

where τ_c is the correlation time of the coherent fluctuations for scales comparable to the clump scale ($\tau_c = 1/k_0^2 D$). The clump lifetime exhibits the logarithmic peaking in the phase-space variables which is characteristic of the exponential separation of trajectories occurring over most of the clump scale. An earlier assertion that particles in a sufficiently small volume stay correlated for a time (τ_{cl}) exceeding the correlation time (τ_c) of the turbulence is now obvious from Eq. (5).

3. THE SOURCE

The clump lifetime τ_{cl} is the decay time of the microscale correlation, which characterizes the evolution of the two-point equation, Eq. (2). In the steady state, the solution of the two-point equation is given approximately by

$\langle g(1)g(2)\rangle = \tau_{cl} S_{12}$. The decay is counteracted by the source, which drives the microscale correlation,

$$S_{12}(\kappa_1, \kappa_2, v_\parallel) = \frac{i|e|}{T_e} \sum_{\substack{k' \\ \omega'}} \exp(ik'y_-)(\omega' - \omega'_{*e}) \langle f(1) \rangle \langle \hat{g}(\kappa_2)\hat{\Phi}(\kappa_1) \rangle_{k', \omega'} \quad (11)$$

For convenience, we work in the space of κ_1 and κ_2 obtained from Fourier transforming,

$$S_{12}(\kappa_1, \kappa_2) = \int d\eta_1 \exp(i\kappa_1 \eta_1) \int d\eta_2 \exp(i\kappa_2 \eta_2) S_{12}(\eta_1, \eta_2)$$

The source is proportional to the rate of relaxation of the average distribution $S = -\langle f \rangle \partial \langle f \rangle / \partial t$. Consequently, associated with the driving of the microscale correlation is the extraction of the expansion-free energy stored in the density gradient of the average distribution. A detailed picture giving insight into the mechanism by which the expansion-free energy is made accessible through the source is provided by an analogy with the physics of discreteness as described in the Balescu–Lenard equation.[1,2] This analogy is suggested when the source is separated into coherent and incoherent components. We recall that

$$\hat{g}_k(\kappa) = \hat{g}_k^{(c)}(\kappa) + \tilde{g}_k(\kappa) \quad (12)$$

where

$$\hat{g}_k^{(c)}(\kappa) = \frac{-(\omega - \omega_{*e})}{\omega - (\kappa v_\parallel / Rq)} \frac{|e|}{T_e} \langle f \rangle \hat{\Phi}_k(\kappa) \quad (13)$$

is the coherent response and $\tilde{g}_k(\kappa)$ is the incoherent fluctuation. Substituting Eqs. (12) and (13) into Eq. (11), the source is written as

$$S_{12} = \sum_{\substack{k' \\ \omega'}} \left[(\omega' - \omega'_{*e})^2 \frac{|e|^2}{T_e^2} \langle f(1) \rangle \langle f(2) \rangle \pi \delta\left(\omega' - \frac{\kappa v_\parallel}{Rq}\right) \langle \hat{\Phi}(1)\hat{\Phi}(2) \rangle_{k', \omega'} \right.$$
$$\left. + \frac{i|e|}{T_e}(\omega' - \omega'_{*e}) \langle f(1) \rangle \langle \tilde{g}(2)\hat{\Phi}(1) \rangle_{k', \omega'} \right] \quad (14)$$

The first term is the coherent component and the second is the incoherent component. In writing the first term, we anticipate that we will require the real part of the source; hence, we retain only the residue contribution $i\pi\delta(\omega - \kappa v_\parallel / Rq)$ from the pole $(\omega - \kappa v_\parallel / Rq)^{-1}$. Since $S_{12} = -\langle f \rangle (\partial \langle f \rangle / \partial t)$, we associate with the right-hand side of Eq. (14) the processes that drive the evolution of $\langle f \rangle$. The equation $\partial \langle f \rangle / \partial t = -S_{12}/\langle f \rangle$ is analogous to the Balescu–Lenard equation and allows us to identify with the coherent and incoherent components of the source the processes of diffusion and drag, respectively. The diffusion is quasilinear diffusion, except that here the coherent response, through Poisson's equation, shields the clump. The drag term represents the friction experienced by the clump as it emits Cherenkov radiation while moving through the plasma.

The analogy with particle collisions is useful in understanding an important cancellation that occurs in the source. To see the cancellation, it is necessary to express the clump-shielded potential $\hat{\Phi}$ in terms of the incoherent potential $\tilde{\Phi}_k = 4\pi q \int dv_\| \tilde{g}_k(\kappa, v_\|)$. The relationship between $\hat{\Phi}$ and $\tilde{\Phi}$ is obtained from Poisson's equation $L_k(\kappa)\hat{\Phi}_k(\kappa) = \tilde{\Phi}_k$ where $L_k(\kappa)$ is the eigenmode operator. We further recall that clumps are resonant, moving at the ballistic velocity $u = \omega/k_\| = \omega Rq/\kappa$. Thus,

$$\langle \tilde{g}(2)\hat{\Phi}(1)\rangle_{k' \atop \omega'} = 2\pi\delta(\omega - k'_\| v_\|) L_k^{-1}(\kappa_1)\langle \tilde{g}(2)\tilde{\Phi}(1)\rangle_{k'}$$

$$\langle \hat{\Phi}(1)\hat{\Phi}(2)\rangle_{k' \atop \omega'} = L_{-k'}^{-1}(\kappa_1) L_k^{-1}(\kappa_2) \int dv_\| \langle \tilde{g}(2)\tilde{\Phi}(1)\rangle_{k'} 2\pi\delta(\omega' - k'_\| v_\|)$$

$$= \frac{2\pi}{|k_\||} L_{-k'}^{-1}(\kappa_1) L_k^{-1}(\kappa_2) \langle \tilde{g}(u')\tilde{\Phi}\rangle_{k'} \quad (15)$$

and

$$S_{12} = \sum_{k' \atop \omega'} \left\{ \left[\frac{|e|}{T_e} \langle f(u)\rangle (\omega' - \omega'_{*e}) 2\pi\delta(\omega' - k_\| u) L_{-k'}^{-1}(\kappa_1) L_k^{-1}(\kappa_2) \right] \right.$$
$$\left. \times \left[\frac{|e|}{T_e}(\omega' - \omega'_{*e}) \frac{\pi}{|k_\||} \langle f(u)\rangle \langle \tilde{g}(u')\tilde{\Phi}\rangle_{k'} + iL_{k'}(\kappa_2)\langle \tilde{g}(u)\tilde{\Phi}\rangle_{k'} \right] \right\} \quad (16)$$

From the imaginary part of $L_{k'}(\kappa)$ comes the drag term's contribution to the source. This consists of the electron and ion dissipation involved in the mode. The electron dissipation, representing the electron–electron drag, is that of inverse Landau damping; the ion dissipation, representing the electron–ion drag, consists of the shear damping (linear) and the nonlinear damping (here, ion Compton scattering) which saturates finite-amplitude turbulence. Since the diffusion term is proportional to the electron dissipation, the diffusion cancels with the electron–electron drag. This cancellation is generic to clump phenomena and reflects the shielding of clumps by collective modes. For quasineutral fluctuations, the shielding of a single clump species, say electrons, is expressed by a shielded quasineutrality condition, $n_e = n_e^{(c)} + \tilde{n}_e = -n_i^{(c)}$. Rewriting the righthand side as an ion-response function times the potential, the imaginary part of this condition yields a relation between electron density and ion dissipation. This velocity-averaged shielding relation is extended to phase-space densities because ballistic, particle-like clumps select from the velocity continuum a single velocity, projecting onto the configuration space. This leaves only the drag with the ions, so that

$$S_{12} = \sum_{k' \atop \omega'} \{ [\langle f(u)\rangle(\omega' - \omega'_{*e}) 2\pi\delta(\omega' - k_\| u) i\varepsilon_{\text{IM}}^{\text{ION}}(k', \omega', \kappa_2)]$$
$$\times L_{-k'}^{-1}(\kappa_2) L_k^{-1}(\kappa_2)\langle \tilde{g}(u')\tilde{n}\rangle_{k'} \} \quad (17)$$

where $i\varepsilon_{IM}^{ION}$ is the ion dissipation. The scaling of the source with ion dissipation reflects emission by an electron clump due to its drag on the ions.

A final form for the source is obtained by expressing the inverse eigenfunction operators L^{-1} in terms of the wave function $\Psi_{k}(\kappa)$ and dielectric response $\varepsilon(k, \omega)$,

$$L^{-1}_{k'\atop\omega'} = \frac{\Psi_{k'\atop\omega'}(\kappa_1)}{N_{k'\atop\omega'}\varepsilon(k', \omega')} \int d\kappa' \Psi_{k'\atop\omega'}(\kappa')$$

where $N_{k'\atop\omega'}$ is the normalization constant. Thus,

$$S_{12} = \sum_{k'\atop\omega'} 2\pi \frac{\langle f(u)\rangle(\omega' - \omega'_{*e})\varepsilon_{IM}^{ION}(k', \omega')|\Psi_{k'\atop\omega'}(\kappa)|^2}{|\varepsilon(k', \omega')|^2} \overline{\langle\tilde{g}(u)\tilde{n}\rangle}_{k'} \quad (18)$$

where

$$\overline{\langle\tilde{g}(u)\tilde{n}\rangle}_{k'} = \frac{1}{|N_{k'\atop\omega'}|^2} \int d\kappa_1 \int d\kappa_2 \Psi_{k'\atop\omega'}(\kappa_1)\Psi_{-k'\atop-\omega'}(\kappa_2)\langle\tilde{g}(u)\tilde{n}\rangle_{k'}$$

For electron phase-space granulations in low-frequency (ω_{*e}) turbulence, the consequences of Eq. (18) and its scaling with ion dissipation are important and account for the significant impact of clumps on the collective modes. As mentioned, the electron dissipation is destabilizing ($\varepsilon_{IM}^{ELEC} > 0$) and the ions are stabilizing ($\varepsilon_{IM}^{ION} < 0$). The ion dissipation includes the linear stabilization mechanisms (shear damping) and amplitude-dependent damping mechanisms such as ion Compton scattering and is large, as it must balance the linear instability and its enhancement by the incoherent excitation. With $\omega < \omega_{*e}$, a large, positive source is thus provided for the electron microscale correlation.

4. THE SPECTRUM BALANCE

The steady-state solution of the two-point phase-space density equation is given approximately by $\langle gg\rangle = \tau_{cl}S$. The incoherent part of the correlation is obtained from the total correlation by extracting the coherent part, and the $\langle g^{(c)}\tilde{g}\rangle$ cross correlation. Hence

$$\langle\tilde{g}\tilde{g}\rangle = (\tau_{cl} - \tau_c)S \quad (19)$$

In the last section the source was expressed in terms of $\overline{\langle\tilde{g}(u)\tilde{n}\rangle}_{k'}$, the projection on the eigenfunctions of the incoherent part of velocity-integrated correlation. The quantity $\tilde{g}(u)$ results from the velocity integration of the incoherent density for the ballistic clump with velocity $v = u = \omega/k_\parallel$. By performing the velocity integrations, Fourier transforms in η and y, and eigenfunction projection on both sides of Eq. (19), it is possible to express Eq. (19) entirely in terms of $\langle \tilde{g}(u)\tilde{n}\rangle_k$. The relationship between $\langle \tilde{g}(u)\tilde{n}\rangle_k$ and $\langle \tilde{g}\tilde{g}\rangle_k$ is expressed by

$$\langle \tilde{n}\tilde{n}\rangle_{k\atop\omega} = \int dv_{\parallel+} \int dv_{\parallel-} 2\pi\delta(\omega - k_\parallel v_{\parallel+}) \int dy_- \exp(-iky_-)$$

$$\times \int d\eta_- \exp(-i\kappa_-\eta_-)\langle \tilde{g}\tilde{g}\rangle_k$$

and

$$\int dv_{\parallel+} 2\pi\delta(\omega - k_\parallel v_{\parallel+})\langle \tilde{g}(u)\tilde{n}\rangle_k = \frac{2\pi}{|k_\parallel|}\left\langle \tilde{g}\left(\frac{\omega}{k_\parallel}\right)\tilde{n}\right\rangle_k = \langle \tilde{n}\tilde{n}\rangle_{k\atop\omega}$$

where the two-time correlation $\langle \tilde{n}\tilde{n}\rangle_{k\atop\omega}$ is obtained approximately from the one-time correlation by operating with the propagator $2\pi\delta(\omega - k_\parallel v_\parallel)$ which represents ballistic propagation. Using these equations and the eigenfunction projections, we obtain the spectrum balance equation,

$$\overline{\langle \tilde{g}(u)\tilde{n}\rangle}_k = \int d\kappa_1 \int d\kappa_2 \frac{\Psi_k(\kappa_1)\Psi_{-k}(\kappa_2)}{|N_k|^2} \int dv_{\parallel-} \int dy_- \exp(-iky_-)$$

$$\times \int d\eta_- \exp(-i\kappa\eta_-)(\tau_{cl} - \tau_c)S_{12} \tag{20}$$

where τ_{cl} and S_{12} are given by Eqs. (10) and (18). The integration over $v_{\parallel-}$ has been given previously.[1] The Fourier transforms are performed with the aid of a polar transformation to variables l and θ: $l\cos\theta = e\eta_-/\Delta\eta\sqrt{2}$; $l\sin\theta = k_0 e y_-$. A Bessel function expansion of $\exp(-i\kappa\eta_- - iky_-)$ then facilitates the θ integration to yield

$$\int dv_{\parallel-} \int d\eta_- \exp(i\kappa_-\eta_-) \int dy_- \exp(iky_-)(\tau_{cl} - \tau_c)$$

$$= \frac{16\pi\Delta\eta^2 Rq}{e^3 k_0^2} \cdot \frac{1}{k_\perp} \int_0^1 dl \cos^{-1} l\, J_1(k_\perp l) l \equiv a(k_\perp) \tag{21}$$

where $k_\perp = \sqrt{(2\kappa_-^2\Delta\eta^2/e^2) + k^2/(k_0^2 e^2)}$. Combining this result with Eq. (20) gives

$$\overline{\langle \tilde{g}(u)\tilde{n}\rangle}_k = 2\pi \int d\kappa_1 \int d\kappa_2 \frac{\Psi_k(\kappa_1)\Psi_{-k}(\kappa_2)}{|N_k|^2} a(k_\perp)\langle f(u)\rangle$$

$$\times \int \sum_{\omega'} dk'(\omega' - \omega'_{*e})\Psi_{k'}(\kappa_1)\Psi_{-k'}(\kappa_2)\frac{\varepsilon_{IM}^{ION}(k', \omega')}{|\varepsilon(k', \omega')|^2}\overline{\langle \tilde{g}(u)\tilde{n}\rangle}_{k'}. \tag{22}$$

This equation describes the detailed steady-state balance in the spectrum between linear destabilization, the enhancement by incoherent emission, and the linear and nonlinear damping mechanisms. It plays an analogous role in the two-point theory to the wave–kinetic equation of weak-turbulence theory. One important difference should be noted. Because $\int dv_\| \tau_{cl}$ is amplitude independent as Eq. (21) indicates, and because the source is proportional to $\langle \tilde{g}(u)\tilde{n}\rangle_{k'}$, the incoherent fluctuation amplitude appears to scale out of the balance. The incoherent emission process is ostensibly independent of fluctuation amplitude above some nominal level and the spectrum is determined only up to an unspecified function $N(k)$. However, because $\varepsilon_{\mathrm{IM}}$ details the balance between destabilization mechanisms and the linear and nonlinear amplitude-dependent damping, the spectrum balance equation does, in fact, depend on the collective-mode amplitude. In the steady state it is possible, by supplying the details of the nonlinear damping process to determine N and completely specify the spectrum. Such a calculation will be outlined in the next section.

For this resonant system, we may expand the dielectric $|\varepsilon(k', \omega')|^2$ in the denominator of Eq. (22) about the eigenfrequency, assuming that the spectrum broadening, or damping, at saturation is not too large,

$$\varepsilon(k', \omega') = [k' - k'_r(\omega')]\frac{\partial \varepsilon}{\partial k'} + i\varepsilon_{\mathrm{IM}}[k'_r(\omega), \omega']$$

where $\varepsilon_r[k_r(\omega), \omega] = 0$. We perform the k' integration, evaluating the residue at the pole corresponding to the eigenmode. The correlation then cancels out of the spectrum balance leaving an equation expressing the relation between the total dissipation $\varepsilon_{\mathrm{IM}}(k, \omega_k)$ and the dissipation in the ions, $\varepsilon_{\mathrm{IM}}^{\mathrm{ION}}(k, \omega_k)$,

$$|\varepsilon_{\mathrm{IM}}(k, \omega_k)| = C(k, \omega_k)|\varepsilon_{\mathrm{IM}}^{\mathrm{ION}}(k, \omega_k)| \tag{23}$$

where

$$C(k, \omega_k) = 2\pi \int d\kappa_1 \int d\kappa_2 \frac{|\Psi_k(\kappa_1)|^2 |\Psi_{-k}(\kappa_2)|^2}{|N(k)|^2} a(k_\perp)\langle f(u)\rangle$$

$$\times (\omega_k - \omega_{*e}) \left.\frac{\partial \varepsilon}{\partial k}\right|_{k=k(\omega)} \tag{24}$$

The integrations over κ_1 and κ_2 are transformed to κ_+ and κ_-, and the integrations then allow the shielding response structure functions to sample the dependence of the clump on the two parallel scales. The evolution or clump decay enters into the fast-scale (κ_-) sampling. The slow-scale (κ_+) sampling reflects the degree to which the mode structure in η shields the clumps.

Since the total dissipation is composed of electron and ion contributions, $\varepsilon_{\mathrm{IM}} = \varepsilon_{\mathrm{IM}}^{\mathrm{ELEC}} - |\varepsilon_{\mathrm{IM}}^{\mathrm{ION}}|$ and $\varepsilon_{\mathrm{IM}} < 0$, as collective resonances must be damped to balance noise emission, we may rewrite Eq. (23) to obtain a saturation condition:

$$\varepsilon_{\text{IM}}^{\text{ION}}(k, \omega_k) = \frac{\varepsilon_{\text{IM}}^{\text{ELEC}}(k, \omega_k)}{1 - C(k, \omega_k)} \quad (25)$$

This expresses the balance in the steady state between the linear electron destabilization of inverse Landau damping enhanced by the incoherent noise emission $[1 - C(k, \omega_k)]^{-1}$ and the linear and nonlinear damping in the ions. A related expression,

$$\varepsilon_{\text{IM}}(k, \omega_k) = \frac{C(k, \omega_k)}{1 - C(k, \omega_k)} \varepsilon_{\text{IM}}^{\text{ELEC}}(k, \omega_k) \quad (26)$$

gives the total dissipative mode responses to the incoherent emission process in terms of a numerical enhancement of the electron dissipation. The width of the frequency spectrum at fixed k is just $\varepsilon_{\text{IM}}/(\partial \varepsilon/\partial \omega)$ and is given by

$$\Delta \omega_k = \gamma_k^{\text{ELEC}} \frac{C(k, \omega_k)}{1 - C(k, \omega_k)} \quad (27)$$

5. ILLUSTRATIONS

We first report the results obtained for stability of the universal mode in the presence of incoherent emission. For comparison, we then consider a different problem, that of trapped electrons in toroidal geometry, and obtain formulas for the frequency and wavenumber spectrum as well as energy flux for the energy-transport problem.

The universal mode is a density-gradient-driven drift mode in a slab geometry with a sheared magnetic field $\mathbf{B} = B_0(\hat{z} + x/L_s \hat{y})$. The linear properties of this mode have been extensively investigated. Destabilization is provided by electrons that resonate with the wave. Ion inertia is stabilizing, and is increasingly effective for stronger shear. It is now well established that the universal mode is linearly stable for all values of wave number and shear.[11,12] The level of thermal fluctuations in an ohmic-discharge plasma is sufficient to trigger incoherent fluctuations, which tap the expansion-free energy as already described. The level of incoherent fluctuations grows, exciting the collective modes that reach finite amplitude. Nonlinear ion Compton scattering provides saturation and the overall damping necessary to maintain the steady state.

In previous sections, we have outlined the calculation of the incoherent spectrum relevent to the universal mode. Using the ballooning representation, the radial eigenvalue problem is reexpressed as an eigenvalue problem in η. The x and y variations of the slab geometry pass over into the η and $y = \phi r/q$ variations of the ballooning representation. Hence, the κ_+ and κ_- integrations of Eq. (24) for the clump-enhanced growth factor

$C(k, \omega_k)$ provide a sampling of radial structure by the radial eigenmode shielding response.

Assuming Pearlstein–Berk mode structure by writing the shielding response as a Gaussian, $\Psi_k(\kappa) = \exp(-\alpha\kappa^2)$ (where α is, in general, complex), the κ_+ and κ_- integrations may be performed using saddle-point contour methods. With normalized wavefunctions, the final answer is relatively insensitive to the detailed value chosen for α, beyond the usual scaling of L_s/L_n, the ratio of shear length to density scale length which reflects the mode width of Pearlstein–Berk structure. The final result for $C(k, \omega_k)$ is

$$C(k, \omega_k) \cong 100(\Delta\eta)^2 \left(\frac{Rq}{L_n}\right) \frac{k\rho_s}{1 + k^2\rho_s^2} \cdot \frac{k_0}{k}\left(\frac{m_e}{2m_i}\right)^{1/2} \qquad (28)$$

The mass ratio dependence arises from a phase-velocity scaling $(\omega/k_\parallel v_{te})$, which reflects the emission process of ballistic electron clumps into modes at frequency ω_k. We note that the incoherent emission process occurs over the entire η extent of the mode, in contrast to another theory of nonlinear destabilization of the universal mode,[13] which relies on effects within a small layer around the rational surface.

Evaluating Eq. (28) for parameter values consistent with the universal mode, we obtain a quantitative measure of the impact of clumps on the universal mode stability. We rewrite the saturation condition, Eq. (25), separating the nonlinear and linear damping in the ion term and writing the nonlinear damping rate at saturation as a function of linear dissipation:

$$\gamma_{nl}^{ION} = \gamma_{C.I.} = \frac{\varepsilon_{IM}^{ELEC}(k, \omega_k)/(\partial\varepsilon/\partial\omega_k)}{1 - C(k, \omega_k)} - \frac{\varepsilon_{IM,l}^{ION}(k, \omega_k)}{\partial\varepsilon/\partial\omega_k} \qquad (29)$$

The nonlinear destabilization is the clump-induced growth rate. When positive, it indicates that incoherent emission has driven the modes to finite amplitude, despite shear damping, and has effectively destabilized the mode. The clump-induced growth rate is expressed in terms of the linear dispersion function. This reflects the fact that it is the response function with its associated dissipative processes that shields the clumps. Furthermore, we assume that the response structure in the nonlinear regime is well approximated by the structure of the linear response. The validity of this assumption is verifiable *a posteriori* in terms of the magnitude of the ratio, $\gamma|_{C.I.}/\omega_{*e}$, and is already evident in Eq. (28) where the small parameter (u/v_{te}) plays a central role.

The evaluation of Eq. (29) is displayed graphically in Fig. 1, where $\gamma|_{C.I.}$ is plotted as a function of $k\rho_s$ and compared with Im $\omega(k)$ from linear theory. It is seen that clumps effectively destabilize an otherwise stable mode. A similar effect has been predicted and observed in simulations of a plasma below the threshold of the ion–acoustic instability.[14] In contrast to our calculation of the steady-state, clump-induced destabilization rate of a finite-amplitude mode, Berman *et al.*[14] determine the nonlinear growth of

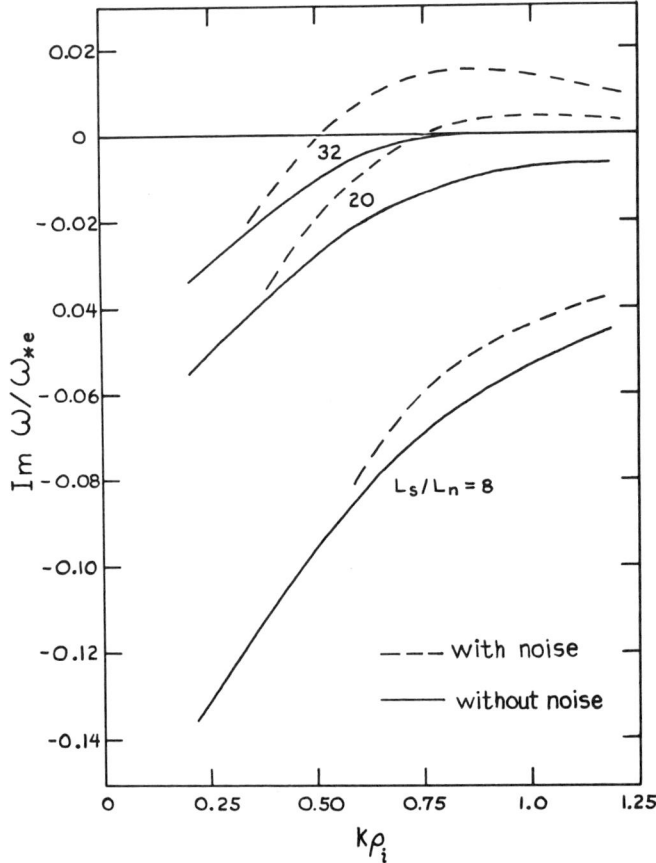

Figure 1. The clump-induced growth rate normalized by ω_{*e} as a function of $k_\perp \rho_s$ for three values of the shear, compared to the linear growth rate in which incoherent fluctuations (noise) are neglected.

the incoherent fluctuation $\langle \tilde{g}\tilde{g} \rangle$ [they solve for $\gamma = d/dt$ in Eq. (6)] away from conditions satisfying the linear dispersion. In the universal mode, the destabilization is not particularly large; for trapped electrons, however, the effect is more significant.

We consider electrons trapped in the magnetic mirrors created by the spiraling of the magnetic field on the toroidal magnetic flux surfaces of tokamak geometry. Such trapped electrons have banana-shaped orbits in a poloidal plane. The collisionless trapped electrons cause an unstable mode. In treating the dynamical equations, considerable simplification is obtained by performing the bounce average, which is possible when the trapped-electron bounce frequency is associated with the fastest time scale in the problem. Further details are found in Ref. 4. The formula for the width of the frequency spectrum is given by Eq. (27), for trapped electrons,

$$C(k, \omega_k) = 2\sqrt{\pi} A(k\rho_s) \frac{\mathcal{S}_k}{\varepsilon_T} \exp\left(\frac{-\omega_k}{\bar{\omega}_d}\right) \left(\frac{\omega_k}{\bar{\omega}_d}\right)^{1/2} \left(1 - \frac{\omega_k}{\omega_{*e}}\right) \qquad (30)$$

where $A(k\rho_s) = (k^2\rho_s^2)^{-1}[1 - J_0(\sqrt{2}k/k_0)]$, $\bar{\omega}_d = \varepsilon_T \omega_{*e}$ is the ballistic frequency of the clump corresponding to the phase velocity $u = \omega/\bar{\omega}_d$, and $\varepsilon_T = L_n/R$. As with the universal mode, phase-velocity dependence enters the relation, underscoring the emission process of ballistic clumps into modes at ω_k. The dependence on the shielding-response wavefunction $\mathcal{S}_k$ indicates that shielding-response structures that experience greater overlap with regions of clump activity allow for more efficient emission. For parameters consistent with the toroidicity-induced mode structure,[15] the frequency broadening is computed to be $\Delta\omega_k/\omega_k \sim 1.1$. This broad line width is indicative of a strong enhancement of the linear growth due to the incoherent emission process.

As mentioned in the previous section, the spectrum may be completely specified from the saturation condition, Eq. (10), by supplying the details of the nonlinear damping process. We consider ion Compton scattering as the process of turbulent energy transfer. The nonlinear ion damping rate is obtained from perturbation theory, as it is in weak–turbulence theory, however, the integral over frequency is performed using the broadened frequency spectrum $\Delta\omega_k/[(\omega - \omega_k)^2 + \Delta\omega_k^2]$ obtained from the two-point theory. The frequency broadening is responsible for broadening the beat resonance, inducing dissipation in the background fluctuations as well as altering the spectrum spatial structure. The net impact of these effects is an enhancement of the nonlinear wave–ion interaction. This may be interpreted as a spreading of the ion-Landau resonance point due to the uncertainty in ω represented by $\Delta\omega_k$. From these considerations, the wavenumber spectrum is determined and found to scale as $N(k_\perp \rho_s) \sim (k_\perp \rho_s)^{-3/2}$ asymptotically.

The fact that incoherent fluctuations are driven by the relaxation of the average distribution according to the drag-induced enhancement of free-energy assessibility implies that the quasilinear prescription of transport is no longer valid. A direct calculation of the flux Γ of energy transport ($\partial E/\partial t + \Gamma \nabla \cdot E = S$) shows that the flux is proportional to the clump-enhanced growth rate

$$\Gamma \propto \frac{\varepsilon_{IM}^{ELEC}}{1 - C(k, \omega_k)} \qquad (31)$$

rather than γ_L, as in conventional quasilinear theory. The enhancement of energy flux is obviously strong where significant broadening of the frequency spectrum occurs.

6. CONCLUSIONS

We have considered herein the incoherent part of the fluctuating density, a constituent of turbulence usually neglected in theoretical treatments. These fluctuations are identified with clumps, granulations in the phase-space density resulting from the properties of the mixing process. We have used the theory of two-point phase-space density correlation as the natural vehicle for treating incoherent fluctuations. The source term has been emphasized and discussed in detail. In particular, we have shown that the driving of the microscale correlation by the source is proportional to the relaxation of the average distribution. Hence, we identify a novel mechanism for the extraction of expansion-free energy. The effect of incoherent fluctuations on the collective modes has been examined. The effect has been described as an emission process that excites the modes. We have considered the steady state in which the modes are damped to balance this emission. We obtain from the net damping in the steady state an analytic formula for the frequency spectrum at fixed k. The theory, then, effectively links the issue of the width of the frequency spectrum with nonlinear stability associated with incoherent fluctuations. Indeed, the width of the frequency spectrum provides a measure of the strength of the nonlinear instability.

We consider two drift-type modes in realistic geometries to obtain a quantitative measure of the impact of clumps. For the universal mode, we find that the mode is destabilized and reaches finite amplitude owing to incoherent emission. The destabilization, however, is a small effect with $\gamma|_{\text{C.I.}}/\omega_{*e} \sim 0.02$ for $L_s/L_n = 32$. For trapped electrons, the impact of incoherent emission can be significant. In this case, with toroidicity-induced mode structure, the frequency spectrum is broad ($\Delta\omega_k/\omega_{*e} \sim 1.1$). Considering the details of ion Compton scattering necessary to achieve saturation, we are able to completely specify the spectrum and show the wavenumber dependence. Finally, the effect on transport is assessed with the result that incoherent emission enhances the transport of energy increasing the flux by a numerical factor $(1 - C)^{-1}$.

ACKNOWLEDGMENT

This work was supported by U.S. Department of Energy Grant No. DE-FG05-80ET-53088.

REFERENCES

1. T. H. Dupree, *Phys. Fluids* **15**, 334 (1972).
2. T. Boutros-Ghali and T. H. Dupree, *Phys. Fluids* **24**, 1839 (1981).

3. P. H. Diamond, P. L. Similon, P. W. Terry, C. W. Horton, S. M. Mahajan, J. D. Meiss, M. N. Rosenbluth, K. Swartz, T. Tajima, R. D. Hazeltine, and D. W. Ross, *Plasma Physics and Controlled Nuclear Fusion Research 1982*, 259 (1983).
4. P. W. Terry and P. H. Diamond, in *Annual Sherwood Controlled Fusion Theory Conference*, 1C24 (1982).
5. E. Mazzucato, *Phys. Rev. Lett.* **48**, 1828 (1982).
6. B. A. Carreras, P. H. Diamond, P. L. Similon, and P. W. Terry, *Bull. Am. Phys. Soc.* **27**, 946 (1982).
7. P. W. Terry and P. H. Diamond, in *Annual Sherwood Controlled Fusion Theory Conference*, 3Q21 (1983).
8. J. H. Misguich and R. Balescu, *Plasma Phys.* **24**, 289 (1982).
9. J. W. Connor, R. J. Hastie, and J. B. Taylor, *Proc. R. Soc. Lond. Ser.* **A365**, 1 (1979).
10. E. A. Frieman and Liu Chen, *Phys. Fluids* **25**, 502 (1982).
11. D. W. Ross and S. M. Mahajan, *Phys. Rev. Lett.* **40**, 324 (1978).
12. K. T. Tsang, P. J. Catto, J. C. Whitson, and J. Smith, *Phys. Rev. Lett.* **40**, 327 (1978).
13. S. P. Hirshman and K. Molvig, *Phys. Rev. Lett.* **42**, 648 (1979).
14. R. H. Berman, D. J. Tetreault, T. H. Dupree, and T. Boutros-Ghali, *Phys. Rev. Lett* **48**, 1249 (1982).
15. L. Chen and C. Z. Cheng, Princeton Plasma Physics Laboratory Report PPPL-1562 (1980).

AUTHOR INDEX

Abdullov, K., 283
Ablowitz, M. J., 90, 274, 283
Abraham, R., 78
Agladze, K. I., 110
Aharnov, A., 230
Albert, J., 224
Altshul, L. M., 296, 308
Anisimov, S. I., 146, 153
Arneodo, A., 104, 110
Arnol'd, V. I., 178, 189, 190, 196, 207, 213, 224, 241, 250

Baesens, C., 205
Balescu, R., 205, 207, 295-309, 317, 325, 328, 333, 341, 351
Barbar, M. N., 230
Belcher, J. W., 286, 291
Belousov, B. P., 96, 117, 125
Benard, B., 171, 235
Benjamin, T., 283
Bennetin, G., 124, 167, 251
Berezovskii, M. A., 153
Berk, H. L., 347
Berman, R. H., 152, 274, 283, 347, 351
Berne, B. J., 251
Berry, M. V., 64
Bialek, J., 64
Bishop, A. R., 125, 167, 284
Blacher, S., 124
Bogolubsky, J., 283
Bona, J., 283
Boozer, A. H., 31, 224, 233
Bountis, T., 50, 58, 64
Boutros-Ghali, T., 152, 283, 298-300, 302, 306, 309, 350, 351
Brown, G., 273, 283
Brumer, P., 251
Bulson, J. F., 89

Campbell, D. K., 125, 284
Cantwell, B., 283
Carreras, B. A., 351
Carrigan, R. A., 56
Cary, J. R., 46, 48, 49, 51-56, 281

Casati, G., 64
Catto, P. J., 351
Cerny, K., 284
Channon, S. R., 34, 37, 42, 190, 191, 207
Chen, L., 351
Cheng, C. Z., 351
Cheung, P. Y., 131-153
Chirikov, B. V., 25, 31, 34, 37, 41, 42, 56, 64, 78, 88, 189, 190, 191, 197, 203, 207, 219, 224, 251
Choi, D. I., 261, 271, 296, 298, 299, 308
Chu, F. Y. F., 284
Cohen, E. G. D., 171-187
Cohen, R. H., 31
Coles, D., 283
Collet, P., 11, 19, 110, 325
Connor, J. W., 351
Crutchfield, J. P., 116, 124, 125
Currie, J. F., 284

Darrow, C. B., 153
Davis, L., 286, 291
De Boor, C., 125
De Fainchtein, R., 65-78
Degtyarev, L. M., 167
Denavit, J., 167
Diamond, P. H., 335-351
Dimotakis, P., 283
Dobrowolny, M., 286, 289, 291
Doolen, G. D., 153, 155-167
Doveil, F., 64
DuBois, D. F., 153, 155-167, 251, 266, 271
Dubin, D. H. E., 49, 252, 255, 270
Dupree, D. H. E., 49, 252, 255, 270
Dupree, T. H., 152, 167, 244, 251, 252, 283, 284, 298, 299, 300, 301, 302, 306, 308, 309, 317, 325, 327, 334, 337, 350, 351

Eckmann, J. P., 19, 325
Eggleston, D. L., 153
Escande, D. F., 64
Espedal, M., 251, 271

Faddeev, L. D., 276, 284

Faehl, F. J., 166
Farmer, J. D., 124, 125, 167
Feigenbaum, M. J., 3, 19, 224, 242, 251
Fesser, K., 167
Fibonacci, L. A., 16, 209, 214, 215, 217, 220, 224
Fied, G. R., 271
Filonenko, N. N., 56, 251
Finn, J. M., 125, 252
Fisher, M. E., 230
Flaschka, H., 78
Ford, J., 50, 64, 207
Frederickson, P., 125
Frieman, E. Q., 351
Frisch, U., 291
Fujisaka, H., 100, 110, 236, 239
Fyfe, D., 271, 291

Galeev, A. A., 152, 251, 264, 271
Galgani, L., 124, 167, 251
Gardner, C. S., 89
Gardner, C. W., 284
Gefen, Y., 230
George, C., 205
Ginzbur, 93-109
Giorgilli, A., 124
Goldman, M. V., 152, 166, 167
Goldstein, M. L., 291
Goldston, R. J., 31, 224
Gottlieb, D., 291
Gould, R. W., 79, 284
Grad, H., 56, 211, 224
Grappin, R., 286, 291
Grassberger, P., 124
Grebogi, C., 127-130, 244, 251
Grecos, A., 205
Greene, J. M., 3-19, 216, 223, 224, 241, 250
Greenside, H. S., 124
Gresillon, 251
Guckenheimer, J., 125
Gurel, O., 251

Hafizi, B., 153, 155-167
Hai, F., 152
Haker, H., 110
Hamilton, I., 209, 210, 212, 251
Hanson, J. D., 56, 125
Hasegawa, A., 89, 242, 243, 247, 251, 255, 270
Hastie, R. J., 351
Hatori, T., 21-31
Hay, R., 223, 224
Hazeltine, R. D., 224, 351
Heiles, C., 207
Helleman, H. G., 50, 58, 64, 207, 251
Henderson, D., 251

Henon, M., 34, 190, 207
Hentschel, H. G. E., 238, 239
Herrara, J. C., 64
Herring, J. R., 252
Hinton, F. L., 224, 255, 270
Hirschman, S. P., 245, 252, 351
Holland, R. R., 284
Holmes, P. J., 207
Hopf, 94
Horton, C. W., Jr., 89, 207, 252, 253-271, 278, 283, 284, 296, 297, 299, 308, 351
Huberman, B. A., 125
Hudson, J. L., 125
Huerre, P., 110
Huson, F. R., 56

Ichikawa, Y. H., 21-31, 79-90
Ikezi, H., 152
Ivanov, M. F., 153

Jacobi, 57
Jeffrey, A., 80, 90
Jeffries, C., 125
Jensen, R. V., 251
Johnson, J. L., 224
Joyce, G., 291

Kac, M., 78
Kadanoff, L. P., 3, 14
Kadomtsev, B. B., 265, 271, 327, 334
Kako, F., 89, 90
Kamimura, T., 21-31, 89, 278, 283, 284
Kaplan, J. L., 112, 120, 125
Karney, C. F. F., 19, 31, 33-42, 190, 191, 207, 223, 224
Karpman, V. I., 284, 296, 308
Kaufman, 242, 249
Kaw, P. K., 56
Kawahara, 107
Keener, J. P., 284
Kerr, W., 167
Khazei, M., 89
Kim, H. C., 153
Kingsep, A. S., 89
Kirkpatrick, S., 230
Kirkpatrick, T. R., 171-187
Kleva, R. G., 224, 251, 252
Koch, H., 19
Koga, S., 110
Kolmogorov, A. N., 6, 78, 190, 207, 213, 217, 218, 221, 224, 235, 241, 286, 323
Koniges, A. E., 252
Konno, K., 80, 89, 90
Kono, M., 90, 327-334

Kortweg-de Vries, 80, 83, 86, 107, 275
Kotschenreuther, M., 252
Kraichnan, R. H., 167, 242, 243, 245, 246, 251, 252, 265, 271
Krinsky, V. I., 110
Krommes, J. A., 49, 224, 241-252, 264, 270, 271, 325
Kruer, W. L., 56, 167
Krumhansl, J. A., 284
Kruskal, M. D., 56, 274, 284
Kubo, R. 230, 242, 243, 251, 325
Kulsrud, R. M., 56
Kumar, P., 284
Kuo-Petravic, G., 224
Kuramoto, Y., 93-110
Kuwahara, M., 86, 89

Landau, 93-109, 247, 249, 342, 346, 349
Langmuir, 131-266
Larcheveque, M., 239
Larichev, V. D., 270, 271
Laval, 251
Lebowitz, J. L., 34, 37, 42, 190, 191, 207, 208
Lee, W. W., 49, 247, 248, 249, 252, 270
Lee, Y. C., 140, 153
Lenard, 302, 308, 328, 333, 341
Leorat, J., 291
Leslie, D. C., 252
Leung, P., 153
Lichtenberg, A. J., 6
Liewer, P. C., 284
Littlejohn, R. G., 43-50
Lomdahl, P., 167
Lonngren, K. E., 84, 89, 90
Lorenz, E. N., 121-125
Lotko, W., 274, 284
Lumley, 283
Lyapunov, 111-124, 155, 243, 244, 258, 267, 324

McCormick, W. D., 125
McDonald, S. W., 127-130
MacKay, R. S., 3, 17, 19, 64, 223, 224, 242, 250
McLaughlin, D. W., 284
McWilliams, J. C., 271
Mahajan, S. M., 351
Mahoney, J., 283
Makhankov, V., 283
Makino, M., 89
Malmberg, J. H., 153
Mandelbrot, B. B., 225, 230
Mangenery, A., 291
Manheimer, W. M., 152
Mankin, J. C., 125
Manneville, P., 100, 110

Martin, P. C., 243-246, 251
Maslov, E., 284
Mather, J. N., 58, 64
Matsumoto, K., 124, 125
Matthaeus, W. H., 285-291
May, R. M., 251
Mazzucato, E., 269, 271, 351
Meiss, J. D., 89, 273-284, 351
Melnikov, V. K., 196, 207
Meneguzzi, M., 291
Micheleson, P., 89
Mikhailovskii, A. B., 270
Miles, J. W., 83, 90
Mima, 242, 243, 247, 251, 255, 270
Misguich, J. H., 295-309, 317, 351
Misra, B., 207
Miura, R. M., 284
Miyamoto, K., 56
Molvig, K., 245, 252, 351
Monin, A. S., 239, 325
Montgomery, D., 257, 271, 285-291
Month, M., 56, 64
Moon, H. T., 110
Morales, G. J., 140, 153
Morel, P., 239
Mori, H., 232, 235-240, 243, 251, 323, 325
Morikawa, G. K., 89
Morrison, P. J., 276, 283, 284
Moser, J., 6, 78, 190, 207, 208, 213, 224, 241
Mozer, F. S., 282

Nagasawa, T., 84, 86, 89
Nagashima, H., 89, 125
Nagashima, T., 86, 113, 124
Nakamura, Y., 84, 86, 90
Nelkin, M., 240, 325
Neumann, J. V., 325
Neville, E. H., 208
Newell, A. C., 93, 110
Nicholson, D. R., 167
Nicolaenko, B., 125, 284
Nicolis, G., 205
Nishida, Y., 84, 86, 89
Nishihara, K., 89
Nogaki, K., 167
Nozaki, K., 88, 89

Oberman, C., 49, 224, 251, 252, 270
Oikawa, M., 90
Olver, P., 284
O'Neal, T. M., 153
Orszag, S. A., 245, 251, 287, 291
Oseledec, V. I., 125
Ott, E., 124, 125, 127-130, 164, 167, 252, 284

Packard, N. H., 124, 125
Papadopoulos, K., 152
Patterson, G. S., 291
Pearlstein, 347
Pelletier, G., 307-309
Percival, I. C., 58, 64
Perdang, J., 124
Pereira, N. R., 167
Perez, J., 125
Peseli, H. L., 89
Petrosky, T. Y., 78, 189-208
Petrov, I. V., 153
Petviashvili, V., 284
Pignaturo, T., 124
Pogutse, O. P., 327, 334
Pomot, C., 307-309
Pouquet, A., 286, 291
Prigogine, I., 190, 191, 198, 200, 205, 207, 208
Procaccia, I., 124, 238, 239

Qian, S. J., 153

Ragdeev, R. Z., 56, 90, 152, 167, 224, 264, 271
Ranch, H. E., 208
Rand, D. A., 124
Rasmussen, J., 89
Rayleigh, 171
Rebenchick, A. M., 153
Rechester, A. B., 8, 25, 31, 224
Redekopp, L. G., 110
Reichl, L. E., 65-78, 89, 205, 207, 271
Resibios, P., 208
Reznik, G. M., 270, 271
Roberts, P. H., 325
Rose, H. A., 153, 155-167, 243, 245, 246, 251, 266, 271
Rosenbluth, M. N., 31, 56, 224, 351
Roshko, A., 273, 283
Ross, D. W., 351
Rossler, O. E., 113-125, 249
Roux, J. C., 124, 125
Rowberg, R., 152
Rudakov, L. I., 89, 166
Ruelle, D., 124, 243, 251
Russell, D. A., 125, 164, 167
Rutherford, P. H., 224
Ryutov, D. D., 56

Saeki, K., 89
Sakagami, H., 89
Sasagawa, F., 230, 231
Sato, T., 89
Satsuma, J., 90
Schieve, W. C., 205

Schmidt, G., 44, 57-64
Scott, A. C., 272, 282
Segur, H., 90, 274, 283
Shapiro, V. D., 152, 167
Shaw, R. S., 125
Shchur, L. N., 153
Shebalin, J., 291
Shepelyansky, D. L., 34, 37, 42
Schevchenko, V. I., 152, 167
Shimada, I., 113, 124
Shraiman, B., 244, 251
Sidont, E., 152
Siggia, E. D., 243, 245, 246, 251
Similon, P., 251, 351
Simoyi, R. H., 125
Slusher, R. E., 274, 284
Smith, J., 331
Snapp, R., 205
Solovev, G. I., 167
Spitzer, L., 56
Stenzel, R. L., 152, 153
Stewartson, K., 93, 110
Strelcyn, J. M., 124, 167, 251
Stringer, T., 223
Stuart, J. T., 93-97, 110
Stupakov, G. V., 56
Sudan, R. N,. 89, 167, 251, 284
Surko, C. M., 274, 284
Suzuki, M., 225-231, 311-325
Swartz, K., 351
Swift, J., 111-125
Swinney, H. L., 124, 125, 273
Szebehely, A. G., 89, 207, 271

Tabor, 284
Tajima, T., 349
Takayoshi, F., 240, 323, 325
Takens, F., 112, 120, 124, 125
Tanikawa, T., 131-153
Taniuti, T., 80, 89, 90
Taylor, J. B., 219, 224, 351
Taylor, R. J., 152
Temerin, M., 284
Tennyson, J. L., 56, 205
Ter Haar, D., 166, 284
Terry, P. W., 252, 253, 255, 269, 271, 335-351
Testa, J., 125
Tetreault, D. J., 152, 251, 283, 299, 309, 351
Thomson, J. J., 167
Thornhill, S. G., 166, 284
Toda, M., 65-67, 78
Toedteimier, 83
Tran, M. Q., 153
Treive, 284

Author Index

Tresser, C., 110
Trivelpiece, 79
Trullinger, S. E., 167, 284
Tsang, K. T., 351
Tsuda, I., 124, 125
Tsukabayashi, I., 84, 86, 90
Tsuzuki, T., 110
Tsytovich, V. N., 166, 271
Turner, J. S., 125

Ueda, S., 90
Ulam, S. M., 325

Van Dyke, 283
Van Moerbeke, P., 78
Veltri, P. L., 291
Vivaldi, F., 19, 34

Wadati, M., 80, 90
Wagner, C. E., 152
Walker, C. H., 78
Waltz, R. E., 270, 271
Washimi, H., 80, 90
Watanabe, S., 86, 90
Watson, G. N., 208
Wayne, C. E., 244, 251
Weatherall, J. C., 167
Wehausen, 283

Weimer, K. E., 246
Wersinger, J. M., 125, 252
Wharton, C., 152
White, R. B., 5, 19, 31, 56, 209-224
Whitson, J. C., 351
Whittaker, E. T., 64, 208
Williams, M., 167
Wilson, K., 12
Winfree, A. T., 110
Woft, A., 111-125
Wong, A. Y., 131-153

Yaglom, A. M., 239, 325
Yajima, N., 79-90
Yakhot, V., 100, 110
Yamada, T., 100, 110
York, E., 125
Yorke, J. A., 112, 120, 124, 125, 127-130
Yoshimura, K., 86, 90

Zabusky, N. J., 274, 284
Zakharov, V. E., 90, 146, 153, 155, 164, 166, 266, 276, 277, 284
Zaslavskii, G. M., 56, 88, 251
Zaslavsky, 224
Zhabotinskii, A. M., 96, 117, 125
Zheng, W. M., 78
Zweben, S. J., 284

SUBJECT INDEX

Accelerating orbit, 26
Accelerator mode, 25, 26, 28, 34, 41
Amplitude turbulence, 101
Anomalous transport, 337
Arnold-Chirikov diffusion coefficient, 189
Attractor, 127

Balescu-Lenard collison term, 328, 333
Basin of attraction, 127
BBGKY hierarchy, 173
BBGKY time-correlation functions, 173
Belousov-Zhabotinsky reaction, 96, 117
Boltzmann equation, 176, 177, 179
Boundary conditions, 173
Bouyancy force, 172, 173
Breather mode, 88
Brownian motion, 228

Cavitons, 131, 137, 142, 146, 151, 155, 157
Chaos, 80, 86, 93, 155, 158, 324
Chaotic attractor, 259
Chaotic orbit, 26
Chapman-Enskog method, 177, 179
Closure approximations, 241, 243, 249, 250
Closure problem, 243
Clump correlations, 307, 328
Clump kinetics, 332
Clump lifetime, 307, 340
Clumps, 141, 150, 166, 295, 302, 316, 324, 327, 331, 335, 336
Coherent wave, 132
Collision operator, 175
Collision operator for fluctuations, 267
Convective instability, 171
Correlation enhancement, 307
Correlation length, 185
Correlations:
 density, 337
 coherent, 303
 incoherent, 304, 307
 long time, 31, 34
 microscale, 336, 340
 momentum, 173, 183
Critical point, 183, 185, 186

Critical slowing down, 183
Cylindrical rolls, 173

Diagram technique, 296
Diffusion, 21, 209
 relative, 306
 turbulent, 235
Diffusion coefficient, 34, 38, 41, 189, 203, 322
Diffusion rate, 25
Direct interaction approximation, 244, 246, 250, 339
Discrete maps, 216
Dissipation, electron and ion, 342
Dissipative forces, 172
Drift waves, 133, 135, 274, 279, 280
 ergodic behavior, 258
 frequency spectrum, 253, 258
 instability, 253
Dynamic alignment, 284, 287

Eigenvectors, 179
Elastic pendulum, 191
Electromagnetic radiation, 143, 150
 emission spectrum, 144
Electron-ion drag, 342
Energy conservation in nonlinear transfer, 267
Energy transport, 349
Enstrophy, 235, 256
Enstrophy cascade, 236, 238
Equipartition of modal energy, 267
Ergodicity, 258, 277
Expansion free energy, 341

Feigenbaum constant, 105
Fibonacci series, 214, 220, 223
Fluctuations, 173
 coherent, 298
 density and velocity, 171
 incoherent, 298, 336
Fokker Planck equation, 201, 314
Fractal, 225
Fractal basin boundaries, 128
Fractal curves, 127
Fractal dimension, 129, 225

Fractional dimensionality, 225
Frequency spectrum, 337, 346

Ginzberg-Landau equation, 93
Global approximation method, 315
Granulations, 327
Guiding center drift, 210
Gyrokinetic equation, 245, 337

Hamiltonian, 52, 54, 59, 66, 68, 156, 191, 212
Hamiltonian mechanics, 43
Hasegawa-Mima equation, 247, 255
Heat flux, 176
Homoclinic orbits, 104
Hopf bifurcations, 94
Hydrodynamic instability, 173
Hydrodynamic modes, 179, 180, 182
Hypercube evolution, 119

Instability, 279
  convective, 171
  drift wave, 253
Instability point, 183, 185
Integrable system, 51
Intermittent transition, 29
Invariants, 44
Inverse scattering transformation, 80
Ion-acoustic solitary waves, 273
Ion-Compton scattering, 349
Island chain, 6, 8, 38, 213

KAM curves, 34
KAM orbits, 6, 57
KAM tori, 58, 59
Kinetic equations, 173, 176, 189, 198, 201, 302, 306, 308
Kinetic operator, 174, 176, 181
Kinetic theory, 171
Kink-antikink pair, 88
Koch curve, 226
Kolmogorov entropy, 217, 221
Kolmogorov inertia range, 323
Korteweg-de Vries equation, 83, 86, 275

Langevin equation, 313
Langmuir turbulence, 156
Langrangian integral, 57, 58
Laser-Poppler velocimetry, 171, 186
Levy-process, 225, 229
Liouville equation, 199
Loop soliton, 80
Lorenz strange attractors, 121
Lyapunov exponents, 112, 114, 117, 123, 157, 164, 258, 324

Lyapunov time, 155, 157

Macroparticles, 327
Magnetic coordinates, 210
Magnetic field lines, 43
Magnetic geometry, 3
Magnetic island, 219
Magnetic surfaces, 3, 5, 15, 209, 213
Maps, area preserving, 5, 15, 21
Markovian approximation, 249, 339
Markovian reduction, 265
Mathieu equation, 61
Melnikov function, 196
MHD modes, 223
Microwave scattering, 171
Mode coupling, 182, 185

Navier-Stokes equation, 177
Noble numbers, 16
Noether's theorem, 46
Noncanonical coordinates, 43

Parametric decay, 132
Pearlstein-Berk mode structure, 345
Period doubling, 61, 100, 131, 133
Periodic orbits, 57
Periodic solutions, 95
Phase instability, 98
Phase turbulence, 98
Poincare plot, 221
Propagator, 298, 305

Quadratic mapping, 34
Quasilinear diffusion, 25
Quasinormal approximation, 262

Radial twist map, 22
Random phase approximation, 34
Rayleigh-Benard cell, 171, 183
Rayleigh number, 172, 180
  critical, 173
Reaction diffusion systems, 93
Regularized longwave equation, 275
Relative diffusion, 320, 322, 324, 340
Renormalization, 3, 12, 17, 253, 264
  propagator, 298, 304
  vertex, 200, 304, 306
Residue, 216
Resonance, 53
Resonance zones, 65, 70-78
Rossby-wave equation, 255
Rössler strange attractor, 113

Subject Index

Saturation condition, 280
Scaled thermal spread, 138, 142
Scaling theory, 227, 229, 311, 316, 320, 322, 324
Self-consistent field approximation, 249
Separatrix, 54
Soliton resonance, 79, 83
Solitons, 79, 83, 273
Spastial phase diffusion, 100
Spectrum balance equation, 344
Spectrum broadening, 133, 137
Spiky structures, 146
Standard map, 22, 25, 33, 41, 63
Statistical coupling, 297, 304
Statistical turbulence theory, 262
Stochastic diffusion, 2, 29, 31
Stochastic domain, 220
Stochastic field, 210
Stochastic instability, 88
Stochastic layer, 177, 199, 202
Stochastic orbit, 7, 15
Stochastic region, 54
Stochastic sea, 33, 41
Stochastic threshold, 210, 223
Strange attractor, 111
Strobe plot, 67
Stuart-Landau equation, 94
Survival probability, 37
Symmetries, 44

Tangent bifurcations, 34
Tangent space, 216
Taylor-Chirikov map, 219
Toda lattice, 65, 67
Toroidal system, 5
Trajectory correlation, 327
Transient phenomena, 311, 313, 317
Trapping statistics, 37
Turbulence, 97, 131, 136
  coherent structures, 273
  drift wave, 253
  fully developed, 235
  MHD, 285
Turbulent diffusion, 235

Universal mode, 346

Variational principles, 44
Vlasov hierarchy equation, 338
Vlasov-Poisson equations, 328
Volume contraction, 256
Vortex motion, 80
Vortices, 273

Wave trapping, 148
Whisker mapping, 197

Zakharov equations, 146, 150, 155, 164

RAYMOND H. FOGLER LIBRARY
DATE DUE